Renewable Energy Systems 2023

Renewable Energy Systems 2023

Editor

Maria Vicidomini

MDPI

Basel • Beijing • Wuhan • Barcelona • Belgrade • Novi Sad • Cluj • Manchester

Editor
Maria Vicidomini
Department of Industrial
Engineering, University of
Naples Federico II
Naples
Italy

Editorial Office
MDPI
St. Alban-Anlage 66
4052 Basel, Switzerland

This is a reprint of articles from the Special Issue published online in the open access journal *Applied Sciences* (ISSN 2076-3417) (available at: https://www.mdpi.com/journal/applsci/special_issues/Renewable_Energy_Systems_2021).

For citation purposes, cite each article independently as indicated on the article page online and as indicated below:

Lastname, A.A.; Lastname, B.B. Article Title. *Journal Name* **Year**, *Volume Number*, Page Range.

ISBN 978-3-7258-0727-7 (Hbk)
ISBN 978-3-7258-0728-4 (PDF)
doi.org/10.3390/books978-3-7258-0728-4

Contents

About the Editor vii

Luca Cimmino and Maria Vicidomini
Renewable Energy Systems 2023
Reprinted from: *Applied Sciences* **2024**, *14*, 1918, doi:10.3390/app14051918 1

Mª Teresa Pintanel, Amaya Martínez-Gracia, Mª Pilar Galindo, Ángel A. Bayod-Rújula, Javier Uche, Juan A. Tejero, et al.
Analysis of the Experimental Integration of Thermoelectric Generators in Photovoltaic–Thermal Hybrid Panels
Reprinted from: *Applied Sciences* **2021**, *11*, 2915, doi:10.3390/app11072915 7

Monika Božiková, Matúš Bilčík, Vladimír Madola, Tímea Szabóová, Ľubomír Kubík, Jana Lendelová, et al.
The Effect of Azimuth and Tilt Angle Changes on the Energy Balance of Photovoltaic System Installed in the Southern Slovakia Region
Reprinted from: *Applied Sciences* **2021**, *11*, 8998, doi:10.3390/app11198998 25

Benjamin Murgas, Alvin Henao and Luceny Guzman
Evaluation of Investments in Wind Energy Projects, under Uncertainty. State of the Art Review
Reprinted from: *Applied Sciences* **2021**, *11*, 10213, doi:10.3390/app112110213 46

Henning Thiesen and Clemens Jauch
Potential of Onshore Wind Turbine Inertia in Decarbonising the Future Irish Energy System
Reprinted from: *Applied Sciences* **2022**, *12*, 2984, doi:10.3390/app12062984 74

Viacheslav G. Kadochnikov and Mikhail V. Dvoynikov
Development of Technology for Hydromechanical Breakdown of Mud Plugs and Improvement of Well Cleaning by Controlled Buckling of the Drill String
Reprinted from: *Applied Sciences* **2022**, *12*, 6460, doi:10.3390/app12136460 90

Aiman Mazhar Qureshi and Ahmed Rachid
Comparative Analysis of Multi-Criteria Decision-Making Techniques for Outdoor Heat Stress Mitigation
Reprinted from: *Applied Sciences* **2022**, *12*, 12308, doi:10.3390/app122312308 106

Nikolay Sergienko, Pavel Kalinin, Ivan Pavlenko, Marek Ochowiak, Vitalii Ivanov, Anton Sergienko, et al.
Synthesis of the Energy-Saving Dry Dual Clutch Control Mechanism
Reprinted from: *Applied Sciences* **2023**, *13*, 829, doi:10.3390/app13020829 120

Heriberto Rodriguez-Estrada, Elias Rodriguez-Segura, Rodolfo Orosco-Guerrero, Cecilia Gordillo-Tapia and Juan Martínez-Nolasco
Novel Multibus Multivoltage Concept for DC-Microgrids in Buildings: Modeling, Design and Local Control
Reprinted from: *Applied Sciences* **2023**, *13*, 2405, doi:10.3390/app13042405 138

KeChrist Obileke, Golden Makaka and Nwabunwanne Nwokolo
Recent Advancements in Anaerobic Digestion and Gasification Technology
Reprinted from: *Applied Sciences* **2023**, *13*, 5597, doi:10.3390/app13095597 174

Francesco Calise, Francesco Liberato Cappiello, Luca Cimmino and Maria Vicidomini
Dynamic Modelling and Energy, Economic, and Environmental Analysis of a Greenhouse Supplied by Renewable Sources
Reprinted from: *Applied Sciences* **2023**, *13*, 6584, doi:10.3390/app13116584 . . . . **192**

Shurong Peng, Lijuan Guo, Haoyu Huang, Xiaoxu Liu and Jiayi Peng
ForecastNet Wind Power Prediction Based on Spatio-Temporal Distribution
Reprinted from: *Applied Sciences* **2024**, *14*, 937, doi:10.3390/app14020937 . . . . **209**

About the Editor

Maria Vicidomini

Maria Vicidomini was born in 1988 and graduated cum laude in Environmental Engineering from the University of Naples Federico II, Italy, in 2013. From 2013 to 2014, she was a fellow researcher at the same institution. She obtained a Ph.D. degree in Industrial Engineering in 2018, and since 2019, she has been a researcher at the University of Naples Federico II. Her research activity has mainly focused on the development of dynamic simulation models for energy, exergy, economic and environmental analysis and to assess impact of innovative systems for distributed polygeneration systems supplied by renewable energy (geothermal, solar, wind energy) and natural gas; building integrated solar thermal systems and internal combustion engines for the production of heating, cooling and power; solar heating and cooling systems; solar desalination systems; hybrid renewable systems based on wind, solar and geothermal energy; electrical storage; and electric vehicles. She has also carried out research activity in cooperation with several international institutions (in Portugal, Iran, Turkey, Denmark, China, Croatia, Poland, Germany and Canada). She was a supervisor of several BsC, MS and PhD students at the University of Naples Federico II. She is also a reviewer of around 15 international journals and has served as Guest Editor for Special Issues in international journals. She is a member of the scientific committee of several international conferences and is also involved in the organizing committee of the CEER 2020 Conferences. She was a conference chair and presenter for several sessions at international conferences.

applied sciences

MDPI

Editorial

Renewable Energy Systems 2023

Luca Cimmino and Maria Vicidomini *

Department of Industrial Engineering, University of Naples Federico II, 80125 Naples, Italy; luca.cimmino@unina.it
* Correspondence: maria.vicidomini@unina.it

1. Introduction

In the last few years, the adoption of renewable technologies and energy efficiency strategies has become increasingly pivotal for reaching the ambitious target of an 80–95% reduction in greenhouse gas emissions by 2050 [1]. In order to find a sustainable balance between the production and consumption of energy, and reach those targets, several challenges still need to be addressed. These problems are strongly related to the increasing demand for energy [2] due to the development of both industrialized and emerging countries [3]. As such, several countries have agreed on the necessity of developing and implementing a novel sustainable energy paradigm in the building, transport, and industrial sectors. For this reason, novel pillars of development for the future and a sustainable energy system have to be varied and flexible, such as the waste to energy paradigm [4], a reduction in energy consumption, the circular economy [5], improved quality of life for people [6], pollution and climate change adaption [7], carbon neutrality with the overall goal to reach zero emissions by 2050 [8], etc. To increase the penetration of renewable energy into society, significant care must be taken with issues related to the management of the excess production of renewable electric energy [9], which leads to several severe issues in terms of grid management [10]. The increasing development of power-to-x technologies [11,12] will be crucial to mitigate the phase shifts between renewable energy production and user demand [13]. In this framework, the reduction in energy waste through feasible waste management approaches is also crucial [14]. Not only this provides a renewable source of energy from low/negative-value wastes in the form of heat, electricity, or fuels, but it also minimizes environmental pollution due to a reduction in the volume of waste produced by leading to an increase in the reuse and recycling of waste materials [15]. The use of electric vehicles powered by renewables will significantly influence the reduction in GHG emissions from the transport sector [16].

To develop and support this new approach to energy use and production, scientists and companies have focused their attention on the use of renewable energy technologies, improved functional energy use, energy efficiency, novel efficient energy conversion systems, etc. In this context, this Special Issue "Renewable Energy Systems 2023" aims at summarizing the most up-to-date advancements and central studies dealing with the integration of renewable technologies into new or existing systems for the production of energy and non-energy outputs. In this editorial, the presented papers within the Issue are summarized, spanning the potential use of renewable energy for power production, heating, cooling, water management, and transport. The contributions of the main works in this Special Issue within the above mentioned framework are presented and discussed, using the method outlined in the following section.

2. The Research Topics Represented in This Special Issue

A total of 10 papers were selected for this Special Issue. The main ideas of these papers are briefly reviewed in the following subparagraphs in order to summarize all the presented research. As such, this editorial has been organized into the following sections:

Citation: Cimmino, L.; Vicidomini, M. Renewable Energy Systems 2023. *Appl. Sci.* **2024**, *14*, 1918. https://doi.org/10.3390/app14051918

Received: 8 February 2024
Accepted: 13 February 2024
Published: 26 February 2024

Section 2.1, power plants based on solar systems; Section 2.2, power plants based on wind systems; Section 2.3, power plants based on biomass sources; and Section 2.4, other topics.

2.1. Power Plants Based on Solar Systems

Calise et al. [17] developed a dynamic simulation study on the operation of a greenhouse coupled with renewable energy technologies. The paper is focused on the design of the system components in order to dynamically investigate the operation of this hybrid renewable energy plant. The plant is considered an optimal solution within the green farm framework. In particular, a 20 kW photovoltaic (PV) field is considered adequate to supply the electric energy demand of the farm, in addition to 28 m^2 solar thermal collectors which are designed to meet the thermal energy demand of the greenhouse when its operating temperature falls lower than a specific set point. A biomass auxiliary heater is also included. The plant is designed to produce electricity for buildings close to the greenhouse, their related irrigation pumps, and thermal energy both for the greenhouse's heating and the domestic hot water and space heating energy demands of users in the nearby buildings. As part of a bio-circular economic approach, the plant also includes a pyrogasifier fueled by wood and agricultural waste. This model is applied to a specific case study; a green farm located in Naples (southern Italy). The operation of the whole plant, the energy components, and the greenhouse are dynamically simulated in TRNSYS 18 software. The greenhouse's thermal behavior is also validated using the values in the literature. Subsequently, a detailed energy, economic, and environmental analysis of the whole plant is performed. The proposed plant is able to reduce its total equivalent CO_2 emissions by 148.66 t/y. In addition, it obtains good profitability, with a simple payback of only 1.7 years. Božiková et al. [18] presented a simplified mathematical model for the evaluation of the energy production of a PV field, as a function of its tilt angle and azimuth angle. The authors state that the developed model is useful for ordinary users of PV systems, considering the model's very small number of input parameters. The key objective of this study is the identification of the optimum locations of PV fields, considering the local weather in southern Slovakia. In their work, an experimental system based on two polycrystalline PV modules is also used. In particular, the first part of this experimental system identifies the tilt angle variation, from 0 to 90. The second module aims to calculate energy production as a function of the azimuth angle. Using the carried-out measurements, regression equations highlighting the relationships between the tilt or azimuth angle and the energy produced by the PV system are generated. Although the presented models are obtained for an area of southern Slovakia, these models, described by a simple second-degree polynomial function, can also be considered applicable to the further areas in Central Europe, if the regression equation coefficients are suitably modified. The presented mathematical model is about 4% more precise than other models used for this weather zone in southern Slovakia. Combining their analysis of the results achieved through their experimental and numerical simulation approaches, the authors found that the effect of the tilt angle and azimuth angle of a PV module on its electricity production is about 18% and 24%, respectively. This mathematical model allows for a simple forecast of PV production in real operating conditions in southern Slovakia for design, dimensioning, and optimization purposes. Pintanel et al. [19] investigated the integration of thermoelectric generators (TEGs) into these photovoltaic and thermal collectors (PVT) in order to increase their power production. They performed an experimental and numerical comparison between a PVT panel with and without TEGs. The authors state that the novelty of the presented work is due to both the numerical and experimental analysis they performed on this kind of system. The aH60 model manufactured by Abora Solar was used to perform the analysis. This includes an isolation layer at the back and an isolation camera monitoring the argon in the front layer of the panel. In the layout with the TEGs, the PVT panel included 19 integrated TEG modules in the bench test. The experimental setup also included an isolated 197 L water tank; an impulsion pump; temperature, irradiation, and mass flow probes; an expansion vessel and an aerothermal vessel; a 500 W microinverter; and the electrical panel of the bench. The 3D numerical model of the system was validated with their experimental results, demonstrating the good agreement between the simulated and experimental results. However, the errors in the results were smaller than 10%; therefore, the authors stated that the integration of thermoelectric materials within the PVT panels should be further researched to improve the contact region between both sides of the TEG devices. Their numerical analysis results indicate that the global efficiency of a conventional PVT panel is 81.9%, and the global efficiency of a PVT+TEG panel is 82.2%

2.2. Power Plants Based on Wind Systems

The efficient utilization of renewable energy sources and, in particular, of wind energy, is one of the key strategies to reach the objectives of the 2030 Agenda and the Paris Agreement and reduce CO_2 emissions. For this reason, it is necessary to focus our research efforts on the developments in wind energy projects. It is in this framework that the work of the Murgas et al. [20] was developed. In particular, this work is a literature review which evaluates the models and approaches used in the evaluation of investments in wind systems under uncertainty. In total, 97 studies were evaluated, with 20 different approaches or models detected and grouped into eight different categories, namely, (1) real options, (2) optimization, (3) stochastics, (4) financial evaluation, (5) probabilistic, (6) estimation, (7) numerical prediction, and (8) others. The real options approach, in which the events featured in the possible outcomes are not known, and, therefore, their probability of occurrence cannot be quantifiable, is typical for wind energy systems. In particular, they found that this approach is used in 32% of studies (31 publications). Of the studies adopting the real options approach, 62.5% of them considered the price of electricity to be a source of uncertainty, as did 15.6% with the feed-in rates-subsidy, and 18.8% with the velocity of the wind. The authors stated that future studies should focus on the evaluation of investments in wind energy projects, investigating the real options approach and public opposition to the projects (including NIMBY projects: not in my backyard). A methodology to represent the synthetic inertia of wind turbines as part of an optimization dispatch model is presented in the work of Thiesen and Jauch [21]. They investigated synthetic inertia because it is a crucial element to guaranteeing the stability of power systems, as it limits the speed with which the grid frequency changes. Frequency-converter-connected generation units such as wind turbines do not have an inherent inertial response. WTs and PVs decrease the synchronous inertia in energy systems. The trend of decreasing system inertia is not reflected adequately in current unit commitment and economic dispatch models. However, variable-speed wind turbines, already in use, are a source of synthetic inertia. For this reason, a unit commitment and economic inertia dispatch model of the all-Island Irish power system is used as case study in this work. This case study is selected because the current all-Island Irish power system is already characterized by high amounts of frequency-inverter-connected penetration. The potential of wind inertia is analyzed and calculated, revealing that the synthetic inertia provided by wind turbines is able to save up to 30.99% of CO_2 emissions, reduce curtailment by up to 39.90%, and reduce system costs by 32.72%.

2.3. Power Plants Based on Biomass Sources

The production of biogas from anaerobic digestion processes that start in wastes is a relevant topic in the current energy framework, and it fulfills the goals of the green circular economy [22]. In the work of Obileke et al. [23], a wide-ranging literature review is presented, focusing on the most recent advancements in anaerobic digestion and gasification technology for the sustainable waste-to-energy paradigm. The methodology followed is a conceptual review, which contributes to a better understanding of the topics discussed. The technologies are investigated and analyzed both separately and as a whole hybrid system. The literature review included the most recent studies dealing with the latest developments of these technologies, the intensification of these processes, the supply of several feedstocks, and the quality of the final biogas or syngas obtained. According to the analysis provided, several factors affect the final yield of biogas or syngas. The most influencing factors on anaerobic digestion are the operating temperature and the pH of the biomass. For the gasification process, these instead are the moisture content of the biomass and the gasifying agent used. These parameters also affect the quality of the gas produced, in terms of its lower heating value (LHV). For instance, in the case of gasification with air as the gasifying agent, the LHV is roughly 4–6 MJ/Nm^3, whereas in case of oxygen it is nearly 17–18 MJ/Nm^3, i.e., roughly four times higher. Of course, the selection of the gasifying agent is also dictated by its cost, therefore an optimal solution must be found. However, it was found that both these technologies are effective for energy recovery, with the need for enhancers and intensifiers to improve the efficiency of conversion. The relevance of this work in the current energy framework lies in its potential for global implementation, offering a sustainable approach to waste-to-energy generation.

2.4. Other Topics

Other relevant works in this Issue propose innovative solutions for the energy inefficiency in other energy sectors. In Rodriguez-Estrada et al. [24], the development of a novel microgrid architecture, the modeling of power converters, and the design of a hierarchical control system, including an innovative energy management system (EMS), is proposed. The work presents a novel multibus multivoltage architecture for the application of direct current microgrids to buildings. The aim is to increase the efficiency and the flexibility of power distribution systems by means of modeling and designing these systems using Petri net theory. The results validated each version's performance in a closed loop, with a feedback state controller as the local control stage for every converter in stand-alone operation. The results obtained can set the stage for the further development of finalized microgrids that improve aspects of the environment and the economy. In Sergienko et al. [25], the authors developed a mathematical model for an innovative clutch control mechanism that featured higher energy efficiency and higher performance in terms of the transmission system. Moreover, in this work, it was numerically proven that an equal loading for the frictional discs of each clutch is not guaranteed when there is a radial movement of the carriage when switching on the clutches. This contributes to the increasingly uneven dynamics of torque and disc wear. The results of the synthesis showed that the carriage movement partially depends on the force acting on the pressure disc. In particular, the minimum carriage movement is defined by the wear resistance of the mechanism, determining an inclination angle of the surface profile of the platter between 20 and 29°. This dual-clutch transmission (DCT) system is mainly proposed for the automotive sector, but the results obtained in this work can be extended to several energy system sectors. Qureshi et al. [26] developed a comparative analysis of numerical models in the field of decision-making processes, aiming to highlight the main limitations of several multi-criteria decision methods (MCDM) in terms of the reliability of their decisions. More specifically, these tools are applied to the selection of urban heat mitigation measurements with specific criteria. This numerical investigation allowed for the evaluation of the reliability of diverse MCDMs that adopt the same decision matrix and different normalization techniques, showing the impact of the analytic hierarchy process (AHP) on the decisions made. Of the techniques analyzed, the weight sum method and preference ranking organization method for enrichment evaluation were revealed to be the most reliable. Kadochnikov et al. [27] provided discussion on a non-standard experimental setup developed in the laboratory of the St. Petersburg Mining University's Well Drilling Department. The technique presented allows one to simulate the well cleaning process of making cuttings by evaluating the variations of the parameters that cause buckling, such as the zenith angle of the well or the rotation frequency. Statistica 13 and MS Office Excel software were used for mathematical analysis. The key finding of this work is that a controlled buckling of the drill string significantly enhances mud plug breakdown and well cleaning, especially at critical zenith angles such as 55 degrees. The relevance of this work lies in its assessment of the enhanced efficiency and safety in petroleum engineering, especially in directional drilling operations.

Funding: This research received no external funding.

Conflicts of Interest: The authors declare no conflicts of interest.

References

1. Seck, G.S.; Hache, E.; Sabathier, J.; Guedes, F.; Reigstad, G.A.; Straus, J.; Wolfgang, O.; Ouassou, J.A.; Askeland, M.; Hjorth, I.; et al. Hydrogen and the decarbonization of the energy system in europe in 2050: A detailed model-based analysis. *Renew. Sustain. Energy Rev.* **2022**, *167*, 112779. [CrossRef]
2. Xing, H.; Husain, S.; Simionescu, M.; Ghosh, S.; Zhao, X. Role of green innovation technologies and urbanization growth for energy demand: Contextual evidence from G7 countries. *Gondwana Res.* **2024**, *129*, 220–238. [CrossRef]
3. Calise, F.; Cappiello, F.L.; Cimmino, L.; D'accadia, M.D.; Vicidomini, M. Dynamic analysis and investigation of the thermal transient effects in a CSTR reactor producing biogas. *Energy* **2023**, *263*, 126010. [CrossRef]
4. Rusănescu, C.O.; Ciobanu, M.; Rusănescu, M.; Dinculoiu, R.L. Pretreatments Applied to Wheat Straw to Obtain Bioethanol. *Appl. Sci.* **2024**, *14*, 1612. [CrossRef]
5. Medaglia, R.; Rukanova, B.; Zhang, Z. Digital government and the circular economy transition: An analytical framework and a research agenda. *Gov. Inf. Q.* **2024**, *41*, 101904. [CrossRef]
6. Maury-Micolier, A.; Huang, L.; Taillandier, F.; Sonnemann, G.; Jolliet, O. A life cycle approach to indoor air quality in designing sustainable buildings: Human health impacts of three inner and outer insulations. *Build. Environ.* **2023**, *230*, 109994. [CrossRef]
7. Ashrafuzzaman, M.; Furini, G.L. Climate change and human health linkages in the context of globalization: An overview from global to southwestern coastal region of Bangladesh. *Environ. Int.* **2019**, *127*, 402–411. [CrossRef] [PubMed]
8. Tostado-Véliz, M.; Arévalo, P.; Kamel, S.; El-Sehiemy, R.A.; Senjyu, T. Renewable-Based Microgrids: Design, Control and Optimization. *Appl. Sci.* **2023**, *13*, 8235. [CrossRef]
9. Liu, G.; Wang, Z.; Liu, X.; Kupecki, J.; Zhao, D.; Jin, B.; Wang, Z.; Li, X. Transient analysis and safety-oriented process optimization during electrolysis–fuel cell transition of a novel reversible solid oxide cell system. *J. Clean. Prod.* **2023**, *425*, 139000. [CrossRef]
10. Calise, F.; Cappiello, F.; Cimmino, L.; Vicidomini, M. Dynamic simulation modelling of reversible solid oxide fuel cells for energy storage purpose. *Energy* **2022**, *260*, 124893. [CrossRef]
11. Calise, F.; Cappiello, F.L.; Cimmino, L.; D'accadia, M.D.; Vicidomini, M. Dynamic simulation and thermoeconomic analysis of a power to gas system. *Renew. Sustain. Energy Rev.* **2023**, *187*, 113759. [CrossRef]
12. Sorrenti, I.; Rasmussen, T.B.H.; You, S.; Wu, Q. The role of power-to-X in hybrid renewable energy systems: A comprehensive review. *Renew. Sustain. Energy Rev.* **2022**, *165*, 112380. [CrossRef]
13. Bellocchi, S.; De Falco, M.; Gambini, M.; Manno, M.; Stilo, T.; Vellini, M. Opportunities for power-to-Gas and Power-to-liquid in CO_2-reduced energy scenarios: The Italian case. *Energy* **2019**, *175*, 847–861. [CrossRef]
14. Mujtaba, M.A.; Munir, A.; Imran, S.; Nasir, M.K.; Muhayyuddin, M.G.; Javed, A.; Mehmood, A.; Habila, M.A.; Fayaz, H.; Qazi, A. Evaluating sustainable municipal solid waste management scenarios: A multicriteria decision making approach. *Heliyon* **2024**, *10*, e25788. [CrossRef] [PubMed]
15. Al Shouny, A.; Issa, U.H.; Miky, Y.; Sharaky, I.A. Evaluating and selecting the best sustainable concrete mixes based on recycled waste materials. *Case Stud. Constr. Mater.* **2023**, *19*, e02382. [CrossRef]
16. Van, T.L.C.; Barthelmes, L.; Gnann, T.; Speth, D.; Kagerbauer, M. Enhancing electric vehicle market diffusion modeling: A German case study on environmental policy integration. *Energy Strat. Rev.* **2023**, *50*, 101244. [CrossRef]
17. Calise, F.; Cappiello, F.L.; Cimmino, L.; Vicidomini, M. Dynamic Modelling and Energy, Economic, and Environmental Analysis of a Greenhouse Supplied by Renewable Sources. *Appl. Sci.* **2023**, *13*, 6584. [CrossRef]
18. Božiková, M.; Bilčík, M.; Madola, V.; Szabóová, T.; Kubík, Ľ.; Lendelová, J.; Cviklovič, V. The Effect of Azimuth and Tilt Angle Changes on the Energy Balance of Photovoltaic System Installed in the Southern Slovakia Region. *Appl. Sci.* **2021**, *11*, 8998. [CrossRef]
19. Pintanel, M.T.; Martínez-Gracia, A.; Galindo, M.P.; Bayod-Rújula, A.; Uche, J.; Tejero, J.A.; del Amo, A. Analysis of the Experimental Integration of Thermoelectric Generators in Photovoltaic–Thermal Hybrid Panels. *Appl. Sci.* **2021**, *11*, 2915. [CrossRef]
20. Murgas, B.; Henao, A.; Guzman, L. Evaluation of Investments in Wind Energy Projects, under Uncertainty. State of the Art Review. *Appl. Sci.* **2021**, *11*, 10213. [CrossRef]
21. Thiesen, H.; Jauch, C. Potential of Onshore Wind Turbine Inertia in Decarbonising the Future Irish Energy System. *Appl. Sci.* **2022**, *12*, 2984. [CrossRef]
22. Mellyanawaty, M.; Marbelia, L.; Sarto Prijambada, I.D.; Rochman, Y.A.Y.; Budhijanto, W. Application of Anaerobic Digestion Model No. 1 on thermophilic anaerobic digestion with microbial immobilization media for biogas production from sugarcane vinasse. *J. Environ. Chem. Eng.* **2024**, 112209. [CrossRef]
23. Obileke, K.; Makaka, G.; Nwokolo, N. Recent Advancements in Anaerobic Digestion and Gasification Technology. *Appl. Sci.* **2023**, *13*, 5597. [CrossRef]
24. Rodriguez-Estrada, H.; Rodriguez-Segura, E.; Orosco-Guerrero, R.; Gordillo-Tapia, C.; Martínez-Nolasco, J. Novel Multibus Multivoltage Concept for DC-Microgrids in Buildings: Modeling, Design and Local Control. *Appl. Sci.* **2023**, *13*, 2405. [CrossRef]

25. Sergienko, N.; Kalinin, P.; Pavlenko, I.; Ochowiak, M.; Ivanov, V.; Sergienko, A.; Pavlova, N.; Basova, Y.; Titarenko, O.; Nazarov, A.; et al. Synthesis of the Energy-Saving Dry Dual Clutch Control Mechanism. *Appl. Sci.* **2023**, *13*, 829. [CrossRef]
26. Qureshi, A.M.; Rachid, A. Comparative Analysis of Multi-Criteria Decision-Making Techniques for Outdoor Heat Stress Mitigation. *Appl. Sci.* **2022**, *12*, 12308. [CrossRef]
27. Kadochnikov, V.G.; Dvoynikov, M.V. Development of Technology for Hydromechanical Breakdown of Mud Plugs and Improvement of Well Cleaning by Controlled Buckling of the Drill String. *Appl. Sci.* **2022**, *12*, 6460. [CrossRef]

applied sciences

MDPI

Article

Analysis of the Experimental Integration of Thermoelectric Generators in Photovoltaic–Thermal Hybrid Panels

Mª Teresa Pintanel [1,*], Amaya Martínez-Gracia [1], Mª Pilar Galindo [1], Ángel A. Bayod-Rújula [2], Javier Uche [1], Juan A. Tejero [2] and Alejandro del Amo [3]

1 Mechanical Engineering Department—Research Institute CIRCE, University of Zaragoza, 50018 Zaragoza, Spain; amayamg@unizar.es (A.M.-G.); pilar.galindo.pg@gmail.com (M.P.G.); javiuche@unizar.es (J.U.)
2 Electrical Engineering Department—Research Institute CIRCE, University of Zaragoza, 50018 Zaragoza, Spain; aabayod@unizar.es (Á.A.B.-R.); jatejero@unizar.es (J.A.T.)
3 Abora Solar SL, 50196 Zaragoza, Spain; adelamo@abora-solar.com
* Correspondence: mtpintanel@unizar.es; Tel.: +34-976-762-145

Abstract: Photovoltaic–thermal panels (PVT) have been widely studied in the last years and have proved to be a technically viable and profitable solution. This work analyses the integration of a set of thermoelectric generators (TEG) inside these panels in order to obtain additional power. The thermoelectric material takes advantage of the temperature gap between the hottest part of the system, the output flow from the collector, and the cold water feeding the solar system. An experimental test bench with a PVT having integrated TEGs and the same PVT in parallel without TEGs was mounted to compare both devices. The corresponding CFD simulation was also carried out to better understand the temperature map in the arrangement. Both experimental and computational results show that the manufacture of the panel with integrated TEGs should be carefully studied before becoming a commercial product. They also gave some guidelines for the improvement of the prototype in this integrated product.

Keywords: photovoltaic–thermal hybrid panels; thermoelectric generators; ansys-fluent; experimental results; numerical simulation

Citation: Pintanel, M.T.; Martínez-Gracia, A.; Galindo, M.P.; Bayod-Rújula, Á.A.; Uche, J.; Tejero, J.A.; del Amo, A. Analysis of the Experimental Integration of Thermoelectric Generators in Photovoltaic–Thermal Hybrid Panels. *Appl. Sci.* **2021**, *11*, 2915. https://doi.org/10.3390/app11072915

Academic Editor: Maria Vicidomini

Received: 23 February 2021
Accepted: 18 March 2021
Published: 24 March 2021

1. Introduction

The search for the diversification of generation sources and the increase in the percentage of participation of renewable energies, together with the saving and rational use of energy, are the axis for sustainable energy planning in a decarbonized future. The climatic change affects all countries of the entire world nowadays. Levels of carbon dioxide (CO_2) and other greenhouse effect gases hit a record in 2020 [1]. Consequently, air and water temperature are increasing, the amount of ice and snow is decreasing, and the ocean level is growing up. According to experts, the climatic change effect in the environment could have fatal consequences. The Paris Agreement [2] is the first global climate change accord that looks to reduce global emissions by changing energy consumption and production policies. The employment of renewable source energy and policies that promote energy efficiency will help achieve the accorded objectives in this agreement.

Total energy consumption in the residential sector supposes 25% of the total consumption in Europe [3]. Furthermore, the residential and tertiary sectors consume 60% of the electricity employed in the EU. The use of solar energy as a source of renewable energy in these sectors is a great alternative. Solar hybrid photovoltaic and thermal panels (PVT) can provide electricity and thermal energy by solar irradiation in a single device. Massive implementation of this technology would contribute to the reduction of energy consumption in buildings. PVT panels can be installed in almost any location on the roof of the construction. The generated electricity can be self-consumed, stored in batteries, or sold to the grid. Thermal production of hybrid panels is in the form of hot water that can be employed for different purposes such as domestic hot water (DHW) production, space heating, and refrigeration by adsorption or employing a heat pump.

A way of increasing the electricity production of PVT panels is the inclusion of thermoelectric generators (TEG) inside it. These devices, through Seebeck's effect, produce electricity from a gradient of temperature. The composition of TEGs is an essential factor in achieving a maximum conversion of thermal energy into electricity [4]. It is possible to find three types of thermoelectric generators in the market: high, medium, and low-temperature ranges [5]. Low-temperature range devices have a range work temperature between 25–125 °C, with a maximum work temperature of 250 °C. However, the efficiency decreases drastically when the generator works above 175 °C. Most commercial low-temperature range TEGs are composed of bismuth tellurium (Bi_2Te_3). The medium temperature range module works between 125–425 °C with a high limit temperature of 600 °C. The inconvenience of this type of TEG is its composition. Medium-range thermoelectric generators are made by tellurium (Te), a limited resource on Earth, and lead (Pb), a toxic material. TEGs that can work at high temperatures are composed of silicon (Si) as the main component. These devices provide high-efficiency parameters at temperatures between 425 °C and 725 °C. Depending on the application, a different type of TEG between these is selected.

TEG technology has some advantages that makes it especially suitable in domestic poligeneration schemes [6,7]. TEG devices are safe because they do not contain mobile parts, and no fluids are involved in their operation. Additionally, TEGs are simple, compact, light, silent, and not much maintenance is required. As for disadvantages, the elevated cost for high-temperature range TEGs and the relatively low efficiency of low-temperature range TEGs if thermal integration is not correct [8].

Integration of TEG modules in different applications has been intensively studied. Different review papers where the different applications can be consulted [5,9]. The automobile industry is focused on finding alternative sources of energy. Different authors [10–12] studied the integration of TEGs in the gas exhaust pipe of vehicles in order to recover thermal energy rejected to the ambient. Additionally, other authors like Cheng et al. [13] investigated the location of TEGs in this way that heat from the combustion cycle is taken advantage Ziolkowsli et al. [14] suggested the employment of TEG modules between the hot core stream and the cold bypass flow at the nuzzle of an aviation turbofan engine to reduce the operation cost of the aircraft. TEGs are also widely used in aerospace applications such as space crafts, satellites, and space probes [15,16]. Some researchers are working on TEG integration for wearable applications. Bahk et al. [17] studied the material optimization of this technology to make it flexible and easy to wear in the human body. More recently, Lee et al. [18] simulated different key parameters of TEGs modules to integrate this technology in human body applications.

In domestic applications, the integration of TEGs has also been considered by experts [9]. Nesarajah and Frey [19] optimized the design of TEGs' distribution for waste heat recovery from exhaust pipes of a condensation boiler. A theoretical simulation of exhaust heat recovery in biomass boilers by integrating TEGs was studied by Alanne et al. [20]. Moreover, Höfberger [21] and Moser et al. [22] implemented experimental prototypes with which it is possible to produce 160 W and 300 W of electrical power respectively in biomass pellets boilers of 11–12 kW of thermal capacity. There are other studies with biomass stoves, such as Sornek's paper [23].

In the solar energy fieldwork, TEG integration has been researched. A critical review of the different configurations for thermoelectric generator applications of solar thermal systems was written by Karthick et al. [24]. Date et al. [25] developed a theoretical and experimental study of electricity generation by TEGs using concentrated solar thermal energy. They concluded that having fewer TEGs, a higher concentration ratio, and a higher temperature difference will produce more electricity than having more TEGs and less temperature gradient. A study of the different organic working fluids on solar flat plate collector and TEG modules was presented [26] in order to determine the best one. A scheme of poligeneration based on solar thermal collectors with thermoelectric generators, a sorption module, a reversible heat pump, and a biomass boiler is currently being studied in different locations by Palomba et al. [27].

In respect of the integration of TEG devices into photovoltaics-thermal hybrid panels, there are models and analyses [28,29]. Still, experimental research has been focused on prototypes that do not constitute a real PVT panel [30]. Kilkis [31] has developed a small prototype of a novel PVT panel composed of photovoltaic cells, a flat plate solar collector, TEG modules, and phase change material (PCM) embodiment. He concluded with the experimental results that this combination of layers allows electrical production even after sunset, depending on the total PCM mass. First experimental results show that this new PVT has 14% more total exergy output (including thermal energy and electricity) than a conventional PVT panel.

Investigation in phase change materials is a high-relevance line of research nowadays. The principle of this material is the storage of thermal energy by the phase change of some specific materials. PCM composites based on the coupling of metal foam (i.e., copper) and paraffin are auspicious material in the field of thermal management [32–35]. Zheng et al. [36] studied and compared the thermal performance of copper foam/paraffine composite material with pure paraffin. They concluded that the mentioned PCM showed a reduction in total melting time of 20.5% in comparison with pure paraffin. This characteristic allows the homogenous distribution of the temperature in the material. The application of PCMs has been presented by Kazemian et al. [37]. In this study, experimental research using both ethylene glycol and phase change material (organic paraffine) as work fluid in PVT panels is analyzed. They concluded that using just PCM for cooling reduces the surface temperature of the PV cells inside the PVT panel by 11.05% compared to a conventional PV unit. For this reason, electrical performance is 4.22% higher than that of a conventional PVT panel. Additionally, Zhou et al. [38] proposed the storage of a moderate amount of thermal energy inside PVT panels by including a PCM to reduce the possibility of freezing the working fluid under extreme ambient conditions.

Salari et al. [39] have simulated with a numerical model a hybrid PVT panel with TEGs. They have concluded that up to 10% of increment in the efficiency of the panel can be obtained in comparison with a standard PVT panel. Nevertheless, their paper does not include experimental data.

In this study, the integration of TEGs modules into a real PVT hybrid panel has been carried out. The installation has been monitored to get experimental results. A 3D numerical model of conventional photovoltaics-thermal panel and PVT with TEG modules was also developed. The comparison between simulated and experimental results has been then made in order to validate the model simulation. The first tests show a concordance between the simulated and experimental results. Thus, they indicate the path to follow up to the definite design that allows an integrated PVT-TEG device with an additional performance that conventional PVT. The innovation presented in this paper is the unique combination of the experimental and computational results in a PVT panel with integrated thermoelectric generators inside it. After the validation with experimental results, detailed information from the numerical results obtained in the three-dimensional simulation allows a better understanding of the found inefficiencies. Thus, it gives light to provide new improvements to this integrated device.

2. Materials and Methods

In this study, an experimental test bench was constructed with the objective of testing and further validating the numerical results of the integrated PVT-TEG panel. The test bench has been designed so that two PVT panels can operate simultaneously. One of the panels is a conventional photovoltaics-thermal hybrid panel of the model aH60 manufactured by Abora Solar [40]. This model of PVT panel includes an isolation layer at the back and an isolation camera of argon in the frontal layer of the panel. Another PVT panel (of the same model) in the bench includes 19 integrated TEG modules. Test bench includes other devices such as an isolated water tank of 197 L; an impulsion pump; temperature, irradiation, and mass flow probes; an expansion vessel and an aerothermal; a 500 W microinverter and the electrical panel of the bench. Additionally, a monitoring system has been installed on the test bench in order to report in real-time production data of both panels. The different magnitudes that are being controlled are shown in Table 1. Thanks to this configuration of the bench, it is possible to compare the energy production of both panels at the same operating conditions.

Table 1. Monitored magnitudes of PVT test bench.

Magnitude	Name of Probe	Description	Scale	Readability	B (%)	P (%)	U (%)
Temperature (°C)	T_{in} T_{out1} T_{out2} T_{amb}	Temperature of the flow inlet of the PVT panels Temperature of the flow that leaves the PVT with TEGs Temperature of the flow that leaves the conventional PVT panel Ambient temperature	−50–250	0.1	0.15	0.033	0.15
Mass Flow (kg/s)	m_{in}	Mass flow incoming the system. If both panels are working, the flowmeter shows the total mass flow entering the system.	1–6	0.5	5	8.33	9.72
Solar Irradiation (W/m^2)	Irr	Incident solar irradiation	0–2000	0.01	2.60	0.005	2.60
Pressure	P	Pressure in the hydraulic installation					
Voltage (V)	V1 V2	Continuous Voltage of PVT with TEGs Continuous Voltage of conventional PVT	0–5	0.00489	0.5	0.0978	0.5095
Current (A)	DC1 DC2	Direct current of PVT with TEGs Direct current of conventional PVT	0–20	0.01955	0.04	0.10	0.11

Also in Table 1, uncertainly parameters of the measured variables according to the installed probes on the test bench are listed. *B* represents the systematic uncertainly of the instrument. This parameter is related to the accuracy of the probe. *P* is the precision and represents the random uncertainty. It has relation with the repeatability of the measured. *U* shows the uncertainty and it is calculated as the sum of squares of *B* and *P*.

Uncertainly analysis of the calculi of performance of panels was additionally implemented [41,42]. Equations (3)–(7) are presented in the following lines. These equations govern the behavior of power and performance of the panels. Uncertainly analysis has been obtained according to these equations and parameters showed in Table 1. Uncertainlity of electrical performance is 0.52% and 1.12% for PVT conventional panel and PVT + TEG panel, respectively.

Because of the working temperatures of the PVT panel, between 40 °C and 100 °C, a low range temperature TEG model has been selected for the inclusion of them inside the hybrid panel. The model TEG2-07025HT-SS manufactured by TEC-Solidstate Power Generation [43] is the elected thermoelectric device. This TEG allows a generous capacity of heat transfer in a small configuration. The module withstands temperatures up to 190 °C on the hot side. The P and N elements incorporated in this module offer excellent heat transfer capability in a 40 × 40 mm configuration. According to the manufacturers, with the usual temperatures in PVTs, some 2W could be expected as a maximum, with a voltage match of 1.0 V and a current match of 2 A (internal resistance = 0.5 Ohm)

The integration of TEG modules inside the PVT panel must be made that the TEG sides are set between two sources of different temperatures. In this way, electricity can be produced. The two selected heat sources are mass inlet flow as the cold side and outlet mass flow as the hot side of the thermoelectrical modules. The hot side of the TEG must have excellent contact with the heat sources, the outlet collector pipe of the PVT panel. This pipe is inside the panel between the absorber and isolation layer. It is the pipe where hot water production is collected before leaving the panel and entering into the water tank. The cold side of the TEG module must be in thermal contact and compressed against a heat exchanger in order to dissipate heat from this surface. With the increase of the temperature gradient between the TEG sides, the electric production will be more significant.

In a conventional PVT panel, inlet and outlet pipes are distanced. To a successful integration of the thermoelectrical modules in the hybrid panel, a modification in the inlet pipe has been made. The inlet pipe has been derivated, and before the water enters the panels, water flows inside a pipe set at the back of the panel. This part is where TEGs are integrated. Copper laminas are employed between TEGs and pipes to maximize the heat exchanger and the thermal contact. Furthermore, clamping plates were installed for a consistent subjection between TEGs modules and copper fins. The 19 TEG modules have been connected electrically in series. The positive pole is located on the side of the hybrid panel, and the negative pole is set on the opposite side.

Solar energy (E_s) is the energy input in the studied PVT with integrated TEG modules system, and it is stated according to

$$E_s = I * A_C \tag{1}$$

where I is solar irradiance in W/m^2 and A_C is the gross area of working PVT collectors measured in m^2.

Furthermore, the useful thermal energy production of PVT collectors (Q_{PVT}) is the result of the product between the solar energy that comes from the sun and the thermal performance of the PVT (η_{th_PVT}) that depends on the model of the panel.

$$Q_{PVT} = \eta_{th\_PVT} * Q_s \tag{2}$$

$$\eta_{th\_PVT} = \eta_0 - a_1 * G - a_2 * G * (T_m - T_{amb}) \tag{3}$$

where η_{th_PVT} [44] represents the solar thermal performance of the PVT panel, η_0 is the optical efficiency of the solar hybrid collector, a_1 is the heat loss coefficient in W/m^2K and a_2 is the temperature dependence of thermal heat loss coefficient in W/m^2K^2. These parameters are provided by the manufacturer of PVTs. Temperatures are given in °C. T_m is the average temperature of the PVT panels, and it is defined by the average between the inlet (T_{in}) and outlet (T_{out}) temperatures of the working fluid. T_{amb} is ambient temperature, and G can be calculated as

$$G = (T_m - T_{amb})/I \tag{4}$$

where I represents global solar irradiance.

The electrical production of a conventional PVT is calculated as

$$W_{PVT} = \frac{P_n}{1000} * [1 - \gamma * (T_{cell} - 25)] \tag{5}$$

where P_n is peak power of the module in W , γ is the temperature coefficient losses given by the manufacturer and T_{cell} is the cell temperature measured in °C. In photovoltaic-thermal panels, T_{cell} can be assumed equal as T_m.

In this paper, the integration of TEG modules on a PVT panel is studied. Therefore, the electrical efficiency of TEGs must be considered as well.

$$\eta_{el\_TEG} = \frac{T_h - T_c}{Th} * \frac{\sqrt{1 + Z * \overline{T}}}{+\frac{T_c}{T_h}} \tag{6}$$

where T_h is the temperature of the hot side of the TEG, T_c Is the temperature of the cold side of the *TEG*, $\overline{T}$ is the average temperature of the TEG and Z is the figure of merit of the module. In Equation (6) temperatures are given in K. According to the manufacturer, the product between the figure of merit and the average temperature under nominal conditions has a value of 0.73.

As well as thermal performance, the electrical performance of a PVT panel (η_{el_PVT}) is related electrical production with the incoming energy from the sun

$$\eta_{el\_PVT} = \frac{W_{PVT}}{E_s} \tag{7}$$

and performance of TEG modules (η_{el_TEG}) is related with the heat flux that go throght the thermoelectrical generators (E_{TEG})

$$\eta_{el\_TEG} = \frac{W_{TEG}}{E_{TEG}} \tag{8}$$

In such a way, the electrical performance of PVT + TEG panel is calculated as the addition between the electrical performance of a conventional PVT panel and the electrical performance of TEGs modules considering gradient temperature in the hybrid panel.

$$\eta_{el_{PVT+TEG}} = \eta_{el_{PVT}} + \eta_{elTEG} \tag{9}$$

In this paper, a three-dimensional numerical simulation of the PVT panel and the PVT panel with TEGs has been performed in the software ANSYS [45]. The 3D analysis is selected due to the serial configuration of the TEGs integrated along the tube. This application works under the finite element method for structural analysis and the finite volume method in the case of fluid studies. This software includes different work modules that allow the study of structural, thermal, electrical, and fluid dynamics aspects. In this case study, module Fluent is employed, which is a computation fluid-dynamics (CFD) simulator.

Definition of the three-D model of the PVT and PVT + TEG panels is the first step that must be done. For that purpose, technical specifications of the different layers that compose the PVT panel has been provided by the fabricant and are shown in Table 2. Due to the high complexity of the study, some assumptions have been set in order to simplify the PVT model:

- The thermal behavior of the small pipes of the water collector is assumed identical inside all tubes. For this reason, only one of the 10 pipes is represented;
- Both models of the panel (PVT and PVT + TEG) have been simulated under steady state conditions;
- Water mass flow inside the simulated pipe is considered to be incompressible and laminar;
- The thermophysical properties of the different materials remain constant during the study period.

Table 2. Material properties of the TEG modules and the different layers that composed PVT panels.

Material	Density (kg/m^3)	Specific Heat (J/kg K)	Conductivity (W/m K)
Glass	2383	576	0.793
Argon	2	521	0.016
EVA	960	2090	0.230
PV cells	1190	1552	148
Backsheet	1200	1250	0.360
Absorber (Aluminum)	2719	871	202
Insulation	100	670	0.033
TEG	93	708	0.920
Water	998	4182	0.600

The employed governing equations in the performed numerical simulation of conventional PVT and PVT + TEGs panels are:

Continuity equation:

$$\nabla \vec{v} = 0 \tag{10}$$

Momentum equation:

$$\nabla\left(\rho \vec{v}\vec{v}\right) = -\nabla P + \nabla\left(\mu \nabla \vec{v}\right) \tag{11}$$

Energy equation:

$$\nabla\left(\vec{v}\rho\vec{v}c_p T\right) = \nabla(k\nabla T) \tag{12}$$

where $\vec{v}$ is the velocity vector, ρ is the density of the fluid, P is the pressure vector, μ is the viscosity of the fluid, c_p is the heat capacity and k represents the conductivity.

Considered boundary conditions are shown below:

- External surfaces of the PVT panels are adiabatic. $\nabla(k\nabla T) = 0$.
- Velocity of the fluid in the solid walls is null. $\vec{v} = 0$.
- In the wall of a solid, velocity and temperature of the fluid are that of the solid element. $-k\frac{\partial T}{\partial n}\Big|_f = -k_s\frac{\partial T_s}{\partial n}\Big|_s$.
- Inlet conditions are: $T = T_{in}$; u = 0; v = 0 and w = $-V_{in}$.
- Outlet condition is $P = 0$.

Table 3 includes the main dimensions of the different bodies that compound the PVT panel simulated in the employed numerical software.

Table 3. Geometrical dimensions of PVT bodies simulated in ANSYS.

PVT Layer	Dimensions (mm) (High × Width × Depth)
Glass cover	3.2 × 100 × 1650
Inert gas	10 × 100 × 1650
Glass	3.2 × 100 × 1650
EVA	0.5 × 100 × 1650
PV cell	0.3 × 100 × 1650
EVA + backsheet	1.5 × 100 × 1650
Thermal absorber	2 × 100 × 1650
Thermal collector	∅ 8 (thickness = 0.6) × 1650
Insulant	10 × 100 × 1650
TEG	3 × 40 × 80

Once the physical 3D model (Figure 1) that is going to be simulated is presented, a mesh must be done. The meshing is the process whereby the discretization of the geometry is performed. The selection of the mesh's method and size must be governed by a balance between quality and processing capacity. The quality of elements can be measured with different throw parameters. The selected parameters to analyze and evaluate mesh quality in the studied model are skewness and orthogonal quality [46].

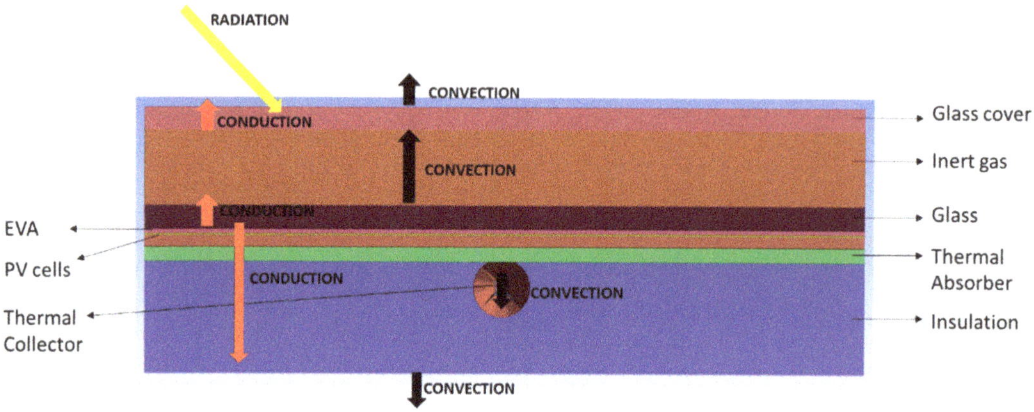

Figure 1. Conceptualization of the layers of a PVT panel and the different forms of heat transfer inside it.

Orthogonal quality is defined as the ratio between the scalar product of the normal vector of each mesh element surface and its centroid vector (that starts from the gravitational center of the element to face center) and the product between the named vectors. When this parameter is closed to one, more orthogonal quality is obtained.

The skewness of a mesh element is related to the angles between its edges. A distorted mesh cell does not generate accurate results. The volume of elements should approximate as possible to a spherical body. Skewness coefficient is calculated as the ratio between the difference of the volumes of the optimal body and the real body and the optimal body volume. An excellent coefficient value of skewness is near 0. Quality analysis of the mesh has been performed before the thermal analysis of the PVT and PVT+TEG models. Solutions obtained from different mesh sizes of the simulated model have been evaluated to select the suitable mesh for this system. In Figure 2, the comparison of the average outlet temperature and the average outlet velocity of the working fluid from the PVT panel with the mesh size can be consulted. Quality analysis of mesh has been performed under these work conditions: solar irradiation of 1000 W/m^2, a fluid inlet temperature of 30 °C, the mass flow of 0.02 kg/s, and an ambient temperature of 30 °C. An increase in the number of mesh elements from 195.000 to 448.000 implies a variation of 0.070% and 0.031% in average outflow velocity and average outflow temperature, respectively.

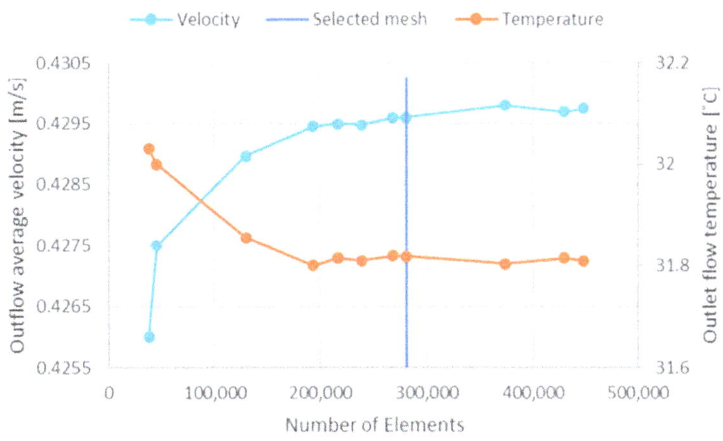

Figure 2. Quality mesh study.

In order to analyze the numerical uncertainly that arises from the space discretization, Richardson's extrapolation method is employed. This method allows the calculation of the exact solution under the assumption that the grid mesh was infinitesimally fine using the data obtained from meshes with reasonable resolution. Exact solution (φ_{exact}) and its associated uncertainly (UC) can be calculated as [47]

$$\varphi_{exact} = \varphi_1 + \frac{\varphi_1 - \varphi_2}{r_{21}^p - 1} \tag{13}$$

$$p = \frac{1}{\ln(r_{21})} * \left[\ln\left(\frac{\varepsilon_{32}}{\varepsilon_{21}}\right) + \ln\left(\frac{r_{21}^p - sign\left(\frac{\varepsilon_{32}}{\varepsilon_{21}}\right)}{r_{32}^p - sign\left(\frac{\varepsilon_{32}}{\varepsilon_{21}}\right)}\right)\right] \tag{14}$$

$$\text{where } r_{21} = \frac{h_2}{h_1};\ r_{32} = \frac{h_3}{h_2};\ \varepsilon_{21} = \varphi_2 - \varphi_1\ and\ \varepsilon_{32} = \varphi_3 - \varphi_2 \tag{15}$$

$$UC = F_s * (\varphi_{exact} - \varphi_1) \tag{16}$$

where φ_i are numerical solutions with i = 1,2 and 3 of the analyzed physical quantity (in this paper outlet flow temperature and outflows velocity) under three different mesh sizes h_1, h_2 and h_3, respectively, where $h_1 < h_2 < h_3$. Richarson's analysis of the employed mesh is summed in Table 4.

Table 4. Parameters employed in the Richardson's method.

	1	2	3
h_i (cm)	0.23	0.31	0.50
Outlet flow temperature (°C)	31.810	31.818	31.856
Outflow velocity (m/s)	0.4298	0.4296	0.4289

Exact temperature value is 31.801 °C what that entails a numerical uncertainly of temperature of 0.0109 °C. In the case of the velocity analysis, exact value achieve a value of 0.4299 m/s and its related uncertainly is 0.0002 m/s. These results reinforce the obtained results in the performed quality analysis.

According to the obtained results, a structured mesh composed of 280.000 elements has been selected in this study. It allows the simulation of the installation with a reasonable computational cost and an adequate quality [46]. On the one hand, the orthogonal quality average value is 0.886 with the least value of 0.79. On the other hand, the skewness average has a value of 0.227 with a maximum value of 0.57. The mesh of the pipe is shown in Figure 3. These mesh elements have been colored according to their quality.

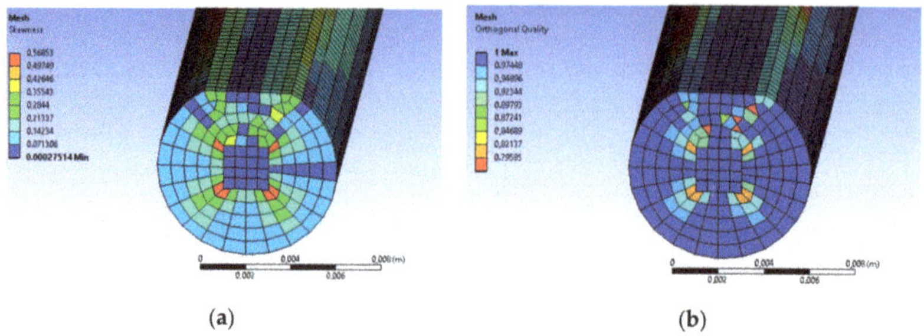

Figure 3. (a) Skewness and (b) orthogonal quality of the tube of the PVT panel simulated.

The PVT and the PVT+TEG models have been validated through experimental data collected on the test bench. Figure 4 (conventional PVT panel) and Figure 5 (PVT + TEG panel) show the comparison between the numerical results obtained from the software Fluent-ANSYS and the experimental data of the outlet flow temperature of both panels. As can be observed, numerical results are nearly equal to the experimental ones.

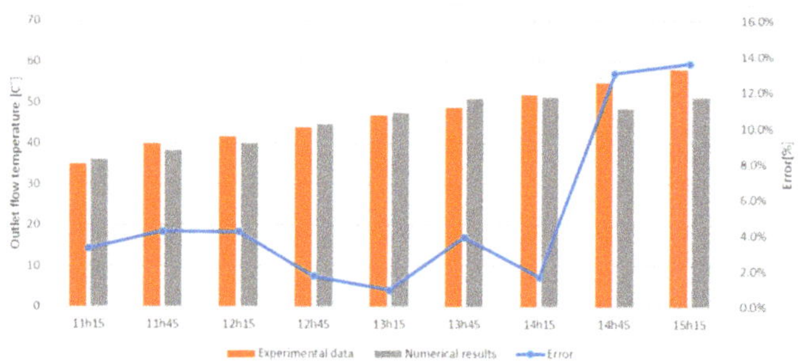

Figure 4. Comparison between experimental and numerical outlet flow temperature of the conventional PVT.

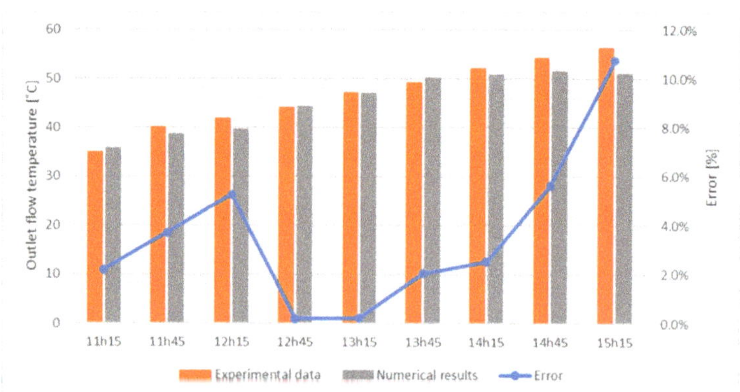

Figure 5. Comparison between experimental and numerical outlet flow temperature of the PVT with integrated TEG modules.

The error between numerical and experimental data does not exceed 13.6%. Except for the test made at 10:45 AM. At this time, error achieves the value of 38.0% and 42.7% in the simulation of a conventional PVT and in that of a PVT + TEG panel, respectively. The mass flow of the working fluid at this time is 0.39 g/s. This value is low. Additionally, it is observed in both Figures 4 and 5 that the increment in the error at 15:15 a.m. At this time, low flow is not the source of error. Analyzing experimental data from both hybrid panels, it can be concluded that the flow is on a transient scenario because impulsion pump goes into operation few minutes before the capture of data. The finding suggests that the elaborated model has limited application with low mass flow experiments and transient situations. Besides, the electrical production of cells embedded on the PVT panels has been validated. The comparison between experimental results and numerical results is shown in Figure 6. In some of the studied moments, the error between real and simulated data is not negligible as it can be observed; its source is related to the incident solar radiation and the ambient temperature. A correction in the calculation has been applied to reduce this error and achieve coherent results in further analysis. The correction factor applied in each range of solar radiation can be seen in Table 5. Electric production of PVT panels calculated with these corrections has a maximum error value of 10% concerning experimental results.

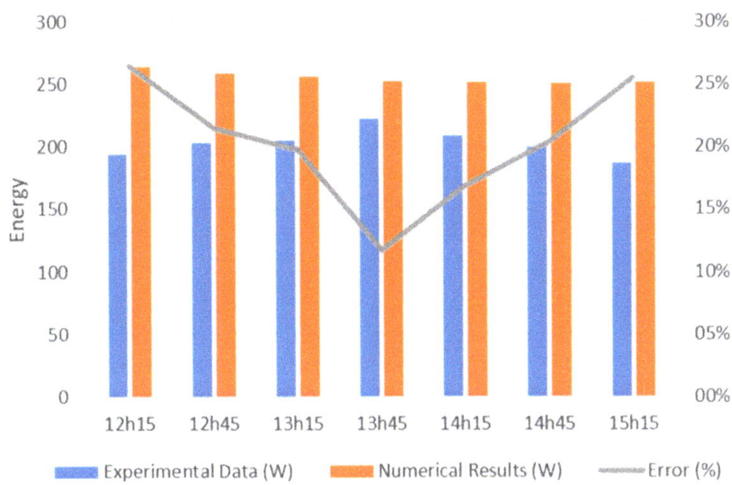

Figure 6. Comparison between experimental and numerical output power of the PVT with integrated TEG modules (without the application of the corrector factor).

Table 5. Correction coefficients for the calculation of electrical energy production of the PVT panel.

Solar Irradiation (W/m^2)	Coefficient
≥1.100	1
1100–900	0.88
≤900	0.78

Experimental data of the electrical production of the integrated TEGs in the PVT panel are currently not available. Monitoring of TEG's sensors is in the implementation phase. Significant values have not been provided at the moment. For this reason, only numerical data are going to be studied in this paper.

3. Results and Discussion

In concordance with the described methodology, four sensibility analyses have been simulated to study the conditions that maximize the performance of the conventional PVT panel and the PVT + TEGs panel. Solar irradiation, inlet mass flow, inlet flow temperature, and ambient temperature are the input parameters of the studied system that has been analyzed.

3.1. Inlet Temperature

Figure 7 shows the thermal and electrical performance of the considered photovoltaics-thermal panels with the variation in the inlet flow temperature. The thermal efficiency of both systems decreases with the increment in the inlet temperature. Additionally, electrical efficiency also displays a reduction. Thermal production of conventional PVT panel is on average 0.2% superior to PVT with integrated TEG modules. The electrical efficiency of TEG modules varies from 0.31% to 0.11%. The highest TEG's efficiency is achieved when the inlet flow temperature is 40 °C. In contrast, the higher thermal efficiency is achieve in both panels at the inlet flow temperature of 10 °C. With the integration of the TEGs in the PVT panel, electrical production increases on average 1.37%.

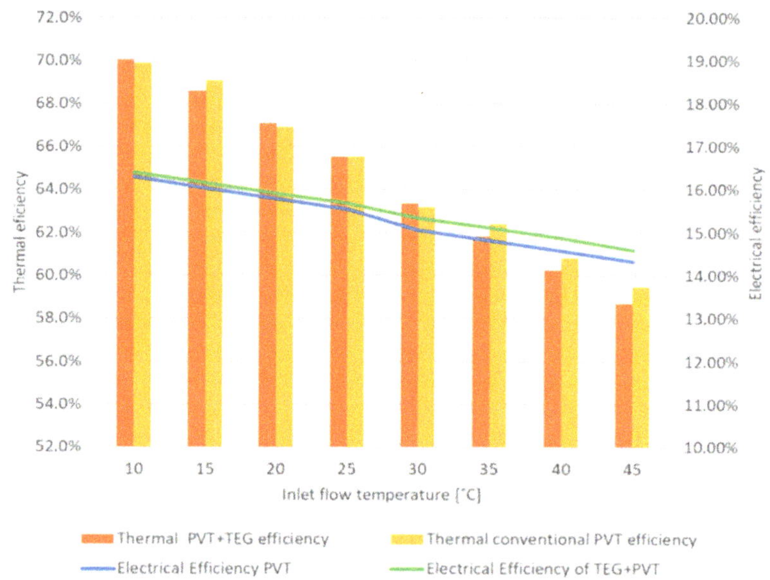

Figure 7. Thermal and electrical efficiency of the studied hybrid panels with the inlet temperature variation.

As it has been commented, as lower is inlet flow temperature of the working fluid of the photovoltaics-thermal panel with TEG modules, the higher is electrical efficiency of the thermoelectric modules. In this optimal situation, the global efficiency of the panel is also better than that obtained with higher inlet temperatures. Global efficiency of 86.5% and 86.2% is achieved in PVT + TEG panel and conventional hybrid panels, respectively, with an inlet flow temperature of 10 °C. In comparison, when the flow inlet temperature is 25 °C global thermal efficiency is 81.2% and 81.1%.

3.2. Ambient Temperature

As it can be consulted in Figure 8, the ambient temperature does not affect to electrical production of PV cells of the hybrid panels. Nevertheless, thermal efficiency is positively related to the variation of the ambient temperature. An increase in hot water production is observed when an increment in the ambient temperature occurs. The thermal efficiency of conventional PVT panels with the ambient temperature of 0 °C and 35 °C is 57.5% and

68.6%, respectively. The difference between thermal production of conventional and PVT + TEG panels is insignificant. The electrical efficiency of TEGs modules increases with the increment in the ambient temperature, reaching a maximum of 0.47% at 35 °C. This result makes sense because TEG efficiency is related to the gradient's increment between sides of the module. These temperatures are related to thermal efficiency too. In this way, it is possible to conclude that an increase in the thermal efficiency of a hybrid panel contributes to the increment in the electrical efficiency of the TEGs modules integrated into the panel. Additionally, an overall electrical efficiency increment is expected as well.

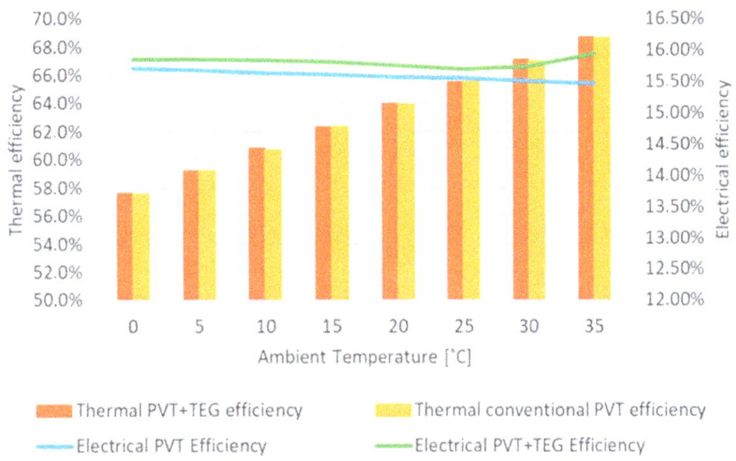

Figure 8. Thermal and electrical efficiency of the studied hybrid panels with the ambient temperature variation.

In low ambient temperature situations, the global electrical efficiency of both panels is nearly equal. Consequently, global efficiency is also similar. Nevertheless, with the increase in this parameter, the thermal efficiency of the collectors and electrical efficiency of TEG modules grow up. In this scenario, the global efficiency of the PVT + TEG panel is 0.64% higher than that of a conventional photovoltaics-thermal panel.

3.3. Inlet Mass Flow

The studied efficiencies of the PVT panels under the parametric analysis that vary the mass flow of the working fluid in the hybrid panels are graphically represented in Figure 9. The reduction of the mass inlet flow under nominal flow supposes an overheating of the PV cells and a reduction in the electrical efficiency of the panels. Electrical efficiency is 15.2% and 15.7%, with a working mass flow of 1 and 5 g/s, respectively. However, this overheating positively affects the TEG modules' efficiency because of the indifference of temperatures between cold and hot sides of them. In particular, the overall electrical efficiency of a PVT + TEG panel is 1.1% higher compared to a PVT panel with a circulating flow of 1 g/s. Despite the overheating produced in the interior of the hybrid panel with the reduction of the mass flow, thermal efficiency does not show an improvement.

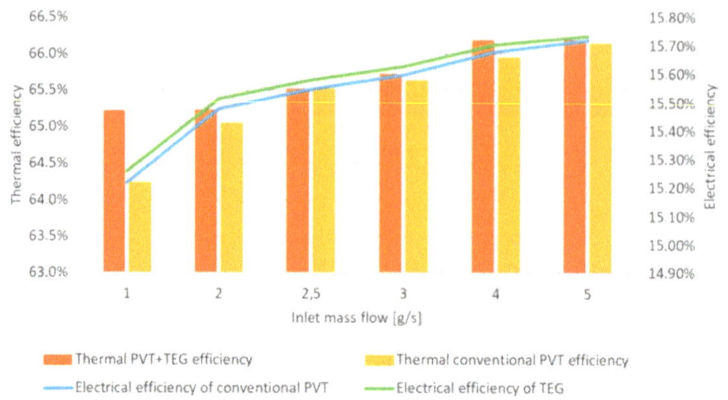

Figure 9. Thermal and electrical efficiency of the studied hybrid panels with the mass flow variation.

In global efficiency terms, an optimal working condition exists when the mass flow has a value of 18 L/h. This parameter is lower than the nominal flow given by the manufacturer. In this situation, the global efficiency of the conventional PVT panel is 81.9%, and the global efficiency of PVT+TEG panel is 82.2%.

3.4. Solar Irradiation

Figure 10 displays the thermal and electrical evolution of the two tested hybrid panels with the variation of the incident solar irradiation. The chart shows that the increase in electrical efficiency of the PVT panels takes place as a consequence of a reduction in solar radiation. In this sensibility analysis, the electrical efficiency of conventional PVT panels varies from 68.9% to 12.1% for incident irradiation in the range 200–1400 W/m^2. The efficiency of the TEG modules and thermal efficiency of the panels keep the opposite trend as PV cells efficiency. With more solar radiation, an increase in thermal production is observed. In the mentioned figure, it is not possible to appreciate the difference between electrical efficiency of the conventional PVT panel and the hybrid panel with integrated TEGs. This is because, on average, this difference has a value of 0.02%.

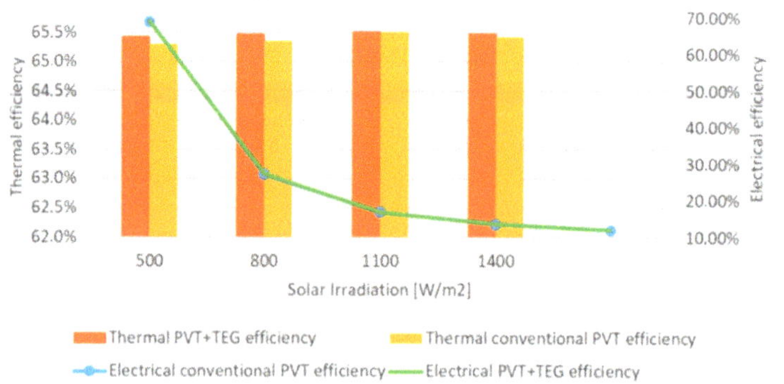

Figure 10. Thermal and electrical efficiency of the studied hybrid panels with the solar radiation variation.

Despite the work with low solar radiation providing good electrical efficiency values in both studied panels, it must take into account that electrical generation will have a low value.

3.5. Thermal Profile of a Photovoltaics-Thermal Panel

In order to identify deficiencies in the proposed configuration for the integration of the thermoelectric generators inside the photovoltaics-thermal hybrid panel, a profile of temperatures has been simulated. The simulation has been developed under standard conditions. These conditions are: solar irradiation of 1100 W/m^2, a fluid inlet temperature of 25 °C, the mass flow of 2.5 g/s, and an ambient temperature of 25 °C. Figure 11 shows the evolution of temperatures through the different layers that composed the hybrid collector. As can be observed, the upper outflow temperature is 7 °C higher than that of the bottom side of the water pipe. The temperature of the top cover glass achieves values around 75 °C. The thermoelectric generator does not have an exceptionally high gradient of temperature. This situation is inappropriate to achieve high-performance rates.

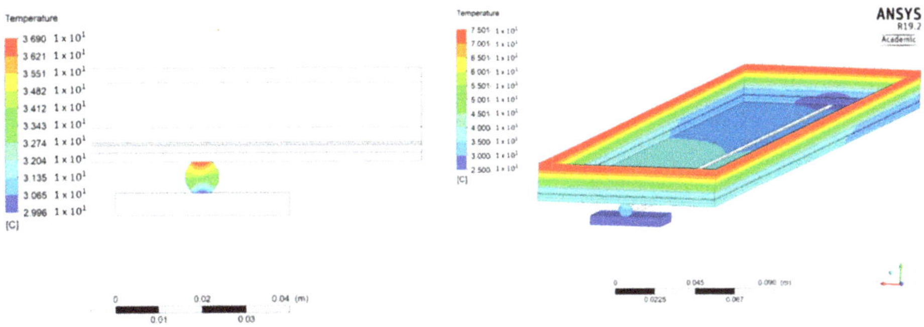

(a) (b)

Figure 11. (**a**) Outflow temperature and (**b**) PVT + TEG thermal profiles of the simulated panel under standard conditions.

Figure 12 displays a detailed profile of the TEG module integrated inside the panel. Temperature distribution on hot side surfaces is not uniform. The contact zone between the pipe and the TEG module is hotter than the distant zones to the pipe. This irregular contact generates problems in the electrical production of the thermoelectric device. If contact is insufficient, the real efficiency of TEG modules is reduced in comparison with theoretical efficiency.

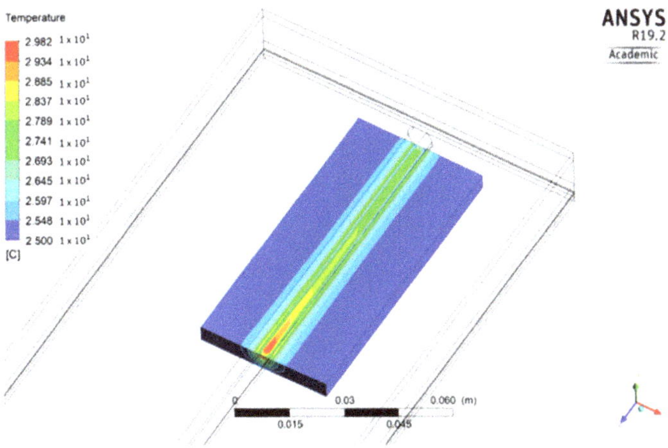

Figure 12. Thermal profile of temperatures from the TEG module integrated inside the photovoltaics-thermal hybrid panel.

4. Conclusions

The research carried out allows the combined simulation of the PVT panel integrated with a TEG in order to make an in-depth analysis of the advantage taken by this relatively easily improved connection. The simple comparison with main output experimental parameters and simulated figures gave reduced errors (less than the 10% in most of the data collected). Thus, the detailed information coming from the CFD allows finding the reasons that the obtained results are yet far from the expected improvement. Therefore, it is worth pointing out that the outcomes of the ongoing results indicate that the integration of thermoelectric materials within the PVT panels should be further researched and improved.

The search for the highest difference of temperature is the first task to deal with. In both experimental and numerical studies the highest recorded gap of temperature does not reach 10 °C. This situation is far from the optimal temperature difference given by the manufacturer of the TEG devices in order to produce the expected amount of electricity. Thus, a new set of experiments is currently being considered. Future research will be surely oriented to determine the optimal integration of thermoelectric generators inside PVT panels by improving the contact region between both sides of the TEG devices. Another possibility is to study the alternative integration of the TEG modules in solar energy installation. However, the TEGs would be placed out of the PVT panels, if insufficienttemperature drop is finally found.

Author Contributions: Conceptualization, A.M.-G., Á.A.B.-R. and A.d.A.; Methodology, M.T.P., A.M.-G. and M.P.G.; Software, M.T.P. and M.P.G.; Validation, Á.A.B.-R., J.A.T., and A.d.A., Formal analysis, M.T.P., A.M.-G. and J.U.; Investigation, J.A.T. and J.U.; Data curation, M.P.G. and J.A.T.; Writing—original draft preparation, M.T.P.; Writing—review and editing, A.M.-G., J.U., and M.P.G.; Visualization, M.T.P.; Supervision, A.M.-G. and Á.A.B.-R.; Funding acquisition, A.M.-G. All authors have read and agreed to the published version of the manuscript.

Funding: This research was funded by the Spanish Ministry of Science, Innovation and Universities through the program "Research Challenges 2018", project "grant number RTI2018-098886-A-I00.

Institutional Review Board Statement: Not applicable.

Informed Consent Statement: Not applicable.

Data Availability Statement: Not applicable.

Acknowledgments: Authors thank the support given to this research by the Spanish Ministry of Science, Innovation and Universities.

Conflicts of Interest: The authors declare no conflict of interest.

References

1. Scripps Institution of Oceanography. The Keeling Curve is a Daily Record of Global Atmospheric Carbon Dioxide Concentration Maintained by Scripps Institution of Oceanography at UC San Diego. Available online: https://keelingcurve.ucsd.edu/ (accessed on 18 January 2021).
2. Paris Agreement. In Proceedings of the Paris Climate Conference (on 22 April 2016, New York) (COP21). December 2015. Available online: https://ec.europa.eu/clima/policies/international/negotiations/paris_en (accessed on 18 January 2021).
3. International Energy Agency (IEA). Energy Consumption by Sector. European Union, 2018. Available online: https://www.iea.org/data-and-statistics/?country=EU28&fuel=Energy%20consumption&indicator=ElecConsBySector (accessed on 18 January 2021).
4. Yusuf, A.; Ballikaya, S. Modelling a Segmented Skutterudite-Based Thermoelectric Generator to Achieve Maximum Conversion Efficiency. *Appl. Sci.* **2020**, *10*, 408. [CrossRef]
5. Jaziri, N.; Boughamoura, A.; Müller, J.; Mezghani, B.; Tounsi, F.; Ismail, M. A comprehensive review of Thermoelectric Generators: Technologies and common applications. *Energy Rep.* **2020**, *6*, 264–287. [CrossRef]
6. Børset, M.T.; Wilhelmsen, Ø.; Kjelstrup, S.; Burheim, O.S. Exploring the potential for waste heat recovery during metal casting with thermoelectric generators: On-site experiments and mathematical modeling. *Energy* **2017**, *118*, 865–875. [CrossRef]
7. Jouhara, H.; Żabnieńska-Góra, A.; Khordehgah, N.; Doraghi, Q.; Ahmad, L.; Norman, L.; Axcell, B.; Wrobel, L.; Dai, S. Thermoelectric generator (TEG) technologies and applications. *Int. J. Thermofluids* **2021**, *9*, 100063. [CrossRef]
8. Chen, J.; Li, K.; Liu, C.; Li, M.; Lv, Y.; Jia, L.; Jiang, S. Enhanced Efficiency of Thermoelectric Generator by Optimizing Mechanical and Electrical Structures. *Energies* **2017**, *10*, 1329. [CrossRef]

9. Kütt, L.; Millar, J.; Karttunen, A.; Lehtonen, M.; Karppinen, M. Thermoelectric applications for energy harvesting in domestic applications and micro-production units. Part I: Thermoelectric concepts, domestic boilers and biomass stoves. *Renew. Sustain. Energy Rev.* **2018**, *98*, 519–544. [CrossRef]
10. Mostafavi, S.A.; Mahmoudi, M. Modeling and fabricating a prototype of a thermoelectric generator system of heat energy recovery from hot exhaust gases and evaluating the effects of important system parameters. *Appl. Therm. Eng.* **2018**, *132*, 624–636. [CrossRef]
11. Cao, Q.; Luan, W.; Wang, T. Performance enhancement of heat pipes assisted thermoelectric generator for automobile exhaust heat recovery. *Appl. Therm. Eng.* **2018**, *130*, 1472–1479. [CrossRef]
12. Orr, B.; Akbarzadeh, A.; Lappas, P. An exhaust heat recovery system utilising thermoelectric generators and heat pipes. *Appl. Therm. Eng.* **2017**, *126*, 1185–1190. [CrossRef]
13. Cheng, K.; Qin, J.; Jiang, Y.; Lv, C.; Zhang, S.; Bao, W. Performance assessment of multi-stage thermoelectric generators on hypersonic vehicles at a large temperature difference. *Appl. Therm. Eng.* **2018**, *130*, 1598–1609. [CrossRef]
14. Ziolkowski, P.; Zabrocki, K.; Müller, E. TEG Design for Waste Heat Recovery at an Aviation Jet Engine Nozzle. *Appl. Sci.* **2018**, *8*, 2637. [CrossRef]
15. Yan, J.; Liao, X.; Yan, D.; Chen, Y. Review of Micro Thermoelectric Generator. *J. Microelectromechanical Syst.* **2018**, *27*, 1–18. [CrossRef]
16. Yang, J.; Caillat, T. Thermoelectric Materials for Space and Automotive Power Generation. *MRS Bull.* **2006**, *31*, 224–229. [CrossRef]
17. Bahk, J.-H.; Fang, H.; Yazawa, K.; Shakouri, A. Flexible thermoelectric materials and device optimization for wearable energy harvesting. *J. Mater. Chem. C* **2015**, *3*, 10362–10374. [CrossRef]
18. Lee, G.; Choi, G.; Kim, C.S.; Kim, Y.J.; Choi, H.; Kim, S.; Kim, H.S.; Lee, W.B.; Cho, B.J. Material Optimization for a High Power Thermoelectric Generator in Wearable Applications. *Appl. Sci.* **2017**, *7*, 1015. [CrossRef]
19. Nesarajah, M.; Frey, G. Optimized Design of Thermoelectric Energy Harvesting Systems for Waste Heat Recovery from Exhaust Pipes. *Appl. Sci.* **2017**, *7*, 634. [CrossRef]
20. Alanne, K.; Laukkanen, T.; Saari, K.; Jokisalo, J. Analysis of a wooden pellet-fueled domestic thermoelectric cogeneration system. *Appl. Therm. Eng.* **2014**, *63*, 1–10. [CrossRef]
21. Höftberger, E.; Moser, W.; Aigenbauer, W.; Friedl, G.; Haslinger, W. Grid Autarchy of Automated Pellets Combustion Systems by the Means of Thermoelectric Generators. Thermoelectr Goes Automot II. 2010, pp. 209–233. Available online: https://www.best-research.eu/files/publications/pdf/I1-2481.pdf (accessed on 18 January 2021).
22. Moser, W.; Friedl, G.; Aigenbauer, S.; Heckmann, M.; Hofbauer, H. A biomass-fuel based micro-scale CHP system with thermoelectric generators. In Proceedings of the Central European Biomass Conference (16 to 19 January 2008, Austria). 2008. Available online: https://www.best-research.eu/files/publications/pdf/236_CP_I-1-21_Paper_Austrian_Bioenergy_Friedl_Moser.pdf (accessed on 18 January 2021).
23. Sornek, K.; Filipowicz, M.; Rzepka, K. The development of a thermoelectric power generator dedicated to stove-fireplaces with heat accumulation systems. *Energy Convers. Manag.* **2016**, *125*, 185–193. [CrossRef]
24. Karthick, K.; Suresh, S.; Hussain, M.M.M.; Ali, H.M.; Kumar, C.S. Evaluation of solar thermal system configurations for thermoelectric generator applications: A critical review. *Sol. Energy* **2019**, *188*, 111–142. [CrossRef]
25. Date, A.; Date, A.; Dixon, C.; Akbarzadeh, A. Theoretical and experimental study on heat pipe cooled thermoelectric generators with water heating using concentrated solar thermal energy. *Sol. Energy* **2014**, *105*, 656–668. [CrossRef]
26. Khanmohammadi, S.; Baseri, M.M.; Ahmadi, P.; Al-Rashed, A.A.; Afrand, M. Proposal of a novel integrated ocean thermal energy conversion system with flat plate solar collectors and thermoelectric generators: Energy, exergy and environmental analyses. *J. Clean. Prod.* **2020**, *256*, 120600. [CrossRef]
27. Palomba, V.; Borri, E.; Charalampidis, A.; Frazzica, A.; Cabeza, L.F.; Karellas, S. Implementation of a solar-biomass system for multi-family houses: Towards 100% renewable energy utilization. *Renew. Energy* **2020**, *166*, 190–209. [CrossRef]
28. Li, D.; Xuan, Y.; Li, Q.; Hong, H. Exergy and energy analysis of photovoltaic-thermoelectric hybrid systems. *Energy* **2017**, *126*, 343–351. [CrossRef]
29. Dimri, N.; Tiwari, A.; Tiwari, G. Thermal modelling of semitransparent photovoltaic thermal (PVT) with thermoelectric cooler (TEC) collector. *Energy Convers. Manag.* **2017**, *146*, 68–77. [CrossRef]
30. Urbiola, E.A.C.; Vorobiev, Y. Investigation of Solar Hybrid Electric/Thermal System with Radiation Concentrator and Thermoelectric Generator. *Int. J. Photoenergy* **2013**, *2013*, 1–7. [CrossRef]
31. Kılkış, B. Development of a composite PVT panel with PCM embodiment, TEG modules, flat-plate solar collector, and thermally pulsing heat pipes. *Sol. Energy* **2020**, *200*, 89–107. [CrossRef]
32. Iasiello, M.; Mameli, M.; Filippeschi, S.; Bianco, N. Simulations of paraffine melting inside metal foams at different gravity levels with preliminary experimental validation. *J. Phys. Conf. Ser.* **2020**, *1599*, 012008. [CrossRef]
33. Ghalambaz, M.; Zhang, J. Conjugate solid-liquid phase change heat transfer in heatsink filled with phase change material-metal foam. *Int. J. Heat Mass Transf.* **2020**, *146*, 118832. [CrossRef]
34. Zhao, C.; Lu, W.; Tian, Y. Heat transfer enhancement for thermal energy storage using metal foams embedded within phase change materials (PCMs). *Sol. Energy* **2010**, *84*, 1402–1412. [CrossRef]
35. Iasiello, M.; Mameli, M.; Filippeschi, S.; Bianco, N. Metal foam/PCM melting evolution analysis: Orientation and morphology effects. *Appl. Therm. Eng.* **2021**, *187*, 116572. [CrossRef]

36. Zheng, H.; Wang, C.; Liu, Q.; Tian, Z.; Fan, X. Thermal performance of copper foam/paraffin composite phase change material. *Energy Convers. Manag.* **2018**, *157*, 372–381. [CrossRef]
37. Kazemian, A.; Hosseinzadeh, M.; Sardarabadi, M.; Passandideh-Fard, M. Experimental study of using both ethylene glycol and phase change material as coolant in photovoltaic thermal systems (PVT) from energy, exergy and entropy generation viewpoints. *Energy* **2018**, *162*, 210–223. [CrossRef]
38. Zhou, F.; Ji, J.; Yuan, W.; Zhao, X.; Huang, S. Study on the PCM flat-plate solar collector system with antifreeze characteristics. *Int. J. Heat Mass Transf.* **2019**, *129*, 357–366. [CrossRef]
39. Salari, A.; Parcheforosh, A.; Hakkaki-Fard, A.; Amadeh, A. A numerical study on a photovoltaic thermal system integrated with a thermoelectric generator module. *Renew. Energy* **2020**, *153*, 1261–1271. [CrossRef]
40. Abora Solar, aH60. Available online: https://abora-solar.com/pdf/AH60SK-ES.pdf (accessed on 18 January 2021).
41. Coleman, H.W.; Steele, W.G. *Experimentation and Uncertainty Analysis for Engineers*, 2nd ed.; John Wiley & Sons Inc.: Toronto, ON, Canada, 1999; ISBN 978-0-470-16888-2.
42. Uche, J.; Acevedo, L.; Usón, S.; Martínez-Gracia, A.; Bayod-Rújula, A.Á. Analysis of a domestic trigeneration scheme with hybrid renewable energy sources and desalting techniques. *J. Clean. Prod.* **2019**, *212*, 1409–1422. [CrossRef]
43. TEG Solidstate Power Generators, TEG2-07025SHT-S. Available online: https://tecteg.com/wp-content/uploads/2014/09/Spec-TEG2-07025HT-SS-rev1-1.pdf (accessed on 18 January 2021).
44. Antonanzas, J.; Del Amo, A.; Martínez-Gracia, A.; Bayod-Rújula, A.; Antonanzas-Torres, F. Towards the optimization of convective losses in photovoltaic–thermal panels. *Sol. Energy* **2015**, *116*, 323–336. [CrossRef]
45. ANSYS. *ANSYS Academic Research Documentation Manual, Release 18.2*; Help System, Guide; ANSYS, Inc.: Canonsburg, PA, USA; Available online: https://www.ansys.com/ (accessed on 18 January 2021).
46. ANSYS. *Lecture 7: Mesh Quality & Advanced Topics*; ANSYS: Canonsburg, PA, USA, 2015.
47. Zhang, H.; Li, Y.; Xiao, J.; Jordan, T. Uncertainty analysis of condensation heat transfer benchmark using CFD code GASFLOW-MPI. *Nucl. Eng. Des.* **2018**, *340*, 308–317. [CrossRef]

applied sciences

MDPI

Article

The Effect of Azimuth and Tilt Angle Changes on the Energy Balance of Photovoltaic System Installed in the Southern Slovakia Region

Monika Božiková [1,*], Matúš Bilčík [1], Vladimír Madola [2], Tímea Szabóová [3], Ľubomír Kubík [1], Jana Lendelová [3] and Vladimír Cviklovič [2]

1 Department of Physics, Faculty of Engineering, Slovak University of Agriculture in Nitra, Tr. A. Hlinku 2, 94976 Nitra, Slovakia; bilcikmatus@gmail.com (M.B.); lubomir.kubik@uniag.sk (Ľ.K.)
2 Department of Electrical Engineering, Faculty of Engineering, Slovak University of Agriculture in Nitra, Automation and Informatics, Tr. A. Hlinku 2, 94976 Nitra, Slovakia; xmadolav@uniag.sk (V.M.); vladimir.cviklovic@uniag.sk (V.C.)
3 Department of Building Equipment and Technology Safety, Faculty of Engineering, Slovak University of Agriculture in Nitra, Tr. A. Hlinku 2, 94976 Nitra, Slovakia; timea.szaboova@uniag.sk (T.S.); jana.lendelova@uniag.sk (J.L.)
* Correspondence: monika.bozikova@uniag.sk; Tel.: +421-37-641-4711

Featured Application: The research results are beneficial for precise prediction of the photovoltaic systems energy production. Presented mathematical models are very valuable for practice, which provide a platform for optimalisation of photovoltaic systems operating conditions in the regions of Central Europe. The research results are focused on the detailed analyzation of the tilt and azimuth angle influence on the photovoltaic energy production. The results are used easily for creation of application for smart devices, because simplified mathematical model equations require the entry of a minimum number of parameters by the user.

Citation: Božiková, M.; Bilčík, M.; Madola, V.; Szabóová, T.; Kubík, Ľ.; Lendelová, J.; Cviklovič, V. The Effect of Azimuth and Tilt Angle Changes on the Energy Balance of Photovoltaic System Installed in the Southern Slovakia Region. *Appl. Sci.* **2021**, *11*, 8998. https://doi.org/10.3390/app11198998

Academic Editor: Alberto Benato

Received: 29 August 2021
Accepted: 22 September 2021
Published: 27 September 2021

Abstract: Energy balance of the photovoltaic system is influenced by many factors. In this article the effect of tilt and azimuth angle changes of the photovoltaic system energy production is analyzed. These parameters have significant impact on the amount of solar radiation which hits on the photovoltaic panel surface and therefore also on the energy absorbed by the module surface. The main aim of research was identification of the optimal position of photovoltaic system installation in the southern Slovakia regions. The experimental apparatus had two setups consisting of polycrystalline photovoltaic modules. The first setup was used for identification of the tilt angle changes in the range (0–90°). The second one was focused on the detection of the azimuth angle effect to the energy production. The measurement results were statistically processed and mathematically analyzed. Obtained dependencies are presented as two-dimensional and three-dimensional graphical relations. Regression equations characterize time relations between the tilt or azimuth angle and the energy produced by the photovoltaic system in Southern Slovakia. Obtained simplified mathematical model was verified by analytical model. Presented models can be used for the dimensioning and optimization of the photovoltaic system energy production.

Keywords: operating conditions; energy production; simplified model; photovoltaics

1. Introduction

Non-renewable energy sources, such as fossil fuel, have been the major source of energy in many regions, including Central Europe region. However, because of the problems associated with the use of these non-renewable energy sources, there is a need for alternative energy sources that are sustainable and non-polluting [1]. Solar energy is a very popular renewable energy source due to its availability. Yearly sum of global irradiation incident on optimally inclined South oriented photovoltaic modules for Central European area is approximately 1300 kWh·m^{-2} [2]. The energy consumption is five times less than the amount of energy captured from the Sun. Based on the presented facts, it is clear that the solar energy can be transformed into electric and thermal energy with positive energetic and economical effect. Nowadays one of the most important reasons for installation of solar systems is their positive ecological aspect and sustainability [3]. Authors [4,5] observed that solar energy offers one of the best solutions to the problem of climate change.

Solar energy can be converted to electricity via photovoltaic (PV) cell. The production of solar photovoltaic energy is increasing annually. For example, the existing solar photovoltaic energy production increased more than 27 times from the production ten years ago; in 2009, it was less than 23 GW. The solar PV installation capacity reached 627 GW in 2019 compared to 512 GW in 2018 [6]. The amount of solar radiation received on a PV module depends on latitude, day of the year, slope or tilt angle, surface azimuth angle, time of the day, and the angle of incident radiation [7,8]. The factors that can be controlled to maximize the amount of radiation flux received upon the PV module are surface azimuth angle and tilt angle by installing a PV module properly [9].

Many researchers [10–13] presented results which declare the fact that for every location on Earth with different radiation characteristics can be found an optimal tilt angle for the best solar energy reception. The output of the PV module is highest when the incident solar ray is perpendicular to the PV module surface [14]. The case study focused on determination of the optimum tilt angle of PV module for each month in Nigeria confirmed its variability during the year. The performance of a PV installation is affected also by azimuth angle. The impact of different horizontal and vertical solar panel orientations on the integration of solar energy in low-voltage distribution grids was detected by authors [15]. The literature [16] presents results for optimum tilt angle and azimuth orientation of solar photovoltaic arrays in order to maximize incident solar irradiance exposed on the array for a specific time period. Especially, the effect of the azimuth angle on the energy production was studied and experimentally evaluated by research [17].

Studies discussed the best performance, design, and simulation for the solar energy systems using optimum tilt angles. There are a number of studies that were carried out in order to find the best performance of solar system areas around the world and others give a comparison between different locations [13].

Optimization of the tilt angle was performed for various locations in European countries including Turkey [18,19], Romania [20], Austria and Germany [21], Italy [22], Greece [16], Cyprus [23], Spain [24]. For Middle Eastern countries such as Oman [25], United Arab Emirates [8], Saudi Arabia [26,27], Egypt [28], Jordan [29,30], and Syria [31]. For Asian Countries, such as Pakistan [32], Thailand [33], Indonesia [34], China [35,36], Taiwan [37], India [38,39], Japan [40], and Bangladesh [41]. For the American continent, such as in Canada [11,42] and the United States of America [43–45], all mentioned studies point to the fact that the tilt angle and azimuth angle change has significant influence on the amount of solar energy absorbed by the surface of the PV modules and so on PV system energy balance. Location of the PV system is very important because of different external operating conditions at the various points on the earth's surface. From the theoretical point of view there exist mathematical models and methods of calculation for comparison of the best tilt angles of solar modules through monthly diffused radiation and actual monthly diffused radiations [46]. The operational parameters of PV module were modeled and discussed in literature [47]. The key formula for getting the PV system energy production is described in [48]. The connection between the energy production and energy consumption in irrigation networks was investigated by [49]. Very important for prediction of PV system energy production is knowledge about the reasons of power losses, which were described in the article [50].

From the practical point of view, the optimal technique to enhance the tilt angle and orientation is solar trackers [51,52]. The active sun trackers follow the path of the Sun, and they optimize the position of the solar module. Tracking systems are used to maximize daily solar energy received by photovoltaic modules [53]. Solar trackers consist of mechanical components that ensure the rotation of the solar module and therefore one of the solar tracker disadvantages is the failure of their mechanical parts and more demanding maintenance than in the case of static solar panels. Trackers are slightly more expensive, need energy for operation, and they are not always applicable because of specific installation conditions.

The information provided above was the reason for further focusing on the research and formulating its goal. The main aim of this research was the creation of the simplified mathematical model. The basis for creating the model was experimental data. The simplified model contains a minimum number of input parameters. It allows the calculation of the energy amount produced by the photovoltaic system in the region of South Slovakia during the calendar year. The model should be easy to use in practice. Due to the applicability of the model equations, it contains only parameters that are easily identifiable in practice for the users. Within the results, the influence of the azimuth angle and the tilt angle on the electrical output of PV system per month was assessed. The created model will be also a part of the application for smart devices which can be used for photovoltaic systems dimensioning. The results of implemented study are valid for Nitra region.

2. Materials and Methods

The definitions of azimuth angle and tilt angle mentioned in literature differ. The tilt angle is the angle of the photovoltaic modules from the horizontal plane for a fixed (non-tracking) mounting [27,54]. Generally, it is recommended that photovoltaic system should be installed with a tilt angle which is equal to the latitude of the site [55,56]. The visualization of tilt angle meaning is in Figure 1 and in detail it is described by Figure 2.

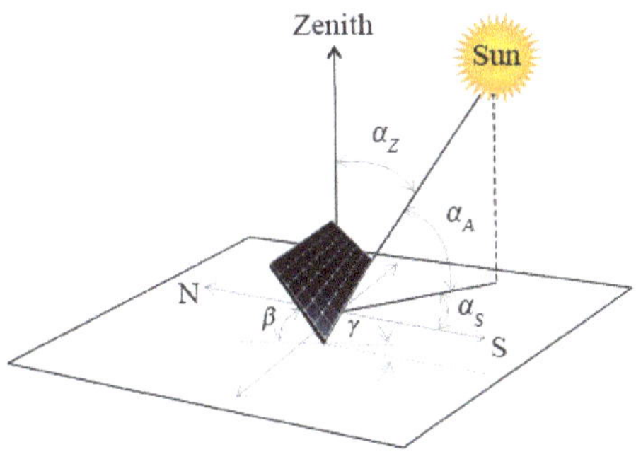

Figure 1. Illustration to the definition of the tilt angle, the solar azimuth angle. and the azimuth orientation.

Figure 2. Experimental PV power plant with PV modules PHONOSOLAR PS240P−20/U.

The azimuth angle indicates the position of the PV modules relative to the South; $-90°$ is East, $0°$ is South and $+90°$ is West [57,58]. The azimuth angle should be South for Northern Hemisphere and North for Southern Hemisphere [59,60]. The next way of azimuth angle definition is at the point of observation. The angle measured between the horizontal plane of the north point and a point on the circle of the horizon intersected by the arc of a vertical plane passing through the zenith and the sun's position at that time [4].

The position of the Sun in the sky at any moment can be defined by two very important angles, the first is the solar altitude angle and the next is the solar azimuth angle. These angles are physical parameters of the position of the Sun with respect to a given place on Earth. Therefore, they are independent of the inclination and orientation of the surface. On Figure 1 the Solar altitude angle α_A is the angle between the horizontal and the line to the Sun ($0° \leq \alpha_A \leq 90°$). The complement of this angle is the zenith angle (α_Z). It is defined by the vertical and the line to the Sun (i.e., the angle of incidence of beam radiation on a horizontal surface); β is the tilt angle and γ is the azimuth angle which determine the azimuth orientation of the solar module.

Solar azimuth angle α_S: angular displacement from South of the projection of beam radiation on the horizontal plane. (In general, $\alpha_S = 0$ is South, $\alpha_S < 0$ is East, and $\alpha_S > 0$ is West).

Once these two angles are established, it will help to define exactly the solar reaching the point on Earth where the solar system is going to be erected. The position of the Sun can be determined also mathematically through equations for the solar altitude angle and the solar azimuth angle (Equations (1) and (2), where α_L is the latitude, δ is the declination and ω is the hour angle.

$$\alpha_A = \arcsin(sin\alpha_L \cdot sin\delta + cos\alpha_L \cdot cos\delta \cdot cos\omega) \quad (1)$$

$$\alpha_S = \arccos\left(\frac{cos\delta \cdot cos\omega - sin\alpha_A \cdot cos\alpha_L}{cos\alpha_A \cdot sin\alpha_L}\right) \quad (2)$$

The next way for their identification is the solar charts. Presented facts are very valuable from the theoretical point of view, but in practice we usually know only orientation represented by azimuth angle γ.

The literature has equations for very precise prediction of the energy and power balance. For example, Equation (3) represents percentage of output power drop (D) and β is the PV module tilt angle shift (°). If we know the output power in selected time range, we can calculate the total energy production of PV system for different value of tilt angle. Equation (3) was obtained for very specific operating conditions described in literature [61].

$$\begin{aligned} D = {} & 1.416674 \cdot 10^{-9} \cdot \beta^6 - 3.59548024 \cdot 10^{-9} \cdot \beta^5 - 5.57767010 \cdot 10^{-6} \cdot \beta^4 + \\ & +1.64493758 \cdot 10^{-5} \cdot \beta^3 - 1.724028491 \cdot 10^{-4} \cdot \beta^2 - 7.11435953 \cdot 10^{-3} \cdot \beta + \\ & +99.37322857 \end{aligned} \quad (3)$$

One of the best ways for the creation of the simplified mathematical model for PV system energy production in the selected location is to perform experimental observations in real operating conditions.

The next part describes the PV power plant and model PV system situated in the area of southern Slovakia. The monitoring of operational parameters started in January 2011 an continues today. The research was carried out during the years 2012–2019. The PV power plant is type grid on and it consists of 432 polycrystalline PV modules PHONOSOLAR PS240P−20/U (Phono Solar, Willich Germany) and the whole area of the PV system is 714.6 m^2 (Figure 2).

The tilt angle of PV modules is 35°, which is in the literature [2] considered to be ideal for Central Europe. The azimuth orientation of the PV modules was South with the azimuth angle 0°. Efficiency of PV module is 14.8%. Power of one PV module is 240 Wp, so the total installed power is 103.68 kWp. For converting DC voltage that is supplied by photovoltaic cells, AC voltage 230 V with frequency 50 Hz serve 3 voltage converters type

Schüco Central Inverter type SGI 33 (Schüco, Prague, Czech Republic) is used. Efficiency of these converters is 97.1%. Monitoring of the PV power plant is solved using the RS485 interface, and the inverters are connected by FTP cables and monitored by Solar-Log 2000 (Solare Datensysteme GmbH, Geislingen-Binsdorf, Germany) (Figure 3). The Solar-Log 2000 offers the option to measure the amount of self-produced power consumption and it presents graphically via the Solar-Log™ WEB. The proposed system is in accordance with the technical recommendations and requirements for the interface between the PV power plant and the electrical grid according to EN 61727.

(a) (b)

Figure 3. (a) Monitoring device: The Solar-Log 2000; (b) Voltage converters type Schüco Central Inverter type SGI 33.

The SGI 33 inverter provides communication via the RS−485 standard. A chip with an integrated Bluetooth 5.2 radiosystem EFR32BG22C224F512GN32-C in a TQFN32 case is used in place of the microcontroller (Silicon Laboratories, Austin USA). It contains two UART modules. The first module is designed for communication with the inverter in the semi-duplex RS−485 mode with hardware flow control of the UART/RS485 converter of the MAX 481 integrated circuit. The second UART module communicates with the galvanically isolated KNX TinySerial Interface 810 and provides information on the operating parameters of the PV system with an intelligent installation element. Block scheme of communication is shown in (Figure 4).

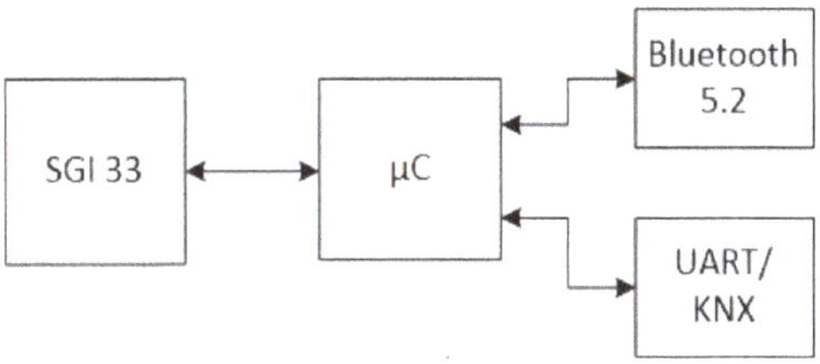

Figure 4. Block scheme of communication.

The model PV solar system was constructed from the identical polycrystalline PV module PHONOSOLAR PS240P−20/U. Each PV module was placed on a separate construction that allowed changes of the tilt angle. The operational data were collected separately for every PV module. The number of PV modules was selected according to number of different azimuth orientation represented by the azimuth angles from −90° to 90° (Figure 5). The tilt angles and azimuth angles were changed separately and manually for every PV module. The solar trackers were not used because approximately 92% of buildings have a sloping roof in southern Slovakia and non-stationary systems cannot be installed. However, roofs differ in the tilt and azimuth orientation.

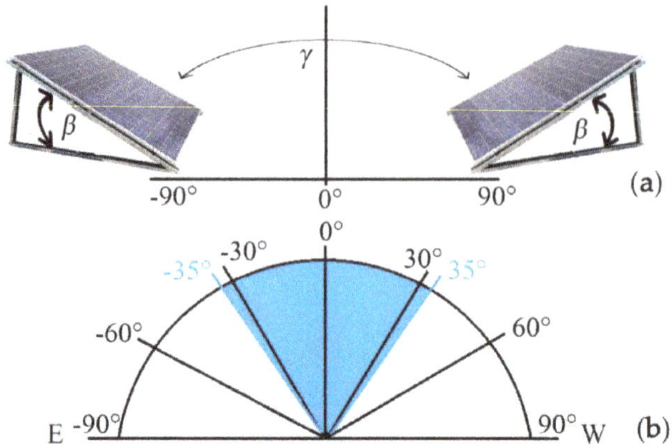

Figure 5. (**a**) The azimuth orientation of the model PV system; (**b**) The recommended orientation (azimuth angle marked in blue color) for southern Slovakia.

The azimuth angle 0° means South, the azimuth angle +90° is West, and −90° is East. The range (−5–−85°) means Southeast (SE) orientation and azimuth angle in the range (5°–85°) means Southwest (SW) orientation of PV module (Figure 5). The results for different azimuth angles (−90°, −80°, −70°, −55°, −50°, −45°, −35°, −25°, −20°, −15°−10°, −5°, 5°, 10°, 15°, 20°, 25°, 35°, 40°, 45°, 50°, 55°, 70°, 80°, 90°) were experimentally detected by comprehensive monitoring of operational conditions (ambient temperature, temperature of PV module surfaces, intensity of solar radiation, wind speed, output electric current and voltage, output electric power, energy production). The tilt angle was changed gradually by 5°. Experimental data were collected, sorted, and numerically and statistically processed. The method of group data analysis was applied on huge data files. The time relations were detected in hourly, daily, and monthly time ranges. The measured data from the selected part model PV system (1 PV module) were compared with data from real PV power plant with the same tilt angle and azimuth angle. The results were converted to a unit of the PV module active area. The difference between each compared value was less than 0.46%. The experimental data from both PV systems with tilt angle 35° were also used for calculation of energy production. The correlation analysis confirmed 99.54% correlation between the values of energy produced by a model PV system and PV power plant.

Then were created two-dimensional graphical dependencies and by regression analysis were detected suitable mathematical equations which describe the PV system energy production as a function of tilt angle and azimuth angle. In the next step polynomial approximation on the three-dimensional graphical dependencies was performed. The verification of mathematical models was made by iteration method. In the next part the PV power plant and the model PV system are described.

Large data files were processed with spreadsheet software Microsoft Excel and Matlab® version R2015b. All data were analysed using analysis of variance (ANOVA). The comparison of the averages was carried out by Duncan's test with a 95% confidence level. The arithmetic averages, medians, and standard error of the arithmetic average were computed from the data. The depth data analysis and the data extraction were applied on the data files obtained from real PV system for creating a mathematical model.

3. Results

The first part of the study was focused on the identification of optimal tilt angle for southern Slovakia region. In this section the time relations for different tilt angles (0°, 15°, 30°, 45°, 60°, 75°, and 90°) were evaluated. At first, data from autonomous model solar system (1 PV module) were detected, then the results were processed and the data were recalculated on the experimental PV power plant. The dependencies of energy produced by the PV system for tilt angle range from 0° to 90° in different calendar months are presented in Figure 6. Especially for tilt angle 35° was measured the amount of produced electrical energy not only by the model PV system but also by the PV power plant. The correlation between the calculated values and the experimentally obtained results was from 99.18% to 99.89% during the year. The output power of the system was the highest for the tilt angle 30° and the optimal energy balance was detected for tilt angle from 15° to 30°. A very small difference of about 0.9% in energy production of PV system was identified for range (30–35°). If the tilt angle is greater than the 35°, the power decreases. The group data analysis and data extraction were carried out for identifying the model equation within the observed experimental dependence. For mathematical description of power energy balance for different tilt angle β was found mathematical Equation (4), which is a polynomial function of the second degree with the coefficient of determination $R^2 = 0.9999$.

$$E_m(\beta) = -0.9\,\beta^2 + 61.929\,\beta + 7514.3 \tag{4}$$

where: $E_m(\beta)$ is the energy produced by the PV system per month (kWh), β is the tilt angle (°). Figure 7 shows a more general relation which represents the result of the evaluation of the PV module tilt angle influence on the electricity production. The simulation in software Matlab® was used for detection of the optimal tilt angle for examined PV system installed in southern Slovakia region. The maximum of energy production was obtained for the tilt angle 34.5°.

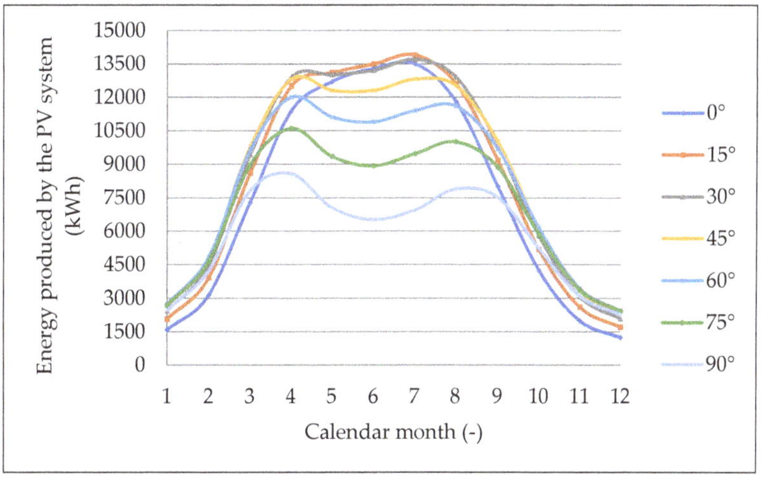

Figure 6. Dependencies of energy produced by the PV system for tilt angle range from 0° to 90°.

A part of the model equation verification has the polynomial function form described by Equation (4). An analytical model was created by using of iterative method for system modelling. Finally, both models (polynomial and analytical) were compared.

A 3rd order differential equation with constant coefficients—Equation (5)—was used to create the analytical model.

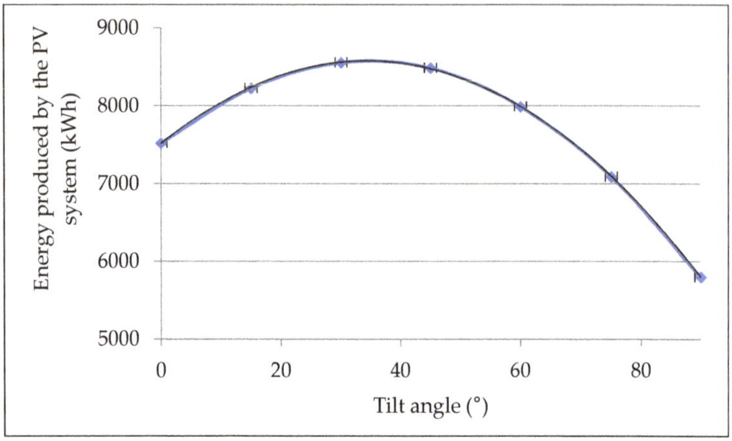

Figure 7. Dependence of the energy produced by the PV system for the different tilt angles.

The differential equation describes the behavior of the output variable. From the physical point of view, it represents the amount of energy produced by the PV system, which is characterized by a transfer function in a complex plain0—Equation (6).

$$\frac{d^3\,E(\beta)}{d\beta^3} + 0.02644 \cdot \frac{d^2\,E(\beta)}{d\beta^2} + 0.007259 \cdot \frac{d\,E(\beta)}{d\beta} + 5.472 \cdot 10^{-11} \cdot E(\beta) = -0.004562 \cdot \beta \tag{5}$$

$$G(s) = \frac{-0.004562}{s^3 + 0.02644 \cdot s^2 + 0.007259 \cdot s + 5.472 \cdot 10^{-11}} \tag{6}$$

This is based on the values of the independent variable, which was the tilt angle of PV modules β. The 95.96% reliability of the transfer function with respect to the experimental data was determined (Figure 8).

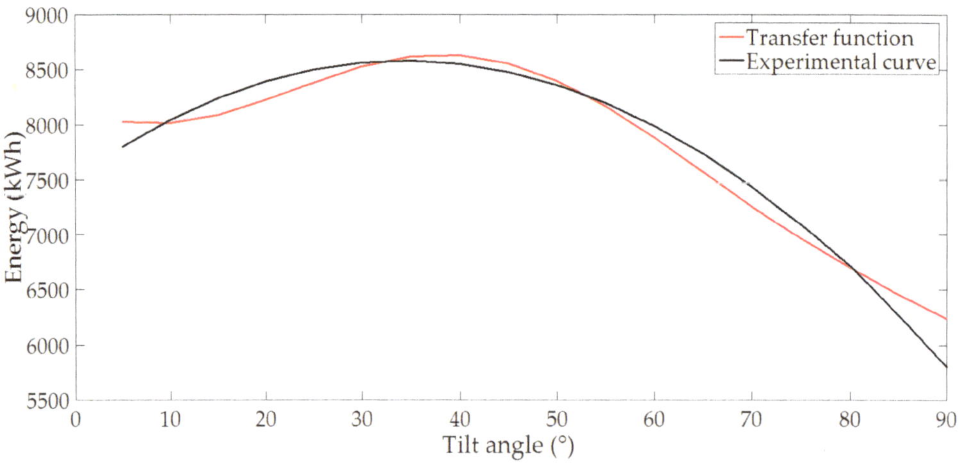

Figure 8. Comparison of the experimental curve and the curve of the analytical model expressed by the transfer function.

General solution of the differential Equation (7) has the form expressed as follows–Equation (7):

$$E(\beta) = \left(7245.99 \cdot e^{-7.53823 \cdot 10^{-9}\beta}\right) + \left(e^{-0.01322\beta} \cdot (-24.2152 \cdot \sin(0.0841679\beta) + 268.311 \cdot \cos(0.0841679\beta))\right) \tag{7}$$

From the global extreme analysis of the model polynomial function described by Equation (6) and analytical model function—Equation (7), the global maxima of the functions were identified for the tilt angle 34,405°.

The influence of the tilt angle on the PV system electricity production during the year was also modelled by three-dimensional graphical dependence shown in Figure 9. On the presented 3D dependence was applied polynomial approximation of second degree, which was used for the x-axis and the y-axis. The x-axis shows the time represented by calendar months; the y-axis shows the individual positions of PV modules which correspond to the tilt angles. The z-axis shows the average values of the electricity produced by the PV system. The final three-dimensional model dependency is shown in Figure 10. It can be described by the characteristics which are presented in Table 1 and for mathematical description of three-dimensional relation was obtained Equation (8). The relation can be used for prediction of PV system energy balance as a function of tilt angle and month. The importance of mathematical modelling dependencies that enable the calculation of the PV system energy production was also declared by the authors [18,47,62].

$$E_m(t,\ \beta) = -1438 + 56.33 \cdot t + 4218 \cdot \beta - 0.8993 \cdot t^2 + 0.8507 \cdot t \cdot \beta - 340.9 \cdot \beta^2 \tag{8}$$

where $E_m(t,\ \beta)$ is the energy produced by the PV system per month (kWh), β is the tilt angle of PV system (°), and t is the time (month).

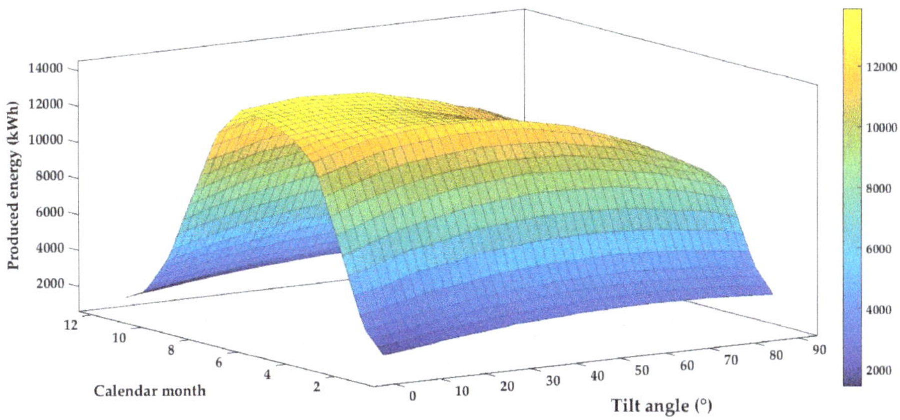

Figure 9. Three-dimensional dependence before polynomial approximation—the influence of tilt angle on the electric energy production of the PV system.

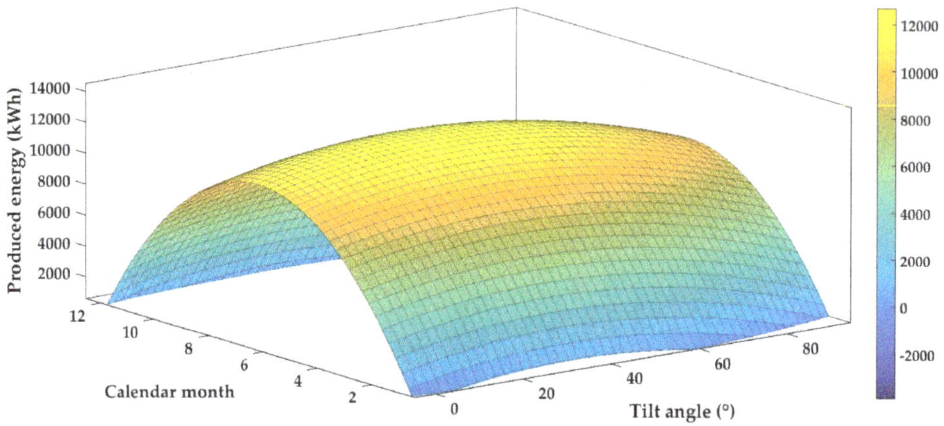

Figure 10. Three-dimensional dependence after polynomial approximation—the influence of tilt angle on the electric energy production of PV system for calendar months.

Table 1. The statistical and regression characteristics of approximated polynomial function for three-dimensional dependence on Figure 10.

Sum of Squares	**1.735×10^8**		
Coefficient of Determination	**0.8723**		
Standard error of arithmeticaverage	1.491×10^3		
Regression Equation	$E_m(t, \beta) = A + B{\cdot}t + C{\cdot}\beta + D{\cdot}t^2 + E{\cdot}t{\cdot}\beta + F{\cdot}\ \beta^2$		
Coefficient	Value	Minimum of the Range	Maximum of the Range
A (kWh)	−1438	−3063	−185.8
B (kWh·month^{-1})	56.33	12.4	100.3
C (kWh)	4218	3785	4652
D (kWh·month^{-2})	−0.8993	−1.315	−0.4835
E (kWh·month^{-1})	0.8507	−2.278	3.979
F (kWh)	−340.9	−371.6	−310.2

The author [2] in the publication presents that any change in the tilt angle of the PV panel is the cause of electric energy production decreasing up to 10%. Measurements on the model PV module showed the significant influence of the tilt angle, where the perpendicular placement of the PV module (90°) reduces the electricity production from 13,200 kWh to 6520 kWh, which represents 49.39%. If the tilt angle of the PV module is in the range from 0° to 30° during the months from April to August, the difference of electric energy production by PV system is minimal. Especially, the energy range was from 12,700 kWh to 13,100 kWh, which means the difference is 3.05%. However, if we compare the arithmetic average representing the energy production by the PV system in different months for different tilt angles, it is obvious that they differ from the total monthly average of 7668.571 kWh. This is especially in the range 5800–8560 kWh, which represents 24.37–11.62%. From the photovoltaic practice point of view, the average influence of the tilt angle on the energy production is approximately 18%. A considerable influence of the tilt angle is evident also from the graphical dependencies presented in Figures 6, 7, 9 and 10. The general mathematical description of the tilt angle influence is represented by Equation (4). This equation allows the simple calculation of electric energy production by the PV system after entering the time value and the position which corresponds to the tilt angle β of the PV panel. These parameters are easy to identify in practice. Based on the model equation, it is possible to determine the energy balance of a PV plant with an installed output of 100 kWp for any tilt angle and calendar month of the year in the southern Slovakia region.

The second part of research was focused on identification of optimal azimuth angle for southern Slovakia region. Azimuth angle change was measured on PV power plant and the model PV system. The results are summarized as graphical dependencies. For clarity, results are presented in two figures. At first, dependencies were created in the range −90–5° shown in Figure 11 and then were made relations for azimuth angles 0–90° in Figure 12. The results were compared with values experimentally obtained from the model PV system for seven different azimuth orientations (−90°, −60°, −30°, 0°, 30°, 60°, and 90°). The difference between values obtained from the PV power plant and model PV system was approximately 0.58%.

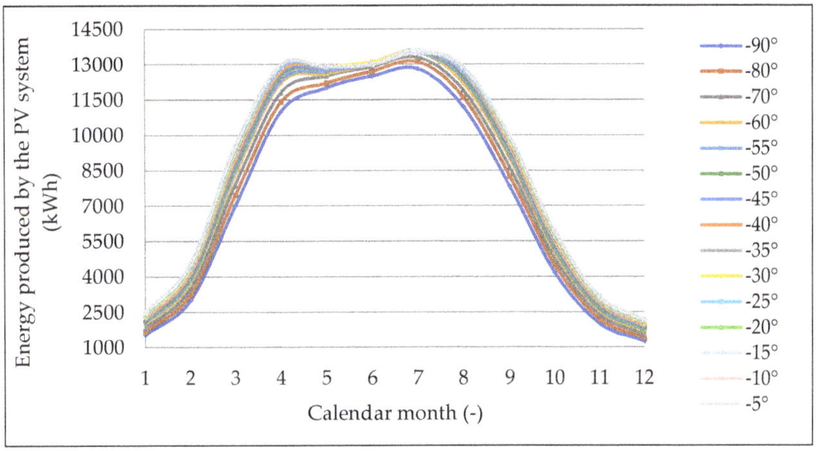

Figure 11. Dependencies of energy produced by the PV system for azimuth angle from −90° to 0° in different calendar months.

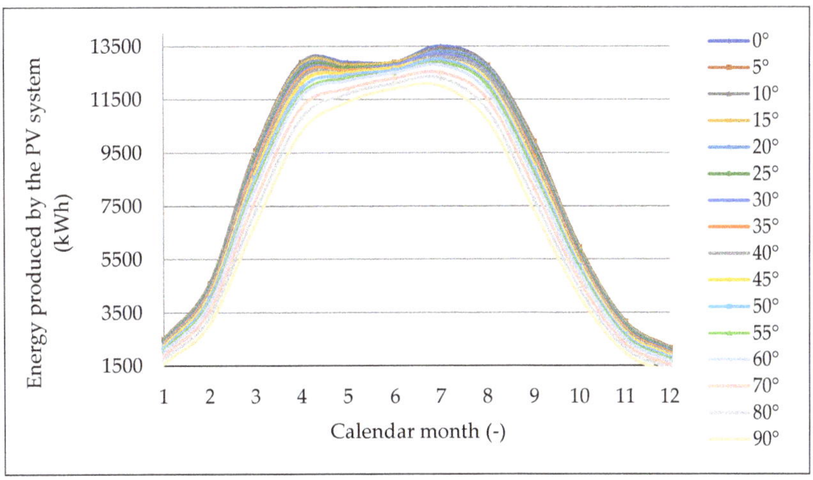

Figure 12. Dependencies of energy produced by the PV system for azimuth angle from 0° to 90° in different calendar months.

The measured data were statistically processed and the selected statistical characteristics for each data group were calculated. Especially, arithmetic average, deviations from arithmetic average, variance σ^2, standard deviations σ, standard error of arithmetic average $\overline{\delta}(t)$, and the median were identified. Selected summary results are presented in Table 2.

Table 2. Average amount of energy ($\overline{E}$) produced by PV system with different azimuth orientation and tilt angle 35°, average energy produced per month ($\overline{E}_m$).

Azimuth angle	−90°	−80°	−70°	−60°	−55°	−50°	−45°	−40°	−35°	−30°	−25°
$\overline{E}$ (kWh)	7190	7480	7750	7990	8090	8190	8280	8350	8420	8470	8520
Median (kWh)	7415	7855	8275	8670	8840	8995	9150	9290	9415	9525	9605
Azimuth angle	−20°	−15°	−10°	−5°	0°	5°	10°	15°	20°	25°	30°
$\overline{E}$ (kWh)	8560	8580	8590	8590	8580	8560	8540	8500	8450	8380	8310
Median (kWh)	9675	9730	9775	9785	9780	9755	9715	9650	9565	9460	9345
Azimuth angle	35°	40°	45°	50°	55°	60°	70°	80°	90°		
$\overline{E}$ (kWh)	8230	8150	8050	7950	7830	7710	7450	7170	6870		
Median (kWh)	9220	9065	8905	8735	8555	8365	7965	7530	7080		

Month	January	February	March	April	May	June
$\overline{E}_m$ (kWh)	2200.32	4094.19	8758.71	12280.65	12567.74	12793.55
$\overline{\delta}$(P) (kWh)	52.38	84.16	143.67	129.42	69.67	52.78
Month	July	August	September	October	November	December
$\overline{E}_m$ (kWh)	13225.81	12325.81	9221.29	5361.61	2778.71	1868.39
$\overline{\delta}$(P) (kWh)	72.67	104.48	101.07	97.96	64.6	51.17

Summary results for energy produced by PV system with different azimuth orientation are presented in Table 3. It is clear that the largest amount of electrical energy is produced for the azimuth orientation from −5° to −35°, which represents the energy range (8420–8590) kWh. In the South orientation 0°, the PV system produced 8580 kWh. In the Southwest orientation was the maximum energy production in range 8560–8450 kWh obtained for azimuth angles from +5° to +20°.

Table 3. The statistical and regression characteristics of approximated polynomial function for three-dimensional dependence in Figure 14.

Sum of Squares	**4.975 × 10^8**		
Coefficient of Determination	**0.9314**		
Standard error of arithmetic average	1.1680 × 10^3		
Regression equation	$E_m(t,\ \gamma) = A + B{\cdot}t + C{\cdot}\gamma + D{\cdot}\ t^2 + E{\cdot}t{\cdot}\gamma + F\ \gamma^2$		
Coefficient	Value	Minimum of the Range	Minimum of the Range
A (kWh)	−2215	−2656	
B (kWh·month^{-1})	−2.104	−7.389	3.18
C (kWh)	4965	4815	5116
D (kWh·month^{-2})	−0.1949	−0.2453	−0.1446
E (kWh·month^{-1})	−0.009663	−0.7277	0.7084
F (kWh)	−396.7	−408	−385.4

The comprehensive evaluation of the results for the azimuth angle that influenced the PV system energy production pointed to the fact that optimum energy is achieved at the azimuth angle −20–15°. However, the average difference is only 3.80% as compared with the orientation in the range (−35–35°), which is presented as acceptable for the azimuth angle for the PV module orientation in literature [16,63–65].

From the results interpretation of the azimuth angle changes simulation on the PV system energy production, it is clear that the time is one of the most important factors. The amount of energy produced by the PV system varies from month to month when changing the azimuth angle of rotation from $-90°$ to $+90°$. The extensive data set will assess the impact in selected model months of the relevant season. The largest amount of electricity is produced in the summer months, specifically in July, where for the above azimuth angle range, the PV system is produced at least 12,000 kWh at +90° orientation and 13,600 kWh at azimuth angle from (−50–15°). It means change of energy production 11.76% in relation to the maximum energy production in July. It follows from the above facts that in the summer months the influence of the azimuth angle is minimal and the changes in the amount of produced energy is on average 12.18%. After analyses of the energy production during the winter months in a similar way, the amount of produced energy in January was 1580 kWh for the azimuth angle +90° to 2520 kWh at the azimuth angle in range (0–5°). It means change of 37.30% from the maximum energy production in January. The influence of the azimuth angle changes in winter is on average 36.55%. It can be compensated for example by optimizing of the PV module tilt angle. From the spring months typical March is the model month, where the amount of produced energy was (6850–9610) kWh; the change in the amount of produced energy in March was affected by an azimuth angle of 28.72%. Minimum energy production was reached in March with azimuth orientation of +90° and maximum energy production with azimuth orientation of 0°. Numerically, the effect of the azimuth orientation changes on energy production is during the spring 19.38%. Similarly, model autumn month October was analysed. The minimum energy production during October was 4090 kWh at +90° and the maximum energy production was 5950 kWh for azimuth orientation 0°. The difference in October was 31.26%, and the average change of energy balance was 31.51% in the autumn season.

The presented results show that in terms of the energy balance, the maximal production was achieved for azimuth angle from $-5°$ to $-10°$ Southeast when the average energy production was 8590 kWh. More generally, the assessment of the effect of the PV module azimuth angle on the amount of produced energy is shown in Figure 13. The final observed dependence can be described by the regression Equation (9). It represents the polynomial function of the second degree with the coefficient of determination $R^2 = 0.998$.

$$E_m(\gamma) = -0.1943 \cdot \gamma^2 - 2.1746 \cdot \gamma + 8568.8 \tag{9}$$

where $E_m(\gamma)$ is the energy produced by the PV system per month (kWh) and γ is the azimuth angle (°).

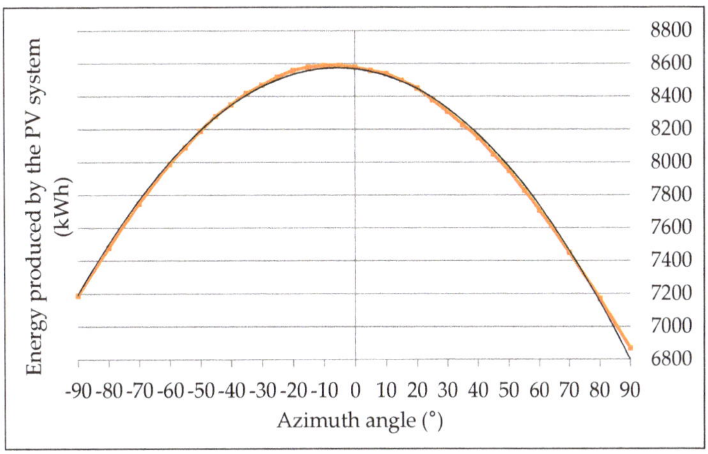

Figure 13. Relation between the PV system energy production and the azimuth angle.

In the case of Figure 13, a purely parabolic trend of functional dependence was detected. Within the analytical model, the first order differential equation with constant coefficients—Equation (10)—was compiled. As in the previous analytical solution, a transfer function in a complex plain described by Equation (11) was used and subsequently a general solution of the differential Equation (10) was found. The coefficient of determination for the transfer function—Equation (11)—was 96.25%,

$$\frac{dE(\gamma)}{d\gamma} + 490.3 \cdot 10^{-6} \cdot E(\gamma) = -0.3879 \cdot \gamma \tag{10}$$

$$G(s) = \frac{-0.3879}{s + 0.0004903} \tag{11}$$

where γ is the azimuth angle (°). From the mathematical description of the parabolic course by polynomial function of the second degree, it is possible to determine the position of the stationary point, which corresponds to an azimuth angle of −5.60°. Due to the information known from PV theory, a calculation was performed for the azimuth angle γ = 0°. The general solution of the differential Equation (10) for the mentioned azimuth angle γ can be expressed in the form of Equation (12):

$$E\ (\gamma) = 8568.8 \cdot e^{-0.0004903\ \gamma} \tag{12}$$

A comparison of the analytical model results with functional dependence is shown on Figure 14.

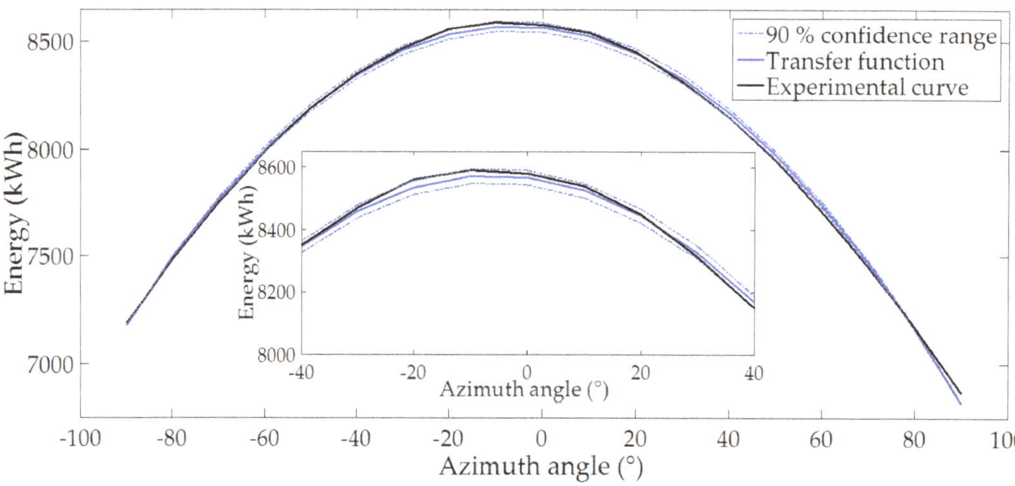

Figure 14. Comparison of relations between the energy and azimuth angle with 90% confidence for azimuth angle range (−90–90°) and detail for range (−40–40°).

Matlab® version R 2015a was used for the mathematical model creation. Three-dimensional dependencies were created by mathematical software. The suitable polynomial approximation was applied to the relation of data files. For the three-dimensional relation, selected statistical parameters and regression coefficients of Equation (13) were calculated, which are summarized in Table 3.

$$E_m(t,\ \gamma) = -3603 + 4966 \cdot t + 192.6 \cdot \gamma - 396.7 \cdot t^2 - 0.05377 \cdot t \cdot \gamma - 6.372\ \gamma^2 \tag{13}$$

where $E_m(t,\ \gamma)$ is the energy produced by the PV system per month (kWh), γ is azimuth angle (°), and t is the time (month).

The final relation for the influence of the azimuth angle on the amount of produced energy by the PV system can be modelled by using three-dimensional graphical dependence, which is shown in Figure 15. The x-axis shows the time represented by calendar months and the y-axis represents the positions which correspond to the azimuth angle of PV module. On the z-axis are plotted average values of produced energy.

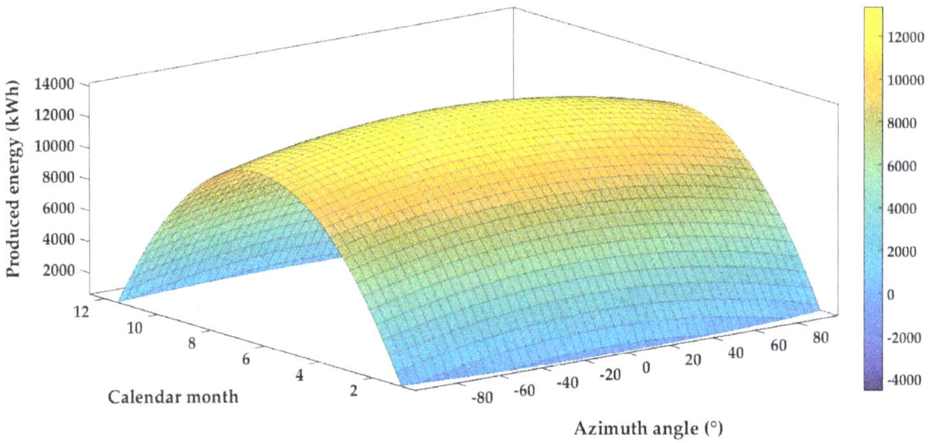

Figure 15. Three-dimensional dependence after polynomial approximation—the influence of azimuth angle on the electric energy production of PV system for calendar months.

4. Discussion

The research results can be divided into two parts. In the first part the effect of the tilt angle changes will be discussed. The second part of presented research aimed to determine the influence of the azimuth orientation on the resulting energy balance of the PV system. The results mentioned in the previous chapter take into account not only experimental data but also mathematical models which were created. The mathematical evaluation of the final function relations was done by software Matlab®.

The polynomial function pointed to the fact that the tilt angle 34.5° is the best for PV module installed in the southern Slovakia region. The obtained result was verified by analytical model, which confirmed that optimal tilt angle of PV module installed in the Southern Slovakia regions is 34,405°, which agrees with the result calculated by using the model polynomial equation. The tilt angle difference between models was 0.095°. Presented results are also in good agreement with the information presented in literature [65] and values of energy production amounts for every month are consistent with the results reported by research in the articles [20,21].

From numerical assessment of results for azimuth angle orientation in the southern Slovakia, it is evident that maximal power and energy production of PV system was obtained for azimuth angle in the range (−20–15°). This fact is in agreement with information presented in the literature [66], but the best energy production is detected if the azimuth angle is −5°, which means the South-East orientation. The average energy losses if the PV modules have South (0°) azimuth orientation were 0.12%. The fact is known also from the literature [13] where general information is presented. Our research confirmed significant influence of azimuth angle and tilt angle on PV system energy production, which is in accordance with results presented by authors [11,17,27,37,53,56].

For verification and evaluation of the experimental results and the created mathematical models special photovoltaic software was also used, which enabled the simulation of the operating conditions of the PV system for the selected area. In this software the simulation of solar radiation and PV system energy production was performed. Software calculates the solar radiation from satellite according to the methodology described in publications [62–64]. The calculation of PV system power is based on Equation (14),

$$P = \frac{G}{1000} \cdot A \cdot eff(I_G, T_m) \tag{14}$$

where P is power of PV module (W), I_G is solar radiation (W.m^{-2}), A is area of PV system (m^2), *eff* is efficiency of PV system (%), and T_m is temperature of PV module (°C) [62]. The total amount of energy PV system production was identified as a product of the PV system power by using Equation (14) and selected time range.

The dependencies found in the experimental work were summarized and made a contribution to a comprehensive basis for the creation of a final mathematical model. The final model will be a platform for creation of smart application. Part of the flowchart is presented in Figure 16. Due to the applicability of the mathematical model in practice and the possibility of its use by the target group, which will be ordinary users, the model had to be simplified. It contains easily identifiable parameters. The created mathematical model allows for the prediction of power or energy of a PV system installed in southern Slovakia region. The equation for the simplified performance model has the form described by Equation (15):

$$P(t) = k_{type} \cdot k_{const} \cdot k_{tilt} \cdot k_{azim} \cdot k_{tcell} \cdot k_{loss} \cdot f_G(t) \cdot P_{inst} \tag{15}$$

where P is power of PV system in kW, k_{type} coefficient for PV module type, k_{const} is coefficient for PV system construction, k_{tilt} coefficient for PV module tilt angle, k_{azim} is coefficient for PV module azimuth angle, k_{tcell} is coefficient for PV cell temperature, k_{loss} coefficient for PV system loss, $f_G(t)$ is the function for solar radiation intensity in kW·m^{-2}, and P_{inst} is installed power of PV system in kW$_{p.}$

The tilt angle of PV panels in relation to the location and azimuth position was investigated experimentally in many studies. Latitude based model for tilt angle optimization for solar systems in the Mediterranean region was detected by [67]. Mentioned article presents also quadratic regression model that allows the prediction of the annual optimal tilt angle. Optimum tilt angle for 1 MW PV system at Sukkur in Pakistan was determined and described in literature [68]. The research results confirmed the tilt angle variations during the year from 0° to 61.1° in the northern Pakistan. It was found that optimal tilt angle for that region is 29.5°. An experimental and mathematical investigation of optimal tilt angle and effects of reflectors on PV energy production was presented by authors [69]. Experimental results show that for gain of optimum power output the tilt angle needs to be changed every month. From experimental study located in Nitte (India), it was found that the PV system produced maximum power output in April for tilt angle 0°; in March optimal tilt angle is 13°, for February 22°, tilt angle for January is 33° and 30° for December. World estimates of PV optimal tilt angles and ratios of sunlight incident upon tilted and tracked PV panels relative to horizontal panels were described in [70].

One of the best ways to optimize the energy balance of PV panels is to use solar trackers that can optimize their position. The use of solar trackers can increase electricity production by around a third, and some claim by as much as 40% in some regions, compared with modules at a fixed angle. In any solar application, the conversion efficiency is improved when the modules are continually adjusted to the optimum angle as the sun traverses the sky. As improved efficiency means improved yield, use of trackers can make quite a difference to the income from a large plant. This is why utility scale solar installations are increasingly being mounted on tracking systems [71]. One of the best technologies for PV module position optimalization are dual axis trackers. Dual axis trackers typically have modules oriented parallel to the secondary axis of rotation. No matter where the Sun is in the sky, dual axis trackers are able to angle themselves to be in direct contact with the Sun [72]. Very innovative idea is presented in literature [51] where authors described design and performance of a new self-powered LCPV solar trackers using bifacial solar cells and concentrating mirror. All mentioned references confirmed that the orientation of the module with respect to the tracker axis is important when modelling performance. However, when installing solar trackers, there are design and economic limitations that do not allow their use. In this case, it is essential to install a PV system with fixed panels into the best operating position and it is appropriate to use the created mathematical model.

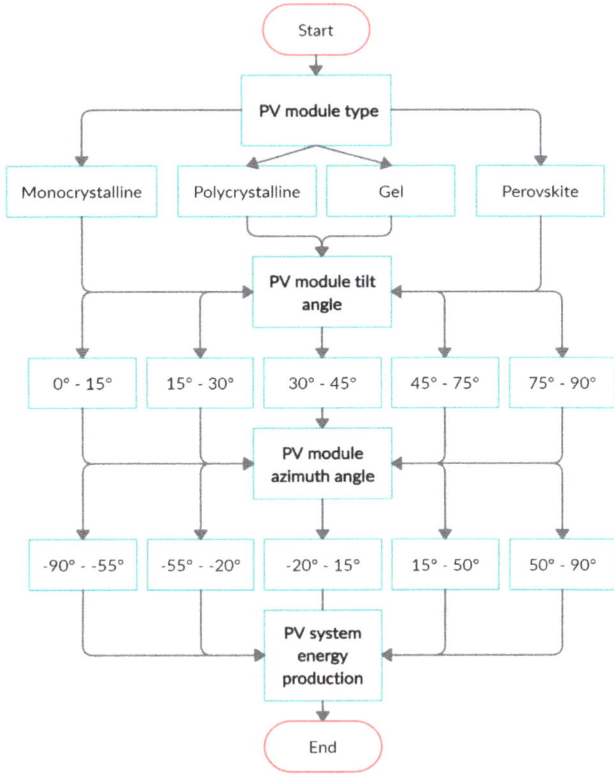

Figure 16. Flowchart of the prepared application for smart devices.

All the above mentioned publications confirm the local variability of the PV panels tilt angle and azimuth orientation. It is in accordance with the results presented in the text. It should be emphasized that the effect of both angle changes on energy production in Central Europe was declared by other experimental results which were obtained during the several years in Czech Republic, specifically in the region of Brno and Prague. Identical methods of processing and evaluation of experimental data were used. It was found that in the region of Nitra in Slovakia the optimal angle of inclination is 34.5°, in the region of Brno in the Czech Republic it is 34.7°, and in the region of Prague the optimal tilt angle is 35°.

5. Conclusions

From the presented results it is a clear well-known fact that the tilt angle and azimuth angle have significant influence on the total energy balance of the PV system. The research also confirmed that is very important to know the optimal tilt angle and azimuth angle orientation for the area in which the PV system is installed. The first benefit of presented research is a simplified mathematical model for energy production expressed through the time relations for tilt angle and azimuth angle which contains a very small number of input parameters which are easily identifiable to ordinary users of PV system. The results of the research presented in the manuscript are localized to the area of Southern Slovakia, but the general forms of model equations are applicable to the area of Central Europe respectively for areas with similar climatic conditions. For a detailed prediction of energy production in certain areas, a modification of regression equation coefficients and differential equations can be applied. The second benefit is that the model can be described by the simple polynomial function of the second degree. Since the model equations were created on the basis of data processing from long-term monitoring of the operating parameters of the PV system, the resulting model, although not based on data obtained by satellite measurements, gives results better corresponding to the real energy production of the PV system at the site. The third benefit is that the mentioned mathematical model gives about 4% more accurate results than models used so far for the locality of southern Slovakia. Experimental methods, numerical simulation, and data analysis find that the effect of PV module tilt angle on the amount of electricity generated by the PV power plant is about 18%. The ideal tilt angle for southern Slovakia region is 34.5°. The effect of azimuth angle changes on the PV system energy production is 24.19% during the year. Maximum energy production was achieved in the selected locality at azimuth angle from the range (−20–15°). All obtained results are a part of final mathematical model for simulation of PV system power and energy balance. This model created the platform for development of application for smart devices. Presented results allow a simple prediction of the energy production of a PV power plant installed in real operating conditions and they can be used in practice for the design, dimensioning, and optimization of PV power plants in the southern Slovakia.

Author Contributions: Conceptualization, M.B. (Monika Božiková) and M.B. (Matúš Bilčík); methodology, M.B. (Monika Božiková) and M.B. (Matúš Bilčík); software, M.B. (Matúš Bilčík) and V.M.; validation, M.B. (Monika Božiková), M.B. (Matúš Bilčík) and V.M.; formal analysis, M.B. (Matúš Bilčík) and V.M.; investigation, M.B. (Monika Božiková), M.B. (Matúš Bilčík) and L'.K.; resources, M.B. (Matúš Bilčík) and T.S.; data curation, M.B. (Monika Božiková), M.B. (Matúš Bilčík), L'.K., V.M., and V.C.; writing—original draft preparation, M.B. (Monika Božiková) and M.B. (Matúš Bilčík), V.C. and V.M.; writing—review and editing, L'.K. and V.C.; visualization, M.B. (Matúš Bilčík) and V.M.; supervision, M.B. (Monika Božiková), V.C., L'.K. and J.L.; project administration, M.B. (Matúš Bilčík) and T.S.; funding acquisition, M.B. (Matúš Bilčík) and T.S. All authors have read and agreed to the published version of the manuscript.

Funding: This publication was funded by the Grant Agency Slovak University of Agriculture (GA SPU), grant number No 03-GASPU-2021.

Institutional Review Board Statement: No applicable.

Informed Consent Statement: No applicable.

Data Availability Statement: The data presented in this study are available in the main text of this document.

Conflicts of Interest: The authors declare no conflict of interest.

References

1. Kamanga, B.; Mlatho, J.S.P.; Mikeka, C.; Kamunda, C. Optimum Tilt Angle for Photovoltaic Solar Panels in Zomba District, Malawi. *J. Sol. Energy.* **2014**, *3*, 1–9. [CrossRef]
2. Libra, M.; Poulek, V. *Photovoltaics: Theory and Practice of Solar Energy Usage*, 1st ed.; Ilsa: Praha, Czech Republic, 2009; 160p. (In Czech)
3. Sadati, S.M.; Qureshi, F.; Baker, D. Energetic and economic performance analyses of photovoltaic, parabolic trough collector and wind energy systems for Multan, Pakistan. *Renew. Sustain. Energy Rev.* **2015**, *47*, 844–855. [CrossRef]
4. Kalogirou, S.A. Solar thermal collectors and applications. *Prog. Energy Combust. Sci.* **2004**, *30*, 231–295. [CrossRef]
5. An Assessment of Solar Energy Conversion Technologies and Research Opportunities. Available online: https://gcep.stanford.edu/pdfs/assessments/solar_assessment.pdf (accessed on 5 June 2021).
6. Renewables 2020 Global Status Report. Available online: https://www.ren21.net/wp-content/uploads/2019/05/gsr_2020_full_report_en.pdf (accessed on 5 June 2021).
7. Xu, R.; Ni, K.; Hu, Y.; Si, J.; Wen, H.; Yu, D. Analysis of the optimum tilt angle for a soiled PV panel. *Energy Convers. Manag.* **2017**, *148*, 100–109. [CrossRef]

8. Jafarkazemi, F.; Saadabadi, S.A. Optimum tilt angle and orientation of solar surfaces in Abu Dhabi, UAE. *Renew. Energy* **2013**, *56*, 44–49. [CrossRef]
9. Antonanzas, J.; Urraca, R.; Martinez-de-Pison, F.J.; Antonanzas, F. Optimal solar tracking strategy to increase irradiance in the plane of array under cloudy conditions: A study across Europe. *Sol. Energy* **2018**, *163*, 122–130. [CrossRef]
10. Hussein, H.M.S.; Ahmad, G.E.; El-Ghetany, H.H. Performance evaluation of photovoltaic modules at different tilt angles and orientations. *Energy Convers. Manag.* **2004**, *45*, 2441–2452.
11. Rowlands, I.H.; Kemery, B.P.; Beausoleil-Morrison, I. Optimal solar-PV tilt angle and azimuth: An Ontario (Canada) case-study. *Energy Policy* **2011**, *39*, 1397–1409. [CrossRef]
12. Salih, S. Effect of Tilt Angle Orientation on Photovoltaic Module Performance. *ISESCO Cent. Promot. Sci. Res.* **2014**, *10*, 19–25.
13. Hafez, A.Z.; Soliman, A.; El-Metwally, K.A.; Ismail, I.M. Tilt and azimuth angles in solar energy applications—A review. *Renew. Sustain. Energy Rev.* **2017**, *77*, 147–168. [CrossRef]
14. Yadav, A.; Chandel, S. Tilt angle optimization to maximize incident solar radiation: A review. *Renew. Sustain. Energy Rev.* **2013**, *23*, 503–513. [CrossRef]
15. Laveyne, J.; Bozalakov, D.; Van Eetvelde, G.; Vandevelde, L. Impact of Solar Panel Orientation on the Integration of Solar Energy in Low-Voltage Distribution Grids. *Int. J. Photoenergy* **2020**, *2020*, 2412780. [CrossRef]
16. Mehleri, E.D.; Zervas, P.L.; Sarimveis, H.; Palyvos, J.A.; Markatos, N.C. Determination of the optimal tilt angle and orientation for solar photovoltaic arrays. *Renew. Energy* **2010**, *35*, 2468–2475. [CrossRef]
17. Dhimish, M.; Silvestre, S. Estimating the impact of azimuth-angle variations on photovoltaic annual energy production. *Clean Energy* **2019**, *3*, 47–58. [CrossRef]
18. Bakirci, K. General models for optimum tilt angles of solar panels: Turkey case study. *Renew. Sustain. Energy Rev.* **2012**, *16*, 6149–6159. [CrossRef]
19. Ertekin, C.; Evrendilek, F.; Kulcu, R. Modeling Spatio-Temporal Dynamics of Optimum Tilt Angles for Solar Collectors in Turkey. *Sensors* **2008**, *8*, 2913. [CrossRef]
20. Stanciu, D.; Stanciu, C.; Paraschiv, I. Mathematical links between optimum solar collector tilts in isotropic sky for intercepting maximum solar irradiance. *J. Atmos. Sol.-Terr. Phys.* **2016**, *137*, 58–65. [CrossRef]
21. Hartner, M.; Ortner, A.; Hiesl, A.; Haas, R. East to west? The optimal tilt angle and orientation of photovoltaic panels from an electricity system perspective. *Appl. Energy* **2015**, *160*, 94–107. [CrossRef]
22. Calabr, E. An algorithm to determine the optimum tilt angle of a solar panel from global horizontal solar radiation. *J. Renew. Energy* **2013**, *1*, 307547. [CrossRef]
23. Ibrahim, D. Optimum tilt angle for solar collectors used in Cyprus. *Renew Energy* **1995**, *6*, 813–819. [CrossRef]
24. De Miguel, A.; Bilbao, J.; Diez, M. Solar radiation incident on tilted surfaces in Burgos, Spain: Isotropic models. *Energy Convers Manag.* **1995**, *36*, 945–951. [CrossRef]
25. Jafari, S.; Javaran, E.J. An Optimum Slope Angle for Solar Collector Systems in Kerman Using a New Model for Diffuse Solar Radiation. *Energy Sources Part A Recovery Util. Environ. Eff.* **2012**, *34*, 799–809. [CrossRef]
26. Tamimi, A.; Sowayan, A. Optimum Tilt Angles of Flat-plate Solar Collectors at Riyadh, Kingdom of Saudi Arabia. *Energy Sources Part A Recovery Util. Environ. Eff.* **2012**, *34*, 1213–1221. [CrossRef]
27. Benghanem, M. Optimization of tilt angle for solar panel: Case study for Madinah, Saudi Arabia. *Appl Energy* **2011**, *88*, 1427–1433. [CrossRef]
28. Elminir, H.K.; Ghitas, A.E.; El-Hussainy, F.; Hamid, R.; Beheary, M.M.; Abdel-Moneim, K.M. Optimum solar flat-plate collector slope: Case study for Helwan, Egypt. *Energy Convers. Manag.* **2006**, *47*, 624–637. [CrossRef]
29. Alatarawneh, I.S.; Rawadieh, S.I.; Tarawneh, M.S.; Alrowwad, S.M.; Rimawi, F. Optimal tilt angle trajectory for maximizing solar energy potential in Ma'anareain in Jordan. *J. Renew. Sus. Energy* **2016**, *8*, 587–598.
30. Shariah, A.; Al-Akhras, M.A.; Al-Omari, I.A. Optimizing the tilt angle of solar collectors. *Renew. Energy* **2002**, *26*, 587–598. [CrossRef]
31. Skeiker, K. Optimum tilt angle and orientation for solar collectors in Syria. *Energy Convers. Manag.* **2009**, *50*, 2439–2448. [CrossRef]
32. Khahro, S.F.; Tabbassum, K.; Talpur, S.; Alvi, M.B.; Liao, X.; Dong, L. Evaluation of solar energy resources by establishing empirical models for diffuse solar radiation on tilted surface and analysis for optimum tilt angle for a prospective location in southern region of Sindh, Pakistan. *Int. J. Electr. Power Energy Syst.* **2015**, *64*, 1073–1080. [CrossRef]
33. Krishna, S.M.; Madhu, M.N.; Mohan, V.; Suresh, M.P.R.; Singh, J.G. A generalized approach for enhaced solar energy harvesting using stochastic estimation of optimum tilt angles: A case study of Bangkok City. *Green* **2015**, *5*, 1–6.
34. Handoyo, E.; Ichsani, D.; Prabowo, P. The Optimal Tilt Angle of a Solar Collector. *Energy Procedia* **2013**, *32*, 166–175. [CrossRef]
35. Shen, Y.; Zhang, J.; Guo, P.; Wang, X. Impact of solar radiation variation on the optimal tilted angle for fixed grid-connected PV array-case study in Beijing. *Glob. Energy Intercon.* **2018**, 460–466.
36. Li, D.H.W.; Lam, T.N.T. Determining the optimum tilt angle and orientation for solar energy collection based on measured solar radiance data. *Int. J. Photoenergy* **2007**, *8*, 085402. [CrossRef]
37. Chang, Y.P. Optimal the tilt angles for photovoltaic modules in Taiwan. *Int. J. Electr. Power Energy Syst.* **2010**, *32*, 956–964. [CrossRef]
38. Jamil Ahmad, M.; Tiwari, G. Optimization of tilt angle for solar collector to receive maximum radiation. *Open Renew. Energy J.* **2009**, *2*, 19–24.

39. Agarwal, A.; Vashishtha, V.K.; Mishram, S.N. Comparative approach for the optimization of tilt angle. *Int. J. Eng. Res. Technol.* **2012**, 1–9.
40. Hiraoka, S.; Fujii, T.; Takakura, H.; Hamakawa, Y. Tilt angle dependence of output power in an 80kWp hybrid PV system installed at Shiga in Japan. *Sol. Energy Mater. Sol. Cells* **2003**, *75*, 781–786. [CrossRef]
41. Ghosh, H.R.; Bhowmik, N.C.; Hussain, M. Determining seasonal optimum tilt angles, solar radiations on variously oriented, single and double axis tracking surfaces at Dhaka. *Renew Energy* **2010**, *35*, 1292–1297. [CrossRef]
42. Siraki, A.G.; Pillay, P. Study of optimum tilt angles for solar panels in different latitudes for urban applications. *Sol. Energy* **2012**, *86*, 1920–1928. [CrossRef]
43. Lave, M.; Kleissl, J. Optimum fixed orientations and benefits of tracking for capturing solar radiation in the continental United States. *Renew Energy* **2011**, *36*, 1145–1152. [CrossRef]
44. Gong, X.; Kulkarni, M. Design optimization of a large scale rooftop photovoltaic system. *Sol. Energy* **2005**, *78*, 362–374. [CrossRef]
45. Moon, S.H.; Felton, K.E.; Johnson, A.T. Optimum tilt angles of a solar collector. *Energy* **1981**, *6*, 895–899. [CrossRef]
46. Tang, R.; Wu, T. Optimum tilt angles for solar collectors used in China. *Appl. Energy* **2004**, *79*, 239–248. [CrossRef]
47. Nguyen, H.; Nguyen, M. Mathematical modeling of photovoltaic cell/module/arrays with tags in Matlab/Simulink. *Environ. Syst. Res.* **2015**, *4*, 1–13. [CrossRef]
48. Duffie, J.A.; Beckman, W.A. *Solar Engineering of Thermal Processes*, 4th ed.; Wiley: New York, NY, USA; 928p.
49. Pardo, M.; Manzano, J.; García-Márquez, D. Energy Consumption Optimization in Irrigation Networks Supplied by a Standalone Direct Pumping Photovoltaic System. *Sustainability* **2018**, *10*, 4203. [CrossRef]
50. Maghami, M.R.; Hizam, H.; Gomes, C.; Mohd, R.; Mohd, A.; Rezadad, S.; Hajighorbani, S. Power Loss Due to Soiling on Solar Panel: A review. *Renew. Sustain. Energy Rev.* **2016**, *59*, 1307–1316. [CrossRef]
51. Poulek, V.; Khudysh, A.; Libra, M. Self powered solar tracker for Low Concentration PV (LCPV) systems. *Sol. Energy* **2016**, *127*, 109–112. [CrossRef]
52. Mishra, J.; Thakur, R.; Deep, A. Arduino based Dual Axis Smart Solar Tracker. *Int. J. Adv. Eng. Manag. Sci.* **2017**, *5*, 239849. [CrossRef]
53. Hailu, G. Fung. Optimum Tilt Angle and Orientation of Photovoltaic Thermal System for Application in Greater Toronto Area, Canada. *Sustainability* **2019**, *11*, 6443. [CrossRef]
54. Alsadi, S.Y.; Nassar, Y.F. Estimation of solar irradiance on solar fields: An analytical approach and experimental results. *IEEE Trans Sustain. Energy* **2017**, *8*, 1601–1608. [CrossRef]
55. Thanailakis, P.T.A. Direct computation of the array optimum tilt angle in constant-tilt photovoltaic systems. *Sol. Cells* **1985**, *14*, 83–94.
56. Mondol, J.D.; Yohanis, Y.G.; Norton, B. The impact of array inclination and orientation on the performance of a grid-connected photovoltaic system. *Renew. Energy* **2007**, *32*, 118–140. [CrossRef]
57. Sidek, M.H.; Azis, N.; Hasan, W.Z. Automated positioning dual-axis solar tracking system with precision elevation and azimuth angle control. *Energy* **2017**, *124*, 160–170. [CrossRef]
58. Mahmoud, Y.; El-Saadany, E.F. A Novel MPPT Technique based on an image of PV modules. *IEEE Trans. Energy Convers* **2017**, *32*, 213–221. [CrossRef]
59. Fouad, M.; Shihata, L.A.; Morgan, E.I. An integrated review of factors influencing the performance of photovoltaic panels. *Renew. Sustain. Energy Rev.* **2017**, *80*, 1499–1511. [CrossRef]
60. Abdallah, R.; Natsheh, E.; Juaidi, A.; Samara, S.; Manzano-Agugliaro, F. A Multi-Level World Comprehensive Neural Network Model for Maximum Annual Solar Irradiation on a Flat Surface. *Energies* **2020**, *13*, 6422. [CrossRef]
61. Shareef, S. The Impact of Tilt Angle on Photovoltaic Panel Output. *J. Pure Appl. Sci.* **2017**, *29*, 112–118.
62. Gracia Amillo, A.M.; Huld, T. Performance comparison of different models for the estimation of global irradiance on inclined surfaces. In *Validation of the model implemented in PVGIS*; JRC Technical Report; Publications Office of the European Union: Brussels, EUR 26075 EN, 2013.
63. Müller, R.; Behrendt, T.; Hammer, A.; Kemper, A. A new algorithm for the satellite-based retrieval of solar surface irradiance in spectral bands. *Remote Sens.* **2012**, *4*, 622. [CrossRef]
64. Müller, R.; Matsoukas, C.; Gratzki, A.; Behr, H.; Hollmann, R. The CM-SAF operational scheme for the satellite based retrieval of solar surface irradiance—A LUT based eigenvector hybrid approach. *Remote Sens. Environ.* **2009**, *113*, 1012–1024. [CrossRef]
65. Poulek, V.; Libra, M. *Solar Energy*, 1st ed.; CUA Praha: Praha, Czech Republic, 2006; 153p.
66. Cenek, M. *Renewable Energy Sources*, 1st ed.; FCC Public: Praha, Czech Republic, 2001; 174p. (In Czech)
67. Hassane, D.; Driss, L. Latitude Based Model for Tilt Angle Optimization for Solar Collectors in the Mediterranean Region. *Energy Procedia* **2013**, *42*, 426–435.
68. Qasir, A.M.; Abdul, Q.R.; Khurshed, A.; Muhammad, F.S.; Shoaib, A.S. Determining Optimum Tilt Angle for 1 MW Photovoltaic System at Sukkur, Pakistan. *Int. J. Photoenergy* **2021**, *2021*, 5552637.
69. Krishna, Y.; Karinka, S.; Fauzan, I.T.D.M.F.; Pai, M. An Experimental and Mathematical investigation of optimal tilt angle and effects of reflectors on PV energy production. *MATEC Web Conf.* **2021**, *335*, 03020. [CrossRef]
70. Mark, Z.J.; Vijaysinh, J. World estimates of PV optimal tilt angles and ratios of sunlight incident upon tilted and tracked PV panels relative to horizontal panels. *Sol. Energy* **2018**, *169*, 55–66.

71. Appleyard, D. Solar trackers: Facing the sun. *Renew. Energy World* **2009**, *12*, 41–53.
72. Barsoum, N. Fabrication of dual-axis solar tracking controller project. *Intell. Control Autom.* **2011**, *2*, 57–68. [CrossRef]

applied sciences

MDPI

Review

Evaluation of Investments in Wind Energy Projects, under Uncertainty. State of the Art Review

Benjamin Murgas [1,*], Alvin Henao [2] and Luceny Guzman [2]

1 Administrative Planning Department, Gobernación de La Guajira, Riohacha 440001, Colombia
2 Industrial Engineering Department, Universidad del Norte, Puerto Colombia 081007, Colombia; henaoa@uninorte.edu.co (A.H.); lguzman@uninorte.edu.co (L.G.)
* Correspondence: n.murgas@laguajira.gov.co

Abstract: The use of renewable energy sources, especially wind energy, has been widely developed, mostly during the last decade. The main objective of the present study is to conduct a literature review focused on the evaluation under uncertainty of wind energy investment using the real options approach to find out whether public opposition (NIMBY projects) has been contemplated, and if so, what have been the flexible strategies applied for its intervention. Overall, 97 publications were analyzed, identifying 20 different models or approaches, which were grouped into eight categories: 1. Real options, 2. Optimization, 3. Stochastics, 4. Financial evaluation, 5. Probabilistic, 6. Estimation, 7. Numerical prediction, and 8. Others. The real options approach, present in 32% of the studies, was the most popular. Twenty-eight types of uncertainties were identified, which were grouped, for better analysis, into nine categories. In total, 62.5% of the studies included the price of electricity as a source of uncertainty; 18.8%, the velocity of wind; and 15.6%, the feed-in rates-subsidy. Both random and non-random techniques were applied to assess the real options and to model the uncertainties. When evaluating real options, the Monte Carlo simulation technique was the most preferred, with 16 (51.6%) applications, followed by non-randomized techniques, decision tree, and dynamic programming, with eight (25.8%) applications each. There is a marked tendency to use stochastic processes to model uncertainty, particularly geometric Brownian motion, which was used in 61.3% (19) of the studies in the sample. When searching for "real options AND (nimby OR public opposition)", no study was found, which shows the possibility of developing research on this aspect to determine its impact on investments in wind energy projects.

Keywords: wind energy; investment appraisal; real options; uncertainty; NIMBY projects; public opposition

Citation: Murgas, B.; Henao, A.; Guzman, L. Evaluation of Investments in Wind Energy Projects, under Uncertainty. State of the Art Review. *Appl. Sci.* **2021**, *11*, 10213. https://doi.org/10.3390/app112110213

Academic Editor: Alberto Benato

Received: 1 September 2021
Accepted: 19 October 2021
Published: 31 October 2021

Publisher's Note: MDPI stays neutral with regard to jurisdictional claims in published maps and institutional affiliations.

1. Introduction

Approximately 80% of the world's energy demand is met with fossil fuels (IEA, 2017a), causing two thirds of the global emissions of CO_2 (https://www.un.org/es/chronicle/article/el-papel-de-los-combustibles-fosiles-en-un-sistema-energetico-sostenible accessed on 19 June 2021).

According to the 2019 report from the International Energy Agency (IEA) (www.eia.gov/aeo accessed on 19 June 2021), world energy consumption is projected to increase by 46.9% in 2050, going from 620 billion BTUs, produced in 2018, to 911 billion BTUs. If the sources of energy were not changed, this would generate a greater demand for fossil fuels, increasing the emission of carbon dioxide (CO_2) affecting the levels of global warming and climate change, greatly risking the survival of humankind [1].

With the aim of promoting a sustainable future for the planet, global strategies been proposed, such as the 2030 Agenda, adopted by the United Nations General Assembly in September 2015, which included a global goal on sustainable energy [2]. In addition, the Paris agreement (2015) established the aim of keeping the global average temperature below 2 °C above pre-industrial levels, and to continue efforts to bring it to 1.5 °C [3].

In 2020, the UN calls for urgent action to achieve zero net greenhouse gas (GHG) emissions by 2050, which involves more than 120 countries, that contribute more than half the global GDP, alongside thousands of companies and investors, cities, regions, and universities. Countries aligned with this goal currently represent two-thirds of the global economy, and 63% of global greenhouse gas emissions. The list includes the United States, the European Union, Japan, South Korea, Canada, South Africa, and China, which have committed to meeting the target by 2060 [4]. Colombia, for its part, seeks to reduce nine million tons of CO_2 by 2030 [5].

Regardless, tackling climate change and the high levels of greenhouse gas emissions continue to be priority problems, meriting urgent solutions on a global scale. This need increases the use of renewable energy sources as a replacement of fossil fuels and, therefore, makes them play an important role in shaping the energy matrix of the future [6–10].

Within renewable energies, wind energy is one of the fastest growing in the world. From 2011 to 2020, its production capacity increased from 220 GW to 733 GW, 233% [11]. However, to achieve the 2050 goal, annual increases in wind power capacity must reach 160 GW in 2025, and rise to 280 GW by 2030, three times the level of 2020, which would give renewable energies a participation of 60% in the world energy total, with 30% coming from wind and solar energy; meaning that it would require going from an investment of 380 billion dollars in 2020 to 1.6 trillion dollars by 2030, according to the IEA [4].

In addition, the growth and development of wind energy creates new challenges and opportunities due to: (i) a greater competitiveness of the energy sector [12–14], (ii) the discontinuity that characterizes the generation of wind energy [15,16], (iii) the need to improve operations and maintenance [17,18], (iv) better conditions as required by the distribution system and grid integration [19,20], (v) the need for a wind energy regulation policy [21], and (vi) the environmental impact associated with power generation [22,23]. The barriers, which prevent a generalized use of wind energy, can be grouped into technological, economic, socio-political, and environmental dimensions, as shown in Table 1 [24].

Colombia projects large investments in wind energy generation, so these barriers can be increased. The areas with the greatest potential for wind energy (La Guajira Department-Upper Guajira region) are in areas that is are legally constituted as indigenous reservations (Decree 2164 [25]), inhabited by Wayuu communities, who act under their own traditions and customs, with total control over the property and use of the territory. In these regions, the communities are frequently opposed to interventions that affect their culture and customs. For these reasons, any type of project requires prior consultations (ILO Convention 169 of 1989, incorporated to Colombian legislation by Law 21 of 1991 [26]), making public opposition something of utmost importance.

This article attempts to review the literature regarding public opposition (NIMBY projects) to determine the flexible strategies implemented to overcome this type of barriers.

Public opposition can stop investment in wind energy generation projects and, consequently, affect their profitability.

Although there are strategies to manage public opposition by involving the communities from the beginning of the project, they may have uncertain behaviors or attitudes during the execution phase, despite planning their participation. We believe that a flexible approach to this type of uncertainty may be appropriate, so this research will explore real options to manage public opposition. Thus, this article is organized as such: the following section contains information on the general strategies used to overcome public opposition in investment projects. Section 3 describes the methodology used to review the state of the art on real options as applied to face different sources of uncertainty. Section 4 shows the results and analyzes them. In particular, Section 4.2 focuses on the uncertainties explored in the articles under the real options approach, and Section 4.4 analyzes public opposition under the same approach, to identify whether public opposition has been considered as uncertainty. Section 5 discusses these results and, finally, the last section shows the conclusions of this article and future lines of research.

Table 1. Barriers to the development of the wind energy sector.

Type of Barrier	Description
Technological	• Integration and grid connection challenges • Lack of support infrastructure • Technology maturity and performance concerns • Harsh offshore natural conditions
Economic and market-based	• High initial capital costs and slow return of investment • Limited sources of funding • Underdeveloped offshore supply chains • Ever-changing policies with impact on profits • Carbon emissions and local air pollutants are not priced or only partially so.
Regulatory, political, and social	• Complex/outdated regulatory frameworks • Inadequate financial policy support • Lack of relevant standards and quality control measures • Lack of qualified and experienced professional labor • Lack of long-term and stable policy goals and a well-coordinated policy mix • Transportation of wind turbine components
Environmental	• Impact on marine fauna • Visual impact • Shadow flicker • Radar interference • Noise • Land use • Public opposition-NIMBY "Not in my backyard"

Source: [24] IRENA, 2019.

2. Strategies to Beat Public Opposition

In order to face public opposition, strategies that involve local communities in the early stages of the project and promote community ownership models have been proposed, such as: (i) working with local planning authorities and keeping them committed during the different stages of development and operation of the site [27], (ii) promoting an equitable distribution of economic benefits and costs, generating additional income from the lease of adjacent land and creating jobs for the people from the area during the installation and operation of the wind park [28], (iii) providing additional services, such as educational programs or visits, improvement and maintenance of the landscape, tourist facilities, and sponsorship of local events, among others [29], (iv) creating citizen participation mechanisms to provide local inhabitants with information and the means to air their apprehensions regarding the project and, potentially, influence its design or even its operation, such as letting local authorities stop plant operation in case of accident or malfunction.

Some authors conclude that material compensations are not fully adequate. Although they may work, their effectiveness is conditioned by their design and implementation. The problem of local opposition is resolved by establishing public policies that integrate elements of territorial, local, and national planning, with spaces for participation and consensus building, and economic compensation systems that strengthen the acceptability of projects, instead of diluting it [30].

3. Methodology

For the development of this research, a search was carried out for relevant literature on investment in wind energy under uncertainty in the electronic databases of Web of Science and Scopus, during the period between 2006 and 2020. The search process was carried out in two stages. In the first, two search criteria were used: (i): "wind energy investment projects with uncertainty", generating 196 publications, and (ii): "economic evaluation of investment in wind power generation with uncertainty, obtaining 24 publications, for a total of 220 publications.

The article selection process was made taking into account the relevance of the article content with the topic of investment in wind energy under uncertainty and the criteria of the most cited articles, with 97 articles being chosen. Subsequently, the models or approaches applied in each one were identified to evaluate wind systems.

Finally, the articles were selected under the real options approach, in which they are subjected to an exhaustive analysis to determine: (i) their purpose, (ii) uncertainties used, (iii) option valuation techniques, (iv) uncertainty modeling techniques, and (v) options used for the evaluation.

In the second stage of the search, the combination of real options and public opposition was explored, using as search criteria: Real options AND (nimby OR public opposition), obtaining only 23 studies related to the issue of public opposition to investment, but none under the real options approach to assess public opposition to investment in wind systems projects.

4. Results Analysis

Due to the boom in the use of wind energy, research on investment in this type of energy generation has increased over the years. A great leap is evident from 2018 onwards, when 57 articles were published, 58.8% of the selected 97. This trend appears in Figure 1.

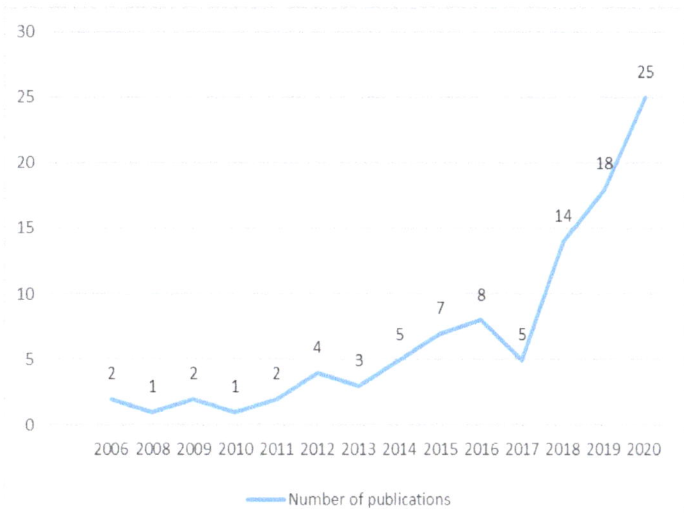

Figure 1. Research on investment in wind energy.

Throughout the literature, there are many models or approaches that was been proposed to evaluate the feasibility of investment in wind power generation, considering the impact of the uncertainty present during the study, siting, and operation phases of a wind farm. Table 2 shows the model proposed in the selected studies, considering the categorization.

Table 2. Models or approaches used in the literature.

Authors	Year	Models or Approaches in the Literature							
		1	2	3	4	5	6	7	Others
Kinias, Tsakalos, & Konstantopoulos [31]	2020	✓							
Detemple & Kitapbayev [32]	2020	✓							
Krömer [33]	2020			✓					
Nasrolahpour, Zareipour, & Rosehart [34]	2020			✓					
Oh & Son [35]	2020			✓					
Ioannou, Angus, & Brennan [36]	2020			✓					
Mehrjerdi & Hemmati [37]	2020			✓					
Liu, He, Liang, Yang, & Xia [38]	2020		✓						
Chowdhury, Pilo, & Pisano [39]	2020		✓						
Tan et al. [40]	2020		✓						
Zhan et al. [41]	2020		✓						
Verleysen, Coppitters, Parente, Paepe, & Contino [42]	2020		✓						
G. Yang, Jiang, & You [43]	2020		✓						
Al, Sirjani, & Daneshvar [44]	2020					✓			
Hübler et al. [45]	2020					✓			
Ge et al. [46]	2020		✓						
H. Yang et al. [47]	2020		✓						
Kong, Li, Liang, Xia, & Xie [48]	2020		✓						
Keck & Sondell [49]	2020							✓	
Stetter, Piel, Hamann, & Breitner [50]	2020				✓				
Abdalla, Smieee, Adma, & Ahmed [51]	2020								17
Henckes, Frank, Küchler, Peter, & Wagner [52]	2020								18
Niromandfam, Movahedi, & Zarezadeh [53]	2020								19
Zhou, Wu, Dong, Tao, & Xu [54]	2020								20
Zhao, Yao, Sun, & Pan [55]	2019	✓							
Maeda & Watts [56]	2019	✓							
Vavatsikos, Arvanitidou, & Petsas [57]	2019	✓							
Askari, Zainal, Ab, Tahmasebi, & Bolandifar [58]	2019	✓							
M. A. Abdulgalil, Khalid, & Alismail [59]	2019			✓					
Junior et al. [60]	2019			✓					
M. Abdulgalil, Khalid, & Alismail [61]	2019			✓					
Yunhao Li, Wang, Gu, Liu, & LI [62]	2019			✓					
Pizarro-alonso, Ravn, & Münster [63]	2019		✓						
Yan, Zhang, Liu, Han, & Li [64]	2019					✓			
Tagliapietra, Zachmann, & Fredriksson [65]	2019						✓		
Thang & Trung [66]	2019		✓						
Quan & Kim [67]	2019		✓						
Borràs, Spelling, Weijde, & Pavageau [68]	2019				✓				
Fuchs, Marquardt, Kasten, & Skau [69]	2019								9
Zhang et al. [70]	2019								8
Ioannou, Angus, & Brennan [71]	2019								16
Ribeiro, Finotti, Perobelli, & Baumgratz [72]	2018	✓							
Finjord, Hagspiel, Lavrutich, & Tangen [73]	2018	✓							
Dalby, Gillerhaugen, Hagspiel, Leth-olsen, & Thijssen [74]	2018	✓							
Yanbin Li, Wu, & Li [75]	2018	✓							
Gazheli & Bergh [76]	2018	✓							
Romanuke [77]	2018	✓							
Y. Yu, Wen, Zhao, Xu, & Li [78]	2018		✓						
Aaboud et al. [79]	2018		✓						
Valinejad et al. [80]	2018		✓						
Jiang et al. [81]	2018		✓						
Deshmukh, Mileva, & Wu [82]	2018		✓						
Z. Li et al. [83]	2018		✓						
Ioannou, Angus, & Brennan [84]	2018				✓				
Esmaieli & Ahmadian [85]	2018								15

Table 2. *Cont.*

Authors	Year	Models or Approaches in the Literature							
		1	2	3	4	5	6	7	Others
Kristiansen, Svendsen, Korpas, & Fleten [86]	2017	✓							
Jannati, Yazdaninejadi, & Talavat [87]	2017					✓			
Chen & Macdonald [88]	2017					✓			
Hamoudi & Maule [89]	2017						✓		
Aquila, Rotela, de Oliveira, & de Queiroz [90]	2017								14
Eryilmaz & Homans [91]	2016	✓							
Kitzing, Juul, Drud, & Krogh [92]	2016	✓							
Pazouki & Haghifam [93]	2016			✓					
Lamadrid, Maneevitjit, & Mount [94]	2016		✓						
Caralis et al. [95]	2016				✓				
Werner & Scholtens [96]	2016				✓				
Sjoerd, Broek, Özdemir, Koutstaal, & Faaij [97]	2016								12
Xiao, Wang, Wang, & Wu [98]	2016								13
Díaz, Gómez-aleixandre, & Coto [99]	2015	✓							
Díaz, Moreno, Coto, & Gómez-aleixandre [100]	2015	✓							
Fang, Li, Wei, & Azim [101]	2015			✓					
Seljom & Tomasgard [102]	2015			✓					
Hong, Lai, Chang, Lee, & Liu [103]	2015		✓						
Siddons, Allan, & Mcintyre [104]	2015				✓				
Rodríguez, del Río, Jaramillo, & Martínez [105]	2015								11
Abadie & Chamorro [106]	2014	✓							
Krogh, Meade, & Fleten [107]	2014	✓							
Correa, Gomes, & Teixeira [108]	2014	✓							
Weibel & Madlener [109]	2014	✓							
Serrano, Burgos, & Riquelme [110]	2014		✓						
Monjas-barroso & Balibrea-Iniesta [111]	2013	✓							
Jin, Botterud, & Ryan [112]	2013			✓					
Kaiser & Snyder [113]	2013						✓		
Reuter, Szolgayová, Fuss, & Obersteiner [114]	2012				✓				
Heinrich, Fuss, Szolgayová, & Obersteiner [115]	2012	✓							
Martinez-cesena, Member, Mutale, & Member [116]	2012	✓							
Ochoa, Betancur, David, Múnera, & Mauricio [117]	2012	✓							
Al-yahyai, Charabi, Al-badi, & Gastli [118]	2012							✓	
Lee [119]	2011	✓							
Dicorato, Forte, Pisani, & Trovato [120]	2011				✓				
Barradale [121]	2010								10
Méndez, Goyanes, & Lamothe [122]	2009	✓							
Muñoz, Contreras, Caamaño, Correia, & Carlo [123]	2009	✓							
Dykes & Neufville [124]	2008	✓							
Magnus, Fleten, Maribu, & Wangensteen [125]	2006	✓							
W. Yu, Sheblé, Lopes, & Matos [126]	2006	✓							

1. Real options
2. Optimization
3. Stochastics
4. Financial evaluation
5. Probabilistic
6. Cost estimate
7. Numerical prediction

OTHERS:

8. Sensibility analysis
9. Qualitative analysis
10. Power purchase agreement (PPA)
11. Error propagation
12. Future-cases
13. Barrier option
14. Value at Risk (VaR)
15. Systems dynamics
16. Key Performance Indicators (KPI)
17. Autoregressive moving average (ARMA)
18. Renewable energy system model
19. Utility function concept
20. Multicriteria decision making (MCDM)

Source: the authors.

Twenty different models or approaches were identified, which were grouped into 8 categories: 1. Real options, 2. Optimization, 3. Stochastics, 4. Financial evaluation, 5. Probabilistic, 6. Cost estimate, 7. Numerical prediction, and 8. Others. Figure 2 presents the established classification and the number of articles that used it.

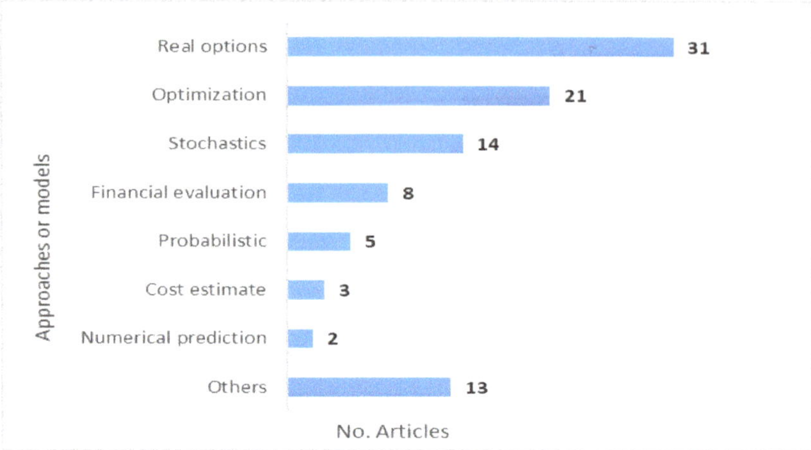

Figure 2. Categories of approaches or models used in the evaluation of investment in wind energy under uncertainty.

Among the established categories, the highest number of publications applied the real options model or approach, with 31 (32%), followed by: optimization, 21 (21.6%); stochastics, 14 (14.4%); financial evaluation, 8 (8.2%); probabilistic, 5 (5.2%); cost estimate, 3 (3.1%); numerical prediction, 2 (2.1%); and others, 13 (13.4%). Below is a description of each category. The category of real options will be discussed later in detail as it is the central purpose of this research.

The optimization category includes research that, with non-random variables, aims to improve efficiency with a lesser amount of resources. They include studies of robust optimizations of capacity distribution of energy systems, robust optimization of wind farm designs, and models that optimize the general design of network expansion planning and programming.

Stochastics covers research that searches for optimization by modeling the parameters of the problem through random variables. There are approaches to evaluate risks of investment in wind farms, to optimize energy generation and wind farms, or to find the optimal design and size of energy storage systems.

The financial evaluation category includes models aimed at the financial and cost evaluation of investment in energy storage systems, investment in wind farms, to evaluate the accuracy of energy cost forecasts, and financial models to establish optimal investment auctions.

Probabilistic includes the models that apply statistical techniques on data obtained through samplings to behaviors that are supposed to be random. It contemplates approaches to determine the optimal allocation of energy storage systems, to probabilistic analysis of wind farm investment designs, to model decisions of the owners and to optimize wind energy conversion processes.

The cost estimation category includes studies that focus on the projection or prediction of costs, considering two phases: determining the cost and the possible real time of the project.

The numerical prediction category includes studies that try to find a description that is as accurate as possible, within a mathematical scheme, of the true meteorological fields which behavior is to be predicted or simulated, being of great interest at the moment of proposing experiments with the models for their validation.

Others, includes approaches with an application, such as Sensitivity Analysis, Qualitative Analysis, Power Purchase Agreements (PPA), Error Propagation, Future Cases, Barrier Option, Value at Risk (VaR), System Dynamics, Parametric Performance Indicators

(KPI), Autoregressive Moving Average (ARMA), Renewable Energy System Model, Utility Function Concept, and Multi-criteria Decision Making (MCDM).

4.1. Studies with a Real Options Approach, Based on the Ends and Purposes of Its Application

This article focuses on conducting a comprehensive analysis on the application of the real options approach to assess the feasibility of investments in wind energy generation projects. One of the main advantages is that it includes, by evaluating flexibility, the adjustments that the decision-maker must make when faced with uncertainty [127,128].

Flexibility creates the possibility of having a range of alternatives, such as: (i) The option to postpone or defer an investment when future prospects are uncertain. It gives the holder the possibility to invest now, or to wait and acquire more information in order to assess the future conditions of the market. In the case of wind energy, it would mean postponing the construction and operation of a wind turbine until demand and prices for the technology are favorable. (ii) The option to modify the operating scale, or expanding the contract, shutting down and restarting, depending on whether market conditions turn favorable or unfavorable. A wind power plant can be expanded, downsized, even closed, depending on market conditions. (iii) The option to quit when the company observes that market conditions are becoming unfavorable. Wind energy projects are subject to changes in regulations, market conditions, and technology. (iv) The option to change, which gives the flexibility to enter and exit depending on market conditions being more favorable, important for the survival of the company. Companies have the option of deciding whether to use the land for agricultural production or to use it for wind power generation. (v) The option to grow, consisting of acquiring or developing a capacity, in anticipation, that will serve to better take advantage of future growth opportunities [58,76]. (vi) The compound option, when, at the end of an investment stage, it is possible to decide whether to stop or defer the start of the next stage. That is, the underlying asset becomes the next option [129].

Of the 97 studies included in Table 2, 31 use real options to evaluate investment in wind systems. Initially, an analysis was made based on the purposes of its application, finding various trends which were grouped into seven categories:

(1) Evaluation of wind energy generation project or investment, (2) Evaluation of energy auctions, portfolios, and energy market investments, (3) Evaluation of renewable energy technologies, (4) Evaluation of the impact of regulatory policies on wind energy projects, (5) Evaluation of wind energy systems, (6) Evaluation of wind resources, and (7) Evaluation of design, size, and location of wind farm projects, shown in Table 3.

Table 3. Investigations according to the purpose of the application in the framework of the real options approach.

Purpose	Authors (Years)
Evaluation of wind energy generation project or investment	[31,56,72,75,77,99,100,106,117,119,123–125].
Evaluation of energy auctions, portfolios, and energy market investments	[31,57,58,86,106,108,122,126].
Evaluation of renewable energy technologies	[32,58,76].
Evaluation of the impact of regulatory policies on wind energy projects	[32,55,73,74,91,92,107,111,126].
Evaluation of wind energy systems	[86,109,114].
Evaluation of wind resources	[116].
Evaluation of design, size, and location of wind farm projects	[109,116].

Source: the authors.

In Figure 3, the purposes of the studies evaluated using the options approach are observed.

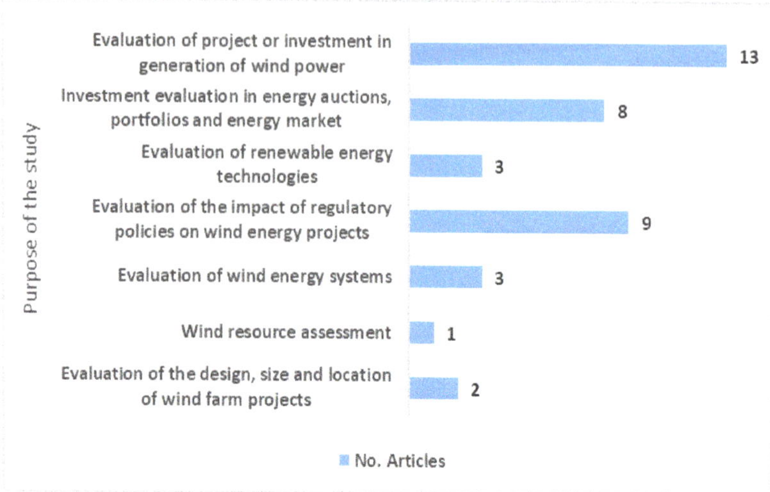

Figure 3. Purpose of the study.

The evaluation of wind energy generation project or investment includes research aimed at optimally evaluating investments from the perspective of the decentralized generation of energy or the viability of projects for the construction of wind farms, is the one of greatest interest by researchers, with the participation of 13 studies.

The evaluation of energy auctions, portfolios, and energy market investments is focused on proposing schemes to measure immersed risks in private markets and in restructuring, to evaluate investment decisions through energy auctions, or to establish portfolios with possible areas for the development of wind farm projects. Eight studies appear.

The evaluation of renewable energy technologies groups the studies focused on optimally supporting the decision to invest in a renewable energy technology, or the adoption of hybrid systems that combine several technologies, such as wind and solar photovoltaic energies, wind and thermal energies, or gas and wind plants. Three studies appear.

The evaluation of the impact of regulatory policies on wind energy projects, seeks to assess the uncertainty caused by the expectation of retroactive changes in wind energy policy, and the flexibilities associated with changing tariffs or subsidies, regulatory frameworks, support schemes such as feed rates and renewable energy certificate trading, and future renewable energy policy. There are nine publications.

The evaluation of wind energy systems, considers the investment required to adopt energy systems integrated by wind energy and hydraulic storage by pumping, or photovoltaic systems, and to expand the network by investing in stages, generating flexibility for the planning of the electrical system. There are three publications.

The evaluation of wind resources integrates studies of wind energy projects planning, considering real options for the evaluation of said resources. There is one publication.

Finally, there are studies aimed at evaluating the design, size, and location of wind farm projects to allow for optimal investment. There are two publications.

4.1.1. Analysis by Group According to Their Application Purposes

Evaluation of Wind Energy Generation Project or Investment

Uncertainties

For the evaluation of wind energy generation projects or investment, uncertainties such as electricity load, energy sales, or energy production [77,99,123,125]; climatic variables or wind conditions [75,99,106,117,125], the price of electricity [31,72,99,117,123–125], carbon price, or other incentives [75,106,124], have been considered. Others have explored things such as market conditions, investment planning, investment costs, capital, or other costs [77,99,119], government policies, and technology levels [75], as uncertainties.

Techniques to Model Uncertainties

To model uncertainties, stochastic processes of the geometric Brownian Movement type–GBM [56,75,106,119,125], the binomial model of Cox, Ross, and Rubinstein [124], reversal of the mean [31,72,123], normal distribution [31], Weibull distribution [77,123], and Box-Jenkins and Ornstein-Uhlenbeck [99,100], have been used.

Used Options

The options that have been used during the evaluation are the option of expansion or closure of facilities [117,124], invest now, postpone, or abandon [72,75,99,123], or the option to switch between alternatives, either capacity [125], or investment [72].

Techniques to Evaluate Used Options

Techniques such as Black-Scholes [119,125], decision trees [106,117,123,124], the binomial model of Cox, Ross, and Rubinstein [124], Monte Carlo simulations [56,75,106,117,123,124], the Wiener process [31], Dynamic programming [56], Monte Carlo (least squares) [99,100], and optimization algorithm [77].

Evaluation of Energy Auctions, Portfolios, and Energy Market Investments

Uncertainties

For the evaluation of energy auctions, portfolios, and energy market investments, one can find energy production [122], electricity price [58,108,122], capital or investment costs [108,122], future deployment of offshore wind energy [86], and electricity demand [58] being considered as uncertainties.

Techniques to Model Uncertainties

To model uncertainties, stochastic processes of the Markov process type [58], Geometric Brownian movement-BGM [57,108,122], the binomial model of Cox, Ross, and Rubinstein [122], and reversal of the mean [86] were used.

Used Options

The options that have been used during the evaluation are the compound type, construction time, either to carry out, postpone, or abandon the investment [86,122], the option of waiting for a better opportunity in market trends to decide when to invest or abandon [108], and the option of building portfolios with possible areas for the development of wind farm projects [57].

Techniques to Evaluate Used Options

For evaluate the options, techniques such as dynamic programming [57,58], the option-games model [58,108], Monte Carlo simulations [57,58,122], decision trees [122], the binomial model of Cox, Ross, and Rubinstein [122], and optimization algorithm [86] have been used.

Evaluation of Renewable Energy Technologies

Uncertainties

For the evaluation of renewable energy technologies, electricity prices were considered as uncertainty [58,76], as were: learning speed [76], electricity demand [58], government policies [32], feed-in rates-subsidies [32,115], alternative technologies [32], and the intermittency of the wind [115].

Techniques to Model Uncertainties

Stochastic processes of the Markov chains type [58], Geometric Brownian Motion-GBM [32,76], and complex formulations [115] were used.

Used Options

The options that have been used are: the option to invest now or wait for when conditions are more favorable [76], to take advantage of the learning effect to anticipate the option to invest and exercise it earlier [76], the option to invest in one or more technologies, whether wind or thermal [58,76], or wind or gas [32].

Techniques to Evaluate Used Options

Techniques such as dynamic programming [58,115], option-games model [58], Monte Carlo simulation [32,58,115], and technological learning curves [76] were used.

Evaluation of the Impact of Regulatory Policies on Wind Energy Projects

Uncertainties

For the evaluation of the impact of regulatory policies in wind energy projects, electricity price [55,73,92,111,126], changes in asset rates [126], energy production [73,111], investments or capital costs [107,111], price index [111], production certificates or feed-in rates-subsidies [32,73,74,107], credit policy [91], wind speed [92], wind turbines hours of use [55], carbon price [55], government policies [32], and alternative technologies [32] were considered as uncertainties.

Techniques to Model Uncertainties

To model uncertainties, stochastic processes of the reversion of the mean type [111,126], Geometric Brownian Movement-GBM [32,55,73,74,92,107], and the Markov process [74,91], have been used.

Used Options

The options used during the evaluation they were the options to invest now, postpone or abandon and the option of capacity generated by support schemes [73,74,92,107], the compound option, associated with changes in wind turbines rates [126], options underlying the change in monthly or annual rate, modeled as an Asian call option on the wind differential [126] the option to invest now, postpone, or abandon [111], to adjust capacity [55,91,111], as generated by regulatory frameworks; and the option to choose between technologies due to policy changes [32].

Techniques to Evaluate Used Options

To evaluate the options, techniques such as decision trees [111], Monte Carlo simulations [32,74,111,126], Black-Scholes [92,107], Monte Carlo (least squares) [55,73,107], dynamic programming [74,91,92], and technological learning curves [74] have been used.

Evaluation of Wind Energy Systems

Uncertainties

For the evaluation of wind energy systems, feed rate-subsidies [115], intermittency of the wind [115], future deployment of wind energy process [86], and wind speed [116] were used as uncertainties.

Techniques to Model Uncertainties

To model uncertainties, tools such as complex formulations [115], stochastic processes of the reversion of the mean type [86], and symmetric distributions [86], were used.

Used Options

The options used for the evaluation are of the construction time type; that is, the option to make the investment for the expansion of the network in stages, where the second stage generates flexibility when planning the electrical system [86,115], and the option to delay the construction of wind energy projects [116].

Techniques to Evaluate Used Options

Techniques such as dynamic programming [115], Monte Carlo simulations [115,116], algorithm optimization [86] and decision trees [116] have been used for the valuation of options.

Evaluation of Wind Resources

Uncertainties

For the evaluation of wind resources, the speed of wind was considered an uncertainty [116].

Techniques to Model Uncertainties

The modeling of uncertainties was carried out using the Weibull distribution and decisions trees [116].

Used Options

The option used was that to delay the construction of wind energy projects to gather more relevant data to evaluate wind resources [116].

Techniques to Evaluate Used Options

The evaluation of options was carried out using Monte Carlo simulations and decisions trees [116].

Evaluation of Design, Size, and Location of Wind Farm Projects

Uncertainties

For the evaluation of the design, size and location of wind farm projects, the uncertainties were solar irradiation, electricity price, capital costs, wind intensity [109], and wind speed [116].

Techniques to Model Uncertainties

To model uncertainties, stochastic processes of the Geometric Brownian Movement type-GBM [109], the Weibull distribution, and decision trees [116] were used.

Used Options

The option used was that to delay the construction of wind energy projects [116].

Techniques to Evaluate Used Options

The evaluation of options has been carried out using Monte Carlo simulations [109,116] and decision trees [116].

4.2. Types of Uncertainties Explored in Publications with a Real Options Approach

Uncertainty is inherent in events where possible outcomes are not known, and, therefore, their probability of occurrence cannot be quantifiable. Regarding projects, uncertainty grows over time, conditioning the occurrence of viability. The management of uncertainty, over the years, has always been a critical variable for decision-makers [130]. In the energy sector, including wind energy, decision-making is generally always conditioned by a level of uncertainty in the data [131].

In the selected articles, with a real options approach, 28 types of uncertainties were identified, which were grouped into nine categories: 1. Power generation, 2. Environmental conditions, 3. Energy price, 4. Costs, 5. Revenues, 6. Regulatory policies, 7. Market, 8. Wind conditions, and 9. Technological progress, as shown in Table 4.

Table 4. Types of uncertainties explored in research with a real options approach.

1. Power Generation	**5. Regulatory Policies**
1.1. Electricity charge	5.1. Production certificates
1.2. Wind energy level	5.2. Changes in asset rates
1.3. Energy production	5.3. Credits policy
1.4. Wind turbine hours of use	5.4. Government policies
1.5. Investment planning	5.5. Feed-in rates-subsidies
1.6. Future deployment of wind energy	
	6. Market
2. Environmental	6.1. Market conditions
2.1. Climate change	6.2. Price index
2.2. Solar irradiation	6.3. Demand for electricity
3. Prices	**7. Wind Conditions**
3.1. Electricity price	7.1. Wind disponibility
3.2. Carbon price	7.2. Wind intensity
3.3. Corn price	7.3. Wind speed
	7.4. Wind intermittency
4. Costs	
4.1. Capital costs	**8. Technological Progress**
4.2. Investment costs	8.1. Alternative technology
	8.2. Technology levels
	8.3. Learning speed

1. Power generation contains the uncertainties that affect the process of generating wind energy, such as electricity charge, wind energy level, energy production, wind turbine hours of use, investment planning, and future deployment of wind energy. Energy production prevails, with three applications, as shown in Figure 4.

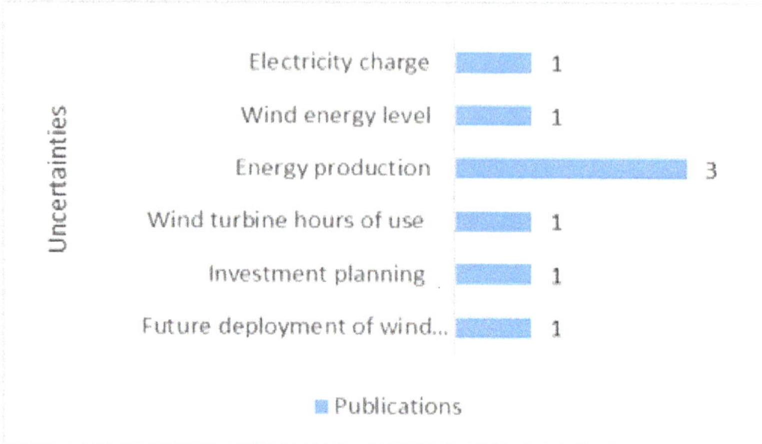

Figure 4. Power generation.

2. Environment refers to the uncertainties present in environmental conditions necessary for the development of projects such as climate change and solar irradiation. Solar irradiation stands out, with two applications, as shown in Figure 5.

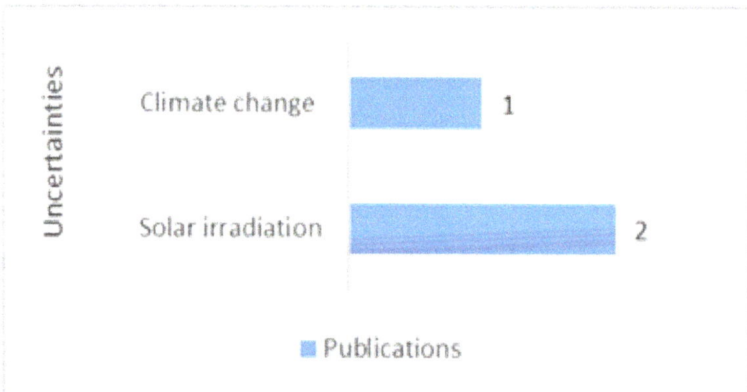

Figure 5. Environment.

3. Energy price considers the uncertainties generated by price fluctuations in the market, in response to the behavior of energy demand and supply, including among others the price of electricity, coal price, and the price of corn. The price of electricity is the source of uncertainty most studied by researchers, being considered in 20 (65.5%) publications, as shown in Figure 6.

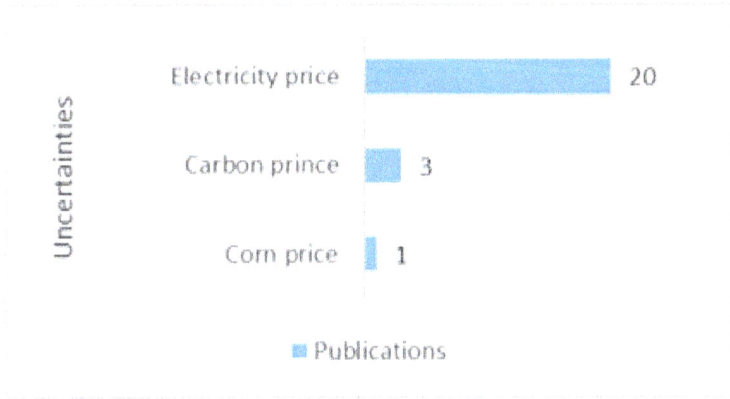

Figure 6. Energy price.

4. Costs studies were integrated that analyze the uncertainties caused by the incidence of the behavior of costs in the profitability of the investment, whether they are capital costs or investment costs. Both costs were studied in four publications, as shown in Figure 7.

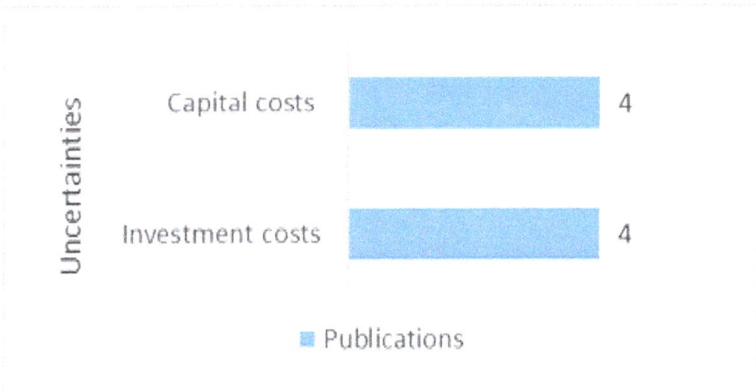

Figure 7. Costs.

5. Regulatory policies include the uncertainties caused by changes implemented in the policies for the generation of wind energy, among which are production certificates, changes in asset tariff, credit policy, government policies, and feed-in tariffs-subsidies. The rates of food and subsidies have been the most studied by researchers, being considered in five publications, as reflected in Figure 8.

Figure 8. Regulatory policies.

6. Market relates sources of uncertainties other than prices and costs which affect their market behavior, such as market conditions, price indices, and electricity demand. Market conditions prevail, with three applications, as shown in Figure 9.

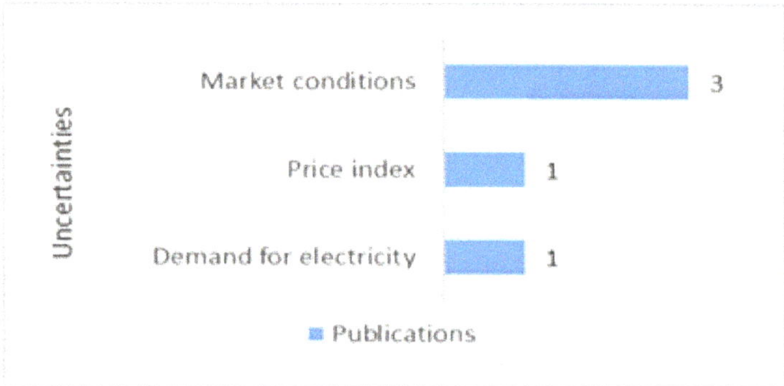

Figure 9. Market.

7. Wind conditions addresses uncertainties caused by variability of the wind characteristics in the generation of wind energy, such as the availability of wind, its intensity, its speed, and its intermittency. After the price of electricity, wind conditions are the source of uncertainty that generates the greatest interest in the studies analyzed, especially wind speed, as it has a great impact on the generation of wind energy. Wind speed was studied in six publications, as shown in Figure 10.

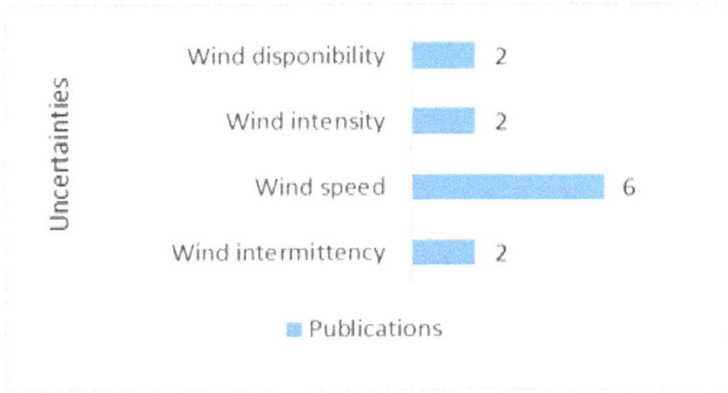

Figure 10. Wind conditions.

8. Technological progress focuses on the analysis of the uncertainties caused by the generation and use of knowledge (R&D) that affect the productivity, such as alternative technology, technology levels, and learning speed. All sources of uncertainties have only been studied in one investigation, as shown in Figure 11.

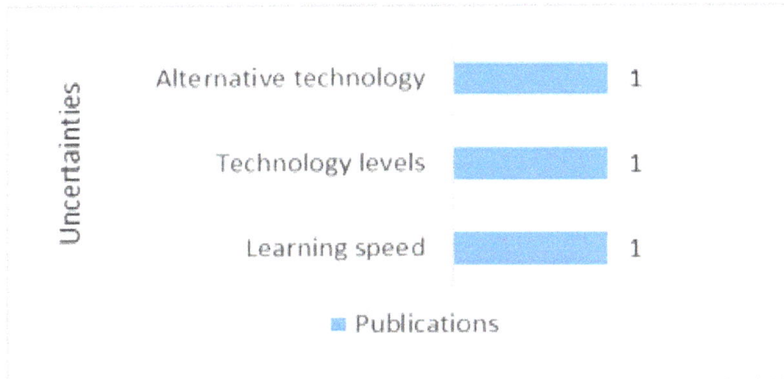

Figure 11. Technological progress.

Table 5 shows, in a disaggregated manner, the type of uncertainty considered in each of the studies with a focus on real options to evaluate investment in wind systems, taking into account the categorization of Table 4. It is observed that 62.5% of the studies treat the price of electricity as a source of uncertainty, 18.8% wind speed, and 15.6% feeding-subsidy rates.

Table 5. Type of uncertainties explored in research with a real options approach.

Authors-Years	1. Power Generation						2. Environ-mental		3. Prices			4. Costs		5. Regulatory Policies					6. Market			7. Wind Conditions				8. Technological Progress			Total
	1.1	1.2	1.3	1.4	1.5	1.6	2.1	2.2	3.1	3.2	3.3	4.1	4.2	5.1	5.2	5.3	5.4	5.5	6.1	6.2	6.3	7.1	7.2	7.3	7.4	8.1	8.2	8.3	
[125]	✓						✓		✓																				3
[124]									✓	✓																			2
[122]			✓						✓			✓																	3
[106]		✓							✓					✓															3
[72]									✓		✓																		2
[119]					✓														✓										2
[31]									✓																				1
[126]									✓						✓							✓							3
[115]																		✓							✓				2
[107]									✓			✓		✓				✓											4
[108]																			✓										1
[91]																✓													1
[73]			✓						✓									✓											3
[74]																		✓											1
[55]				✓					✓	✓																			3
[32]																	✓	✓								✓			3
111]			✓						✓				✓							✓									4
[92]									✓															✓					2
[56]									✓				✓																2
[99]									✓			✓	✓																3
[123]									✓															✓					2
[116]																								✓					1
[117]									✓													✓	✓	✓	✓				5
[109]								✓	✓			✓											✓						4
[100]									✓				✓											✓					3
[75]							✓			✓							✓										✓		4
[76]									✓																			✓	2
[57]																			✓										1
[58]									✓												✓								2
[86]						✓																							1
[77]																								✓					1
Total	1	1	3	1	1	1	1	2	20	3	1	4	4	2	1	1	2	5	3	1	1	2	2	6	2	1	1	1	74

Source: the author.

4.3. Evaluation of Real Options and Modeling of Uncertainties

There are multiple techniques used to assess the real options in the studies analyzed. They are random or numerical and non-random methods or by scenario analysis. Table 6 presents the classification and the studies that used each of the techniques. Within the non-randomized methods, the following techniques were identified:

(1) Decision trees: initially introduced by Cox, Ross, and Rubinstein (1979) as a binomial model to value American-type options. It was later adapted to simulate the evolution of uncertainty in discrete scenarios of multiple interrelated options [132]. It allows modeling the evolution of the price of the underlying asset under uncertainty in discrete scenarios, assuming that the underlying asset follows a multiplicative binomial process [133].

(2) Dynamic programming: an optimization method by way of dividing the problem into superimposed subproblems and optimal substructures, especially when the subproblems are not independent. It is based on the principle of optimum as enunciated by Bellman in 1957: "In an optimal decision sequence, every subsequence must also be optimal", and it allows combining different types of real options with various possible scenarios [134].

(3) Partial Differential Equations: a set of equations, initially used for the valuation of financial options [135,136], later adapted to evaluate specific real options under fixed assumptions [132]. The results of the Black-Scholes model can be obtained from a binomial model for n periods, where n tends to infinity [137].

(4) Option-Games model: a valuation tool that combines the real option approach with game theory, with the aim of quantifying the values of flexibility, allowing for better investment decisions to be taken [138].

(5) Sensitivity analysis: measures the impact that variations in one of the independent variables have on the model [31].

Among the random or numerical models, we find:

(1) Monte Carlo simulation: a numerical method used to evaluate options when there are no closed formulas such as Black-Scholes [137]. Its purpose is to easily value real options for complex projects, since it does not require the formulation of cash flow through differential equations or trees [139]. It creates a distribution of project values from all given sources of uncertainty [140]. An advantage of this method is that it

offers the distribution of the volatility factor, which is key in evaluating the sensitivity of the value of the real options of a project [141].

(2) Least squares Monte Carlo: combines the Monte Carlo simulation with least squares regression, which helps reduce the number of scenarios while still producing an accurate assessment [142]. At any time, the holder of an American option can optimally compare the reward of exercising that option immediately, with the expected reward of not exercising it yet [143].

(3) Optimization algorithms: includes traditional algorithms, such as gradient-based methods and quadratic programming, evolutionary algorithms, heuristic or meta-heuristic algorithms, and various hybrid techniques. Optimization problems tend to be non-linear with complex objectives [144].

Figure 12 shows the techniques used for the valuation of Real Options. They are grouped in two blocks. In the upper one, there are the non-random techniques, with the highest prevalence of Decision Tree and Dynamic Programming with 8 (25.8%) applications each. In the lower one, there are random techniques, among which Monte Carlo simulation is the most preferred by researchers, with 16 (51.6%) applications, which is due to the fact that most of the selected studies contemplate continuous uncertainties such as, for example, the price of energy and wind conditions.

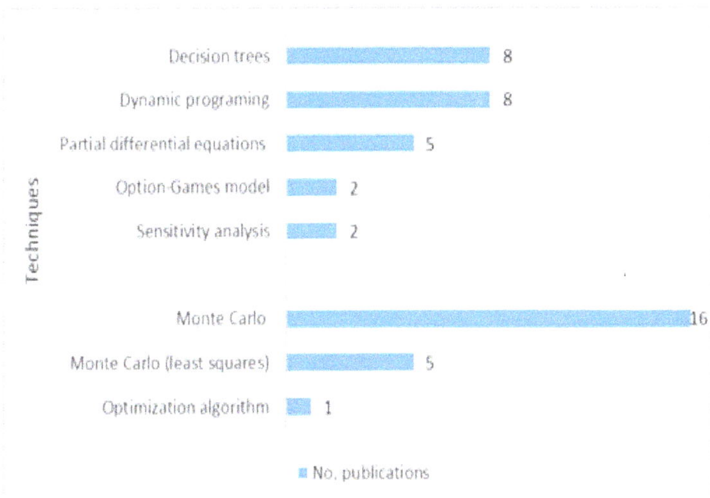

Figure 12. Real options valuation techniques.

Similarly, for the modeling of the sources of uncertainties, a great variety of techniques have been applied. Table 6 shows the different techniques applied in the selected studies. There are random techniques or models, such as:

(1) Geometric Brownian Motion, used in 61.3% (19) of the studies, is a stochastic process in continuous time, generated from a transformation of a standard Wiener process, with the particularity of not allowing asset prices to take negative values [145]. Geometric Brownian Motion appears to be better suited to long-term energy-related investments [146].

(2) Reversal of the mean indicates that values, such as prices, may move away from the mean or intrinsic value, but, over time, they will eventually return to those mean values [147].

(3) The Weibull distribution is the probability function that best describes the wind speed path, thanks to the orthogonal composition of two correlated Gaussian functions. The Weibull density function consists of two parameters: one refers to the

maximum speed and the other indicates the degree of dispersion of the samples (Muñoz et al., 2009).

(4) The Box-Jenkins and Ornstein-Uhlenbeck model is the basis of all modern time series analysis theory, and consists of the analysis of probabilistic or stochastic properties of economic time series where the endogenous variable (Yt) is explained by past values or lags of itself and by stochastic error terms [148].

(5) The ARIMA distribution is an econometric methodology based on dynamic models using time series data, made up of three components: the autoregressive (AR), the integration (I), and the moving average (MA), represented by the parameters p, d, and q, respectively. The model includes the values of the series, prediction of errors, and a random term [149].

(6) The normal distribution is a mathematical model that allows determining probabilities of occurrence for different values of the variable. The graph of the normal distribution has the shape of a bell, and for this reason it is also known as the Gaussian bell, whose central elements are the mean and the variance [150].

(7) Markov chains are discrete stochastic processes used to study the evolution of certain systems in repeated trials, in which the probability of an event occurring depends only on the immediately preceding event. Transition probabilities are used to describe the way the system passes to the next state.

There are also non-random uncertainty modeling techniques, the binomial model of Cox, Ross, and Rubinstein, used to model non-random uncertainties in discrete time, and learning curves, used to model technological uncertainty. This concept was generated from a study by Wright (1936), where he analyzes repetitive production in the construction of airplanes, finding, as a hypothesis, that "the man-hours necessary to complete a production unit would decrease in a constant percentage each time the production doubles" [151], showing that the repetitive action of a process increases the experience and learning on the part of the operator, managing to reduce the production time per unit [152].

Figure 13 shows the techniques used to model the uncertainties studied. They are grouped in two blocks. In the upper part, there are the random techniques, and in the lower part, the non-random ones. It can be seen that the researchers opt for random techniques, preferably Geometric Brownian Motion (GBM), which is consistent with the type of uncertainty considered in the studies, which are mostly continuous, such as electricity prices and wind conditions.

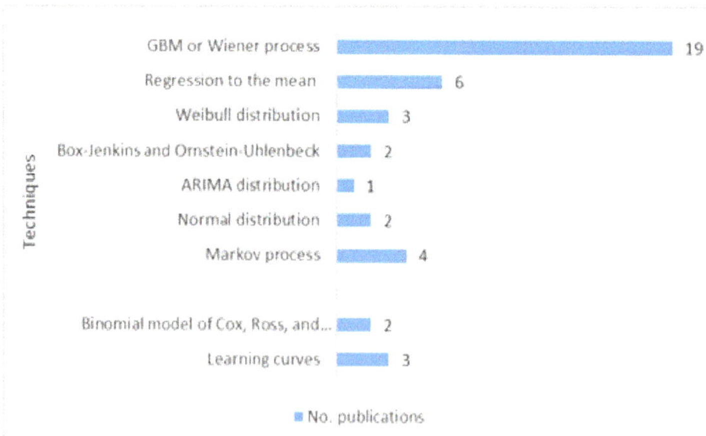

Figure 13. Uncertainty modeling techniques.

Both in evaluating real options and in modeling uncertainties, a standard pattern has not been followed when developing the models, regarding the type and number of techniques to be used, which indicates that, without exception, techniques are combined according to the purpose and scope of each particular study, as seen in Table 6.

Table 6. Techniques to evaluate real options and to model uncertainties.

Author-Year	Technique of Assessment								Modeling Technique								
	Not Random					Random			Random							Not Random	
	1	2	3	4	5	1	2	3	1	2	3	4	5	6	7	1	2
[125]			✓						✓								
[124]	✓	✓				✓										✓	
[106]	✓	✓				✓			✓								
[72]										✓							
[58]				✓		✓									✓		
[31]					✓				✓	✓				✓			
[123]	✓	✓				✓			✓	✓	✓						
[116]	✓	✓				✓					✓						
[117]	✓	✓				✓							✓				
[111]	✓	✓				✓				✓							
[126]						✓			✓	✓							
[115]						✓									✓		
[32]						✓			✓								
[56]						✓			✓								
[109]						✓			✓								
[75]						✓			✓								
[57]						✓			✓								
[122]	✓	✓				✓			✓							✓	
[119]			✓		✓				✓								
[107]			✓				✓		✓								
[108]				✓					✓								
[91]															✓		
[73]							✓		✓								
[74]						✓			✓						✓		✓
[78]							✓		✓								
[92]			✓						✓								
[99]							✓					✓					
[100]							✓					✓					✓
[76]			✓						✓								✓
[86]	✓	✓								✓				✓			
[77]								✓			✓						
	8	8	5	2	2	16	5	1	19	6	3	2	1	2	4	2	3

Valuation of options
- **Non-random**
  1. Decision trees
  2. Dynamic programing
  3. Partial differential equations
  4. Option-Games model
  5. Sensitivity analysis
- **Random**
  1. Monte Carlo
  2. Monte Carlo (least squares)
  3. Optimization algorithm

Modeling of uncertainties
- **Random**
  1. GBM or Wiener process
  2. Regression to the mean
  3. Weibull distribution
  4. Box-Jenkins and Ornstein-Uhlenbeck
  5. ARIMA distribution
  6. Normal distribution
  7. Markov process
- **Non-random**
  1. Binomial model of Cox, Ross, and Rubinstein
  2. Learning curves

Source: the authors.

4.4. Public Opposition under Real Options

The literature search did not find studies that consider uncertainties related to public opposition analyzed under the real options approach; however, some studies that focus their purpose on public opposition to investment in wind energy projects [153–155], to the assembly of transmission lines [156], and to road pricing [157] were identified.

The aspects that generated the greatest public opposition in the execution of the projects were: centralized planning processes [154], implementation of a single technology [156,158], lack of communication prior to decision-making processes [154], environmental impact [155], consultation after decision-making for project execution [154], institutional factors [155], electromagnetic fields [156], network problems [156] and authorization procedures and legal framework [156].

In some studies, solutions to counteract public opposition were proposed, such as: improving the legal framework and procedures [156], implementing different technological options [156,158], participatory models for decision-making [153,154,156], the implementation of financial incentive policies, including permanent tax deductions and investment support or subsidies [157,158], participatory financial investment in projects [158], subsidizing the implementation of Renewable Energy Technology systems at the community level, according to customs and conditions of the housing [158], strengthening information processes and prior consultations [155,156], controling and monitoring the environmental impact [155], and building up institutional capital [153,154].

Surveys [153,154,158] and panels [157] were the preferred methodologies when developing these research.

Socio-economic, situational, and family restrictions have a great impact on the decision of the participants to change their position regarding the execution of projects.

5. Discussion

The purpose of this article is to review the literature on the evaluation of investment in wind systems under the real options approach, to determine the inclusion of public opposition (NIMBY projects) as a source of uncertainty, and to identify the flexible strategies implemented to overcome these types of barriers. As a result of the review, no studies related to the subject of public opposition (NIMBY projects) were found under the real options approach; however, some aspects that generate opposition were identified, which could be valued as real options, such as: the creation of participatory public policies; participatory models for decision-making; financial incentive schemes, including tax deductions and permanent investment support or subsidies; associative schemes for participatory financial investment in projects; implementation of technology systems of Renewable Energies at the community level, according to customs and housing conditions; information processes and prior consultations; environmental monitoring and control systems; and building up an institutional capital.

For the assessment of these opposition generators under the real options approach, some models proposed in studies related to regulatory policies in wind energy projects can be adapted, such as [32,55,73,74,91,92,107,111,126]; models that use the decision tree technique to assess real options: [106,111,116,117,123,124], or those that use dynamic programming, such as [58,91].

Regarding the uncertainties explored in research with a real options approach, most were found to be random variables with continuous states that are discretized over time (price, cost, and wind speeds). However, there are several studies that assume uncertainties as random variables with discrete states at discrete times, modeled by Markov chains, which could be assimilated to model public opposition because they have similar characteristics: climate policy [115], policy changes related to the current support scheme [74], and long-term stochastic uncertainties in the restructured energy market, such as demand and the price of fuel [58].

Regarding the options to use, you can make use of options to invest now, postpone, or abandon [91,92,111,116], and the same is true for the compound options, the one associated with rate change [126], or the one to either carry out, postpone, or abandon the investment before continuing with the next stage [86,107,115,122].

From the analysis of public opposition, it follows that, in order to propose flexible strategies, the perception of the community must first be known; then, it must be determined what alternatives are available from the beginning and when these would be presented, based on consumer preferences.

Given that, despite an exhaustive search in the literature, no studies that included uncertainties, such as public opposition, modeled with current techniques was found, and, therefore, it is not known how people will react, so there is a gap that would be interesting to address, to assess flexibility when this type of uncertainty arises through the real options approach

6. Conclusions

This article provides a review of models or approaches used in the evaluation of investment in wind systems, under uncertainty. In total, 97 publications were analyzed, finding that this type of research has been increasing. The real options approach is the one most used by researchers, applied in 31 publications, representing 32%.

Twenty-eight sources of uncertainty were identified, which were grouped into nine categories: 1. Power generation, 2. Environmental conditions, 3. Energy price, 4. Costs, 5. Income, 6. Energy policies regulation, 7. Market, 8. Wind conditions, and 9. Technological progress. The price of electricity is the source of uncertainty that generates the most interest in the researchers, considered in 20 publications.

Random and non random techniques or models are applied, both to value real options and in uncertainty modeling.

There are a series of barriers to investment in wind power generation projects, within these are public opposition-NIMBY "Not in my backyard", but no studies were found under the real options approach that would develop this theme, which justifies the possibility of exploring what the management of flexibility may be in this case.

Given the importance of investment projects in renewable energy, including wind energy, flexibility is a key factor in the business decision-making process, both in the planning and operating phases. The organizations that have some level of managerial flexibility (administrative and operational) are able to adjust their current decisions to the changes and conditions generated in the future, as caused by the existence of uncertainty. Thus, flexibility becomes an alternative to face uncertainty, because it represents rights and opportunities that any organization can execute when the underlying conditions of a respective project are either favorable or unfavorable. When the value of flexibility is positive, it increases the value of investment cash flows, generating greater social welfare.

The effective use of renewable energy sources, including wind energy, is the best strategy to contribute to meeting the objectives of the 2030 Agenda and the Paris agreement, aimed at reducing CO_2 emissions. For this reason, it is necessary to intensify in the future the level of research on investment in renewable energy projects, under the approach of real options, which allows further assessment of the uncertainty of the projects to establish flexible strategies and establish efficient controls on high risk of the project investment. According to the findings observed in this analysis, the following lines of future research identified others:

1. Within a framework of Real Options, investigate the decision of communities to invest in wind energy [72].
2. Delve into how the application of the real options approach has the potential to increase the expected value of the investment in wind energy, by addressing the impact of uncertainty in the evaluation of the wind resource and contemplating the flexibility within the design and in the project planning [113].
3. Deepen the impact of the uncertainty caused in the future development of regulatory policies for the generation of renewable energy [114].
4. Investigate the impact of additional, sequential, or staged investment options at the optimal investment time and size.

Future investigations related to public opposition:

1. Develop studies that consider the real options approach to evaluate the impact of public opposition on investment in renewable energy projects.
2. Study on how to create institutional capital for wind energy and other renewable resources. This implies that more participatory planning practices and inclusive politics of the communities are needed [153].
3. Research aimed at establishing the determinants for indigenous and Afro-descendant communities within their uses and custom to accept the location of renewable energy projects [155].

Author Contributions: Conceptualization, A.H.; methodology, B.M., L.G. and A.H.; validation, A.H. and L.G.; formal analysis, B.M.; investigation, B.M.; resources, B.M.; data curation, A.H. and L.G.; writing—original draft preparation, B.M.; writing—review and editing, A.H. and L.G.; supervision, A.H., L.G.; funding acquisition, B.M. All authors have read and agreed to the published version of the manuscript.

Funding: This research was funded by the departmento de La Guajira-Colombia, through the project: Master's and Doctoral Studies inside and outside the Country for Professionals in the departamento de La Guajira- BPIN 2013000100032, with resources of Direct Allocation of Royalties, granted through the agreement of Condonable Educational Credit PFAN 019 of 2015, executed by the "Universidad de La Guajira"; and the "Universidad del Norte".

Institutional Review Board Statement: Not applicable.

Informed Consent Statement: Not applicable.

Data Availability Statement: Not applicable.

Conflicts of Interest: The authors declare no conflict of interest.

References

1. Harjanne, A.; Korhonen, J.M. Abandoning the concept of renewable energy. *Energy Policy* **2019**, *127*, 330–340. [CrossRef]
2. Mentis, D.; Howells, M.; Rogner, H.; Korkovelos, A.; Arderne, C. Lighting the World: The first application of an open source, spatial electrification tool (OnSSET) on Sub-Saharan Africa Lighting the World: The first application of an open source, spatial electrification tool (OnSSET) on Sub-Saharan Africa. *Environ. Res. Lett.* **2017**, *12*, 085003. [CrossRef]
3. Lehne, K.-H. Informe Especial: Emisiones de Gases de Efecto Invernadero en la UE: Se Notifican Correctamente, Pero es Necesario Tener un Mayor Conocimiento de las Futuras Reducciones. 2019. Available online: https://www.eca.europa.eu/es/Pages/DocItem.aspx?did=51834 (accessed on 12 July 2021).
4. GWEC. *Global Wind Report | Gwec*; Globle Wind Energy Counc.: Brussels, Belgium, 2021; p. 75.
5. Ministerio de Minas y Energía. *La Transición Energética de Colombia*; Ministerio de Minas y Energía: Bogotá, Colombia, 2020.
6. Kordmahaleh, A.A.; Naghashzadegan, M.; Javaherdeh, K.; Khoshgoftar, M. Design of a 25 MWe Solar Thermal Power Plant in Iran with Using Parabolic Trough Collectors and a Two-Tank Molten Salt Storage System. *Int. J. Photoenergy* **2017**, *2017*. [CrossRef]
7. Deutch, J. Decoupling Economic Growth and Carbon Emissions. *Joule* **2017**, *1*, 3–5. [CrossRef]
8. Zafar, R.; Mahmood, A.; Razzaq, S.; Ali, W.; Naeem, U.; Shehzad, K. Prosumer based energy management and sharing in smart grid. *Renew. Sustain. Energy Rev.* **2018**, *82*, 1675–1684. [CrossRef]
9. Fernandes, B.; Cunha, J.; Ferreira, P. The use of real options approach in energy sector investments. *Renew. Sustain. Energy Rev.* **2011**, *15*, 4491–4497. [CrossRef]
10. Burke, M.J.; Stephens, J.C. Energy Research & Social Science Political power and renewable energy futures: A critical review. *Energy Res. Soc. Sci.* **2018**, *35*, 78–93.
11. IRENA. Renewable Capacity Statistics 2021. 2021. Available online: https://www.irena.org/publications/2021/March/Renewable-Capacity-Statistics-2021 (accessed on 25 July 2021).
12. Bolinger, M.; Wiser, R. Wind power price trends in the United States: Struggling to remain competitive in the face of strong growth. *Energy Policy* **2009**, *37*, 1061–1071. [CrossRef]
13. Mengal, A.; Uqaili, M.A.; Harijan, K.; Memon, A.G. Competitiveness of wind power with the conventional thermal power plants using oil and natural gas as fuel in Pakistan. *Energy Procedia* **2014**, *52*, 59–67. [CrossRef]
14. Astariz, S.; Iglesias, G. Enhancing wave energy competitiveness through co-located wind and wave energy farms. A review on the shadow effect. *Energies* **2015**, *8*, 7344–7366. [CrossRef]
15. Albadi, M.H.; El-Saadany, E.F. Overview of wind power intermittency impacts on power systems. *Electr. Power Syst. Res.* **2010**, *80*, 627–632. [CrossRef]
16. Notton, G.; Nivet, M.L.; Voyant, C.; Paoli, C.; Darras, C.; Motte, F.; Fouilloy, A. Intermittent and stochastic character of renewable energy sources: Consequences, cost of intermittence and benefit of forecasting. *Renew. Sustain. Energy Rev.* **2018**, *87*, 96–105. [CrossRef]
17. Baagøe-Engels, V.; Stentoft, J. Operations and maintenance issues in the offshore wind energy sector. *Int. J. Energy Sect. Manag.* **2016**, *10*, 245–265. [CrossRef]
18. Peeters, C.; Guillaume, P.; Helsen, J. Vibration-based bearing fault detection for operations and maintenance cost reduction in wind energy. *Renew. Energy* **2018**, *116*, 74–87. [CrossRef]
19. Ayodele, T.R.; Jimoh, A.; Munda, J.L.; Agee, J.T. Challenges of grid integration of wind power on power system grid integrity: A review. *Int. J. Renew. Energy Res.* **2012**, *2*, 618–626.
20. Yuan, X.; Cheng, S.; Wen, J. Prospects analysis of energy storage application in grid integration of large-scale wind power. *Dianli Xitong Zidonghua Autom. Electr. Power Syst.* **2013**, *37*, 14–18.
21. Saidur, R.; Islam, M.R.; Rahim, N.A.; Solangi, K.H. A review on global wind energy policy. *Renew. Sustain. Energy Rev.* **2010**, *14*, 1744–1762. [CrossRef]
22. Leung, D.Y.C.; Yang, Y. Wind energy development and its environmental impact: A review. *Renew. Sustain. Energy Rev.* **2012**, *16*, 1031–1039. [CrossRef]
23. Saidur, R.; Rahim, N.A.; Islam, M.R.; Solangi, K.H. Environmental impact of wind energy. *Renew. Sustain. Energy Rev.* **2011**, *15*, 2423–2430. [CrossRef]
24. IREA; IRENA. Future of Wind Deployment, Investment, Technology, Grid Integration and Socio-Economic Aspects. 2019. Available online: https://www.irena.org/-/media/files/irena/agency/publication/2019/oct/irena_future_of_wind_2019.pdf (accessed on 26 July 2021).
25. Presidencia de la República de Colombia. Decreto 2164 de 1995. *D. Of.* **1995**, *1995*, 12. Available online: https://www.mininterior.gov.co/la-institucion/normatividad/decreto-2164-de-1995 (accessed on 19 June 2021).
26. Congreso de Colombia. Ley 21 de 1991: Convenio 169. Convenio sobre Pueblos Indígenas y Tribales en Países Independientes. *D. Of.* 1991, p. 14. Available online: http://www.urosario.edu.co/jurisprudencia/catedra-viva-intercultural/Documentos/ley_21_91.pdf (accessed on 8 June 2021).
27. Aitken, M.; Haggett, C.; Rudolph, D. Wind farms community engagement: Good practice review. *Edinb. Clim.* **2014**, 1–103. Available online: https://core.ac.uk/download/pdf/84000113.pdf (accessed on 15 June 2021).

28. Ledec, G.C.; Rapp, K.W.; Aiello, R.G. *Greening the Wind*; World Bank Publications: Washington, DC, USA, 2011. [CrossRef]
29. Munday, M.; Bristow, G.; Cowell, R. Wind farms in rural areas: How far do community benefits from wind farms represent a local economic development opportunity? *J. Rural Stud.* **2011**, *27*, 1–12. [CrossRef]
30. Hernando, A.; Razmilic, S.; Jiménez, S.; Katz, R.; Martínez, D. Puntos de referencia: [Editorial]. *Rev. ADM* **2017**, 282–284.
31. Kinias, I.; Tsakalos, I.; Konstantopoulos, N. Investment evaluation in renewable projects under uncertainty, using real options analysis: The case of wind power industry. *Investig. Manag. Financ. Innov.* **2020**, *14*, 96–103. [CrossRef]
32. Detemple, J.; Kitapbayev, Y. The value of green energy under regulation uncertainty. *Energy Econ.* **2020**, *89*, 104807. [CrossRef]
33. Krömer, S. Model risk regarding monthly wind energy production for the valuation of a wind farm investment. *Int. J. Energy Sect. Manag.* **2019**, *13*, 862–884. [CrossRef]
34. Nasrolahpour, E.; Zareipour, H.; Rosehart, W.D. Battery investment by a strategic wind producer: A scenario-based decomposition approach. *Electr. Power Syst. Res.* **2020**, *182*, 106255. [CrossRef]
35. Oh, E.; Son, S. Theoretical energy storage system sizing method and performance analysis for wind power forecast uncertainty management. *Renew. Energy* **2020**, *155*, 1060–1069. [CrossRef]
36. Ioannou, A.; Angus, A.; Brennan, F. Stochastic fi nancial appraisal of offshore wind farms. *Renew. Energy* **2020**, *145*, 1176–1191. [CrossRef]
37. Mehrjerdi, H.; Hemmati, R. Wind-hydrogen storage in distribution network expansion planning considering investment deferral and uncertainty. *Sustain. Energy Technol. Assess.* **2020**, *39*, 100687. [CrossRef]
38. Liu, R.; He, L.; Liang, X.; Yang, X.; Xia, Y. Is there any difference in the impact of economic policy uncertainty on the investment of traditional and renewable energy enterprises?—A comparative study based on regulatory effects. *J. Clean. Prod.* **2020**, *255*, 120102. [CrossRef]
39. Chowdhury, N.; Pilo, F.; Pisano, G. Optimal Energy Storage System Positioning and Sizing with Robust Optimization. *Energies* **2020**, *13*, 512. [CrossRef]
40. Tan, J.; Wu, Q.; Hu, Q.; Wei, W.; Liu, F.; Eb, C.H.P. Adaptive robust energy and reserve co-optimization of integrated electricity and heating system considering wind uncertainty. *Appl. Energy* **2020**, *260*, 114230. [CrossRef]
41. Zhan, S.; Hou, P.; Enevoldsen, P.; Yang, G.; Zhu, J.; Eichman, J.; Jacobson, M.Z. Electrical Power and Energy Systems Co-optimized trading of hybrid wind power plant with retired EV batteries in energy and reserve markets under uncertainties. *Electr. Power Energy Syst.* **2020**, *117*, 105631. [CrossRef]
42. Verleysen, K.; Coppitters, D.; Parente, A.; De Paepe, W.; Contino, F. How can power-to-ammonia be robust ? Optimization of an ammonia synthesis plant powered by a wind turbine considering operational uncertainties. *Fuel* **2020**, *266*, 117049. [CrossRef]
43. Yang, G.; Jiang, Y.; You, S. Planning and operation of a hydrogen supply chain network based on the off-grid wind-hydrogen coupling system. *Int. J. Hydrogen Energy* **2020**, *45*, 20721–20739. [CrossRef]
44. Al, A.; Sirjani, R.; Daneshvar, S. New hybrid probabilistic optimisation algorithm for optimal allocation of energy storage systems considering correlated wind farms. *J. Energy Storage* **2020**, *29*, 101335.
45. Hübler, C.; Piel, J.; Stetter, C.; Gebhardt, C.G.; Breitner, M.H.; Rolfes, R. In fl uence of structural design variations on economic viability of offshore wind turbines: An interdisciplinary analysis. *Renew. Energy* **2020**, *145*, 1348–1360. [CrossRef]
46. Ge, L.; Zhang, S.; Bai, X.; Yan, J.; Shi, C.; Wei, T. Optimal Capacity Allocation of Energy Storage System considering Uncertainty of Load and Wind Generation. *Math. Probl. Eng.* **2020**, *2020*, 2609674. [CrossRef]
47. Yang, H.; Yu, Q.; Liu, J.; Jia, Y.; Yang, G.; Ackom, E.; Dong, Z.Y. Optimal Wind-Solar Capacity Allocation with Coordination of Dynamic Regulation of Hydropower and Energy Intensive Controllable Load. *IEEE Access* **2020**, *8*, 110129–110139. [CrossRef]
48. Kong, L.; Li, Z.; Liang, L.; Xia, Y.; Xie, J. A capacity-investment model of wind power with uncertain supply-price under high penetration rate. *J. Clean. Prod.* **2020**, *178*, 917–926. [CrossRef]
49. Keck, R.; Sondell, N. Validation of uncertainty reduction by using multiple transfer locations for WRF—CFD coupling in numerical wind energy assessments. *Wind Energy Sci.* **2020**, *5*, 997–1005. [CrossRef]
50. Stetter, C.; Piel, J.; Hamann, J.F.H.; Breitner, M.H. Competitive and risk-adequate auction bids for onshore wind projects in Germany. *Energy Econ.* **2020**, *90*, 104849. [CrossRef]
51. Abdalla, O.H.; Smieee, L.; Adma, M.A.A.; Ahmed, A.S. Two-stage robust generation expansion planning considering long- and short-term uncertainties of high share wind energy. *Electr. Power Syst. Res.* **2020**, *189*, 106618. [CrossRef]
52. Henckes, P.; Frank, C.; Küchler, N.; Peter, J.; Wagner, J. Uncertainty estimation of investment planning models under high shares of renewables using reanalysis data. *Energy* **2020**, *208*, 118207. [CrossRef]
53. ANiromandfam, A.; Movahedi, A.; Zarezadeh, E. Virtual energy storage modeling based on electricity customers' behavior to maximize wind profit. *J. Energy Storage* **2020**, *32*, 101811. [CrossRef]
54. Zhou, J.; Wu, Y.; Dong, H.; Tao, Y.; Xu, C. Proposal and comprehensive analysis of gas-wind-photovoltaic- hydrogen integrated energy system considering multi-participant interest preference. *J. Clean. Prod.* **2020**, *265*, 121679. [CrossRef]
55. Zhao, X.; Yao, J.; Sun, C.; Pan, W. Impacts of carbon tax and tradable permits on wind power investment in China. *Renew. Energy* **2019**, *135*, 1386–1399. [CrossRef]
56. Maeda, M.; Watts, D. The unnoticed impact of long-term cost information on wind farms' economic value in the USA—A real option analysis. *Appl. Energy* **2019**, *241*, 540–547. [CrossRef]
57. Vavatsikos, A.P.; Arvanitidou, A.; Petsas, D. Wind farm investments portfolio formation using GIS-based suitability analysis and simulation procedures. *J. Environ. Manag.* **2019**, *252*, 109670. [CrossRef]

58. Askari, M.T.; Zainal, M.; Ab, A.; Tahmasebi, M.; Bolandifar, E. Modeling optimal long-term investment strategies of hybrid wind-thermal companies in restructured power market. *J. Mod. Power Syst. Clean Energy* **2019**, *7*, 1267–1279. [CrossRef]
59. Abdulgalil, M.A.; Khalid, M.; Alismail, F. Optimal Sizing of Battery Energy Storage for a Grid-Connected Microgrid Subjected to Wind Uncertainties. *Energies* **2019**, *12*, 2412. [CrossRef]
60. Junior, P.R.; Fischetti, E.; Araújo, V.G.; Peruchi, R.S.; Aquila, G.; Rocha Luiz, C.; Lacerda, L.S. Wind Power Economic Feasibility under Uncertainty and the Application of ANN in Sensitivity Analysis. *Energies* **2019**, *12*, 2281. [CrossRef]
61. Abdulgalil, M.; Khalid, M.; Alismail, F. Optimizing a Distributed Wind-Storage System Under Critical Uncertainties Using Benders Decomposition. *IEEE Access* **2019**, *7*, 77951–77963. [CrossRef]
62. Li, Y.; Wang, J.; Gu, C.; Liu, J.; Li, Z. Investment optimization of grid-scale energy storage for supporting different wind power utilization levels. *J. Mod. Power Syst. Clean Energy* **2019**, *7*, 1721–1734. [CrossRef]
63. Pizarro-alonso, A.; Ravn, H.; Münster, M. Uncertainties towards a fossil-free system with high integration of wind energy in long-term planning. *Appl. Energy* **2019**, *253*, 113528. [CrossRef]
64. Yan, J.; Zhang, H.; Liu, Y.; Han, S.; Li, L. Uncertainty estimation for wind energy conversion by probabilistic wind turbine power curve modelling. *Appl. Energy* **2019**, *239*, 1356–1370. [CrossRef]
65. Tagliapietra, S.; Zachmann, G.; Fredriksson, G. Estimating the cost of capital for wind energy investments in Turkey. *Energy Policy* **2019**, *131*, 295–301. [CrossRef]
66. Van Thang, V.; Trung, N.H. Evaluating efficiency of renewable energy sources in planning micro-grids considering uncertainties. *J. Energy Syst.* **2019**, *3*, 14–25. [CrossRef]
67. Quan, N.; Kim, H.M. Greedy robust wind farm layout optimization with feasibility guarantee feasibility guarantee. *Eng. Optim.* **2019**, *51*, 1152–1167. [CrossRef]
68. Mora, E.B.; Spelling, J.; van der Weijde, A.H.; Pavageau, E.M. The effects of mean wind speed uncertainty on project finance debt sizing for offshore wind farms. *Appl. Energy* **2019**, *252*, 113419. [CrossRef]
69. Fuchs, C.; Marquardt, K.; Kasten, J.; Skau, K. Wind Turbines on German Farms—An Economic Analysis. *Energies* **2019**, *12*, 1587. [CrossRef]
70. Zhang, Y.; Wang, L.; Wang, N.; Duan, L.; Zong, Y.; You, S.; Marechal, F.; Yang, Y. Balancing wind-power fluctuation via onsite storage under uncertainty: Power-to-hydrogen-to-power versus lithium battery. *Renew. Sustain. Energy Rev.* **2019**, *116*, 109465. [CrossRef]
71. Ioannou, A.; Angus, A.; Brennan, F. Informing parametric risk control policies for operational uncertainties of o ff shore wind energy assets. *Ocean Eng.* **2019**, *177*, 1–11. [CrossRef]
72. Ribeiro, A.; Finotti, F.; Perobelli, C.; Baumgratz, A. Counterfactual comparisons of investment options for wind power and agricultural production in the United States: Lessons from Northern Ohio. *Energy Econ.* **2018**, *74*, 299–309.
73. Finjord, F.; Hagspiel, V.; Lavrutich, M.; Tangen, M. The impact of Norwegian-Swedish green certificate scheme on investment behavior: A wind energy case study. *Energy Policy* **2018**, *123*, 373–389. [CrossRef]
74. Dalby, P.A.O.; Gillerhaugen, G.R.; Hagspiel, V.; Leth-olsen, T.; Thijssen, J.J.J. Green investment under policy uncertainty and Bayesian learning. *Energy* **2018**, *161*, 1262–1281. [CrossRef]
75. Li, Y.; Wu, M.; Li, Z. A Real Options Analysis for Renewable Energy Investment Decisions under China Carbon Trading Market. *Energies* **2018**, *11*, 1817. [CrossRef]
76. Gazheli, A.; Van Den Bergh, J. Real options analysis of investment in solar vs. wind energy: Diversi fi cation strategies under uncertain prices and costs. *Renew. Sustain. Energy Rev.* **2018**, *82*, 2693–2704. [CrossRef]
77. Romanuke, V. Wind Farm Energy and Costs Optimization Algorithm under Uncertain Parameters of Wind Speed Distribution. *Stud. Inform. Control* **2018**, *27*, 155–164. [CrossRef]
78. Yu, Y.; Wen, X.; Zhao, J.; Xu, Z.; Li, J. Co-Planning of Demand Response and Distributed Generators in an Active Distribution Network. *Energies* **2018**, *11*, 354. [CrossRef]
79. Aaboud, M.; Aad, G.; Abbott, B.; Abeloos, B.; Abidi, S.H.; AbouZeid, O.S.; Abraham, N.L.; Abramowicz, H.; Abreu, H.; Abreu, R.; et al. Search for dark matter and other new phenomena in events with an energetic jet and large missing transverse momentum using the ATLAS detector. *J. High Energy Phys.* **2018**, *2018*, 1–53. [CrossRef]
80. Valinejad, J.; Marzband, M.; Funsho Akorede, M.; DElliott, I.; Godina, R.; Matias, J.C.; Pouresmaeil, E. Long-Term Decision on Wind Investment with Considering Different Load Ranges of Power Plant for Sustainable Electricity Energy Market. *Sustainability* **2018**, *10*, 3811. [CrossRef]
81. Jiang, X.; Nan, G.; Liu, H.; Guo, Z.; Zeng, Q.; Jin, Y. Optimization of Battery Energy Storage System Capacity for Wind Farm with Considering Auxiliary Services Compensation. *Appl. Sci.* **2018**, *8*, 1957. [CrossRef]
82. Deshmukh, R.; Mileva, A.; Wu, G.C. Renewable energy alternatives to mega hydropower: A case study of Inga 3 for Southern Africa. *Environ. Res. Lett.* **2018**, *13*, 064020. [CrossRef]
83. Li, Z.; Wang, C.; Li, B.; Wang, J.; Zhao, P.; Zhu, W.; Yang, M.; Ding, Y. Probability-interval-based optimal planning of integrated energy system with uncertain wind power. *IEEE Trans. Ind. Appl.* **2018**, *56*, 4–13. [CrossRef]
84. Ioannou, A.; Angus, A.; Brennan, F. A lifecycle techno-economic model of offshore wind energy for di ff erent entry and exit instances. *Appl. Energy* **2018**, *221*, 406–424. [CrossRef]
85. Esmaieli, M.; Ahmadian, M. The effect of research and development incentive on wind power investment, a system dynamics approach. *Renew. Energy* **2018**, *126*, 765–773. [CrossRef]

86. Kristiansen, M.; Svendsen, H.G.; Korpas, M.; Fleten, S.-E. Multistage grid investments incorporating uncertainty in offshore wind development. *Energy Procedia* **2017**, *137*, 468–476. [CrossRef]
87. Jannati, J.; Yazdaninejadi, A.; Talavat, V. Simultaneous Planning of Renewable/Non-Renewable Distributed Generation Units and Energy Storage Systems in Distribution Networks. *Trans. Electr. Electron. Mater.* **2017**, *18*, 111–118. [CrossRef]
88. Chen, L.; Macdonald, E. Wind Farm Layout Sensitivity Analysis and Probabilistic Model of Landowner Decisions. *J. Energy Resour. Technol.* **2017**, *139*, 031202. [CrossRef]
89. Hamoudi, H.; Maule, I. Photovoltaic and wind cost decrease estimation: Implications for investment analysis. *Energy* **2017**, *137*, 1054–1065.
90. Aquila, G.; Rotela, P.; de Oliveira, E.; de Queiroz, A. Wind power feasibility analysis under uncertainty in the Brazilian electricity market. *Energy Econ.* **2017**, *65*, 127–136. [CrossRef]
91. Eryilmaz, D.; Homans, F.R. How does uncertainty in renewable energy policy affect decisions to invest in wind energy? *Electr. J.* **2016**, *29*, 64–71. [CrossRef]
92. Kitzing, L.; Juul, N.; Drud, M.; Krogh, T. A real options approach to analyse wind energy investments under different support schemes. *Appl. Energy* **2016**, *188*, 83–96. [CrossRef]
93. Pazouki, S.; Haghifam, M. Optimal planning and scheduling of energy hub in presence of wind, storage and demand response under uncertainty. *Int. J. Electr. Power Energy Syst.* **2016**, *80*, 219–239. [CrossRef]
94. Lamadrid, A.J.; Maneevitjit, S.; Mount, T.D. The Economic Value of Transmission Lines and the Implications for Planning Models. *Energy Econ.* **2016**, *57*, 1–15. [CrossRef]
95. Caralis, G.; Chaviaropoulos, P.; Ruiz, V.; Diakoulaki, D.; Kotroni, V.; Lagouvardos, K.; Gao, Z. Lessons learnt from the evaluation of the feed-in tariff scheme for offshore wind farms in Greece using a Monte Carlo approach. *J. Wind Eng. Ind. Aerodyn.* **2016**, *157*, 63–75. [CrossRef]
96. Werner, L.; Scholtens, B. Firm Type, Feed-in Tariff, and Wind Energy Investment in Germany: An Investigation of Decision Making Factors of Energy Producers Regarding Investing in Wind Energy Capacity. *J. Ind. Ecol.* **2016**, *21*, 402–411. [CrossRef]
97. Sjoerd, A.; Van Den Broek, M.; Özdemir, Ö.; Koutstaal, P.; Faaij, A. Business case uncertainty of power plants in future energy systems with wind power. *Energy Policy* **2016**, *89*, 237–256.
98. Xiao, Y.; Wang, X.; Wang, X.; Wu, Z. Trading wind power with barrier option. *Appl. Energy* **2016**, *182*, 232–242. [CrossRef]
99. Díaz, G.; Gómez-aleixandre, J.; Coto, J. Dynamic evaluation of the levelized cost of wind power generation. *Energy Convers. Manag.* **2015**, *101*, 721–729. [CrossRef]
100. Díaz, G.; Moreno, B.; Coto, J.; Gómez-aleixandre, J. Valuation of wind power distributed generation by using Longstaff—Schwartz option pricing method. *Appl. Energy* **2015**, *145*, 223–233. [CrossRef]
101. Fang, X.; Li, F.; Wei, Y.; Azim, R. Reactive Power Planning under High Penetration of Wind Energy using Benders Decomposition. *IET Gener. Transm. Distrib.* **2015**, *9*, 1835–1844. [CrossRef]
102. Seljom, P.; Tomasgard, A. Short-term uncertainty in long-term energy system models—A case study of wind power in Denmark. *Energy Econ.* **2015**, *49*, 157–167. [CrossRef]
103. Hong, Y.; Lai, Y.; Chang, Y.; Lee, Y.; Liu, P. Optimizing Capacities of Distributed Generation and Energy Storage in a Small Autonomous Power System Considering Uncertainty in Renewables. *Energies* **2015**, *8*, 2473–2492. [CrossRef]
104. Siddons, C.; Allan, G.; Mcintyre, S. How accurate are forecasts of costs of energy? A methodological contribution. *Energy Policy* **2015**, *87*, 224–228. [CrossRef]
105. Rodríguez, O.; del Río, J.A.; Jaramillo, O.A.; Martínez, M. Wind Power Error Estimation in Resource Assessments. *PLoS ONE* **2015**, *10*, e0124830. [CrossRef]
106. Abadie, L.M.; Chamorro, J.M. Valuation of Wind Energy Projects: A Real Options Approach. *Energies* **2014**, *7*, 3218–3255. [CrossRef]
107. Krogh, T.; Meade, N.; Fleten, S. Renewable energy investments under di ff erent support schemes: A real options approach. *Eur. J. Oper. Res.* **2012**, *220*, 225–237.
108. Correa, M.D.; Gomes, L.L.; Teixeira, L.E.B. Investors' asymmetric views and their decision to enter Brazil's wind energy sector. *Pesqui. Oper.* **2014**, *34*, 319–345.
109. Weibel, S.; Madlener, R. Cost-Effective Design of Ringwall Storage Hybrid Power Plants: A Real Options Analysis. *Energy Procedia* **2014**, *61*, 2196–2200. [CrossRef]
110. Serrano, J.; Burgos, M.; Riquelme, J. *M Optimization of Wind Farm Turbine Layout Including Decision Making Under Risk*; IEEE: Piscataway, NJ, USA, 2014.
111. Monjas-barroso, M.; Balibrea-Iniesta, J. Valuation of projects for power generation with renewable energy: A comparative study based on real regulatory options. *Energy Policy* **2013**, *55*, 335–352. [CrossRef]
112. Jin, S.; Botterud, A.; Ryan, S.M. Impact of Demand Response on Thermal Generation Investment with High Wind Penetration. *IEEE Trans. Smart Grid* **2013**, *4*, 2374–2383. [CrossRef]
113. Kaiser, M.J.; Snyder, B.F. Modeling offshore wind installation costs on the US Outer Continental Shelf. *Renew. Energy* **2013**, *50*, 676–691. [CrossRef]
114. Heinrich, W.; Szolgayová, J.; Fuss, S.; Obersteiner, M. Renewable energy investment: Policy and market impacts. *Appl. Energy* **2012**, *97*, 249–254.

115. Heinrich, W.; Fuss, S.; Szolgayová, J.; Obersteiner, M. Investment in wind power and pumped storage in a real options model. *Renew. Sustain. Energy Rev.* **2012**, *16*, 2242–2248.
116. Martinez-cesena, E.A.; Member, S.; Mutale, J.; Member, S. Wind Power Projects Planning Considering Real Options for the Wind Resource Assessment. *IEEE Trans. Sustain. Energy* **2012**, *3*, 158–166. [CrossRef]
117. Ochoa, M.; Betancur, H.; David, J.; Múnera, G.; Mauricio, Ó. L a Valoración de Proyectos Bajo el Enfoque de Opciones. 2012. Available online: https://www.redalyc.org/pdf/205/20523152009.pdf (accessed on 17 July 2021).
118. Al-yahyai, S.; Charabi, Y.; Al-badi, A.; Gastli, A. Nested ensemble NWP approach for wind energy assessment. *Renew. Energy* **2012**, *37*, 150–160. [CrossRef]
119. Lee, S. Using real option analysis for highly uncertain technology investments: The case of wind energy technology. *Renew. Sustain. Energy Rev.* **2011**, *15*, 4443–4450. [CrossRef]
120. Dicorato, M.; Forte, G.; Pisani, M.; Trovato, M. Guidelines for assessment of investment cost for offshore wind generation. *Renew. Energy* **2011**, *36*, 2043–2051. [CrossRef]
121. Barradale, M.J. Impact of public policy uncertainty on renewable energy investment: Wind power and the production tax credit. *Energy Policy* **2010**, *38*, 7698–7709. [CrossRef]
122. Méndez, M.; Goyanes, A.; Lamothe, P. Real Options Valuation of a Wind Farm. February 2009. Available online: http://www.realoptions.org/papers2009/46.pdf (accessed on 24 July 2021).
123. Muñoz, J.I.; Contreras, J.; Caamaño, J.; Correia, P.F.; Carlo, M. Risk Assessment of Wind Power Generation Project Investments based on Real Options. In *2009 IEEE Bucharest PowerTech*; IEEE: Piscataway, NJ, USA, 2009; pp. 1–8.
124. Dykes, K.; de Neufville, R. Real options for a wind farm in Wapakoneta, Ohio: Incorporating uncertainty into economic feasibility studies for community wind. In Proceedings of the Paper submitted for the World Wind Energy Conference of 2008, Kingston, ON, Canada, 24–26 June 2007.
125. Magnus, K.; Fleten, S.; Maribu, K.M.; Wangensteen, I. Optimal investment strategies in decentralized renewable power generation under uncertainty. *Energy* **2007**, *32*, 803–815.
126. Yu, W.; Sheblé, G.B.; Lopes, J.A.P.; Matos, M.A. Valuation of switchable tariff for wind energy. *Electr. Power Syst. Res.* **2006**, *76*, 382–388. [CrossRef]
127. Aponte, R.; Muñoz, F.; Álzate, L. La Evaluación Financiera De Proyectos Y Su Aporte En La Generación De Valor Corporativo. *Cienc. Y Poder Aéreo* **2017**, *12*, 144–155. [CrossRef]
128. Santos, L.; Soares, I.; Mendes, C.; Ferreira, P. Real options versus traditional methods to assess renewable energy projects. *Renew. Energy* **2014**, *68*, 588–594. [CrossRef]
129. González, Y.; Zuluaga, M.; Maya, C. *El Valor de la Propiedad Industrial Aproximación a un Método de Valoración Financiera de Activos Intangibles*; Universidad EAFIT: Medellín, Colombia, 2010.
130. Attoh-Okine, N.O.; Ayyub, B.M. *Applied Research in Uncertainty Modeling and Analysis*; Springer: New York, NY, USA, 2005.
131. Conejo, A.J.; Carrión, M.; Morales, J.M. *Decision Making under Uncertainty in Electricity Markets*; Springer: New York, NY, USA, 2010; Volume 1.
132. Dixit, R.K.; Pindyck, R.S. *Investment under Uncertainty*; Princeton University Press: Princeton, NJ, USA, 1994.
133. Cuervo, F.I.; Botero, S.B. Aplicación de las opciones reales en la toma de decisiones en los mercados de electricidad. *Estud. Gerenc.* **2014**, *30*, 397–407. [CrossRef]
134. Guerequeta, R.; Vallecillo, A. Técnicas de Diseño de Algorismos. 2000. Available online: https://archive.org/details/TecnicasDiseoAlgoritmosGuequeretaYVallecillo (accessed on 28 July 2021).
135. Black, F.; Scholes, M. The pricing of options and corporate liabilities. *J. Political Econ.* **1973**, *81*, 637–654. [CrossRef]
136. Merton, R.C. Theory of rational option pricing. *Bell J. Econ. Manag. Sci.* **1973**, *4*, 141–183. [CrossRef]
137. Mascareñas, J. *Opciones Reales y Valoración de Activos: Como Medir la Flexibilidad Operativa en la Empresa*; Prentice Hall: Hoboken, NJ, USA, 2004.
138. Ferreira, N.; Kar, J.; Trigeorgis, L. Option Games. *Harv. Bus. Rev.* **2009**, *87*, 101–107.
139. Kozlova, M. Real option valuation in renewable energy literature: Research focus, trends and design. *Renew. Sustain. Energy Rev.* **2017**, *80*, 180–196. [CrossRef]
140. Boyle, P.P. Options: A Monte Carlo approach. *J. Financ. Econ.* **1977**, *4*, 323–338. [CrossRef]
141. Kodukula, P.; Papudesu, C. *Project Valuation Using Real Options: A Practitioner's Guide*; J. Ross Publishing: Fort Lauderdale, FL, USA, 2006.
142. Morrison, S. Nested Simulation for Economic Capital. Barrie + hibbert. 2009. Available online: www.barrhibb.com (accessed on 22 July 2021).
143. Longstaff, F.A.; Schwartz, E.S. Valuing American Options by Simulation: A Simple Least-Squares Approach. *Rev. Financ. Stud.* **2001**, *14*, 113–147. [CrossRef]
144. Yang, X.-S. *Nature-Inspired Algorithms and Applied Optimization*; Springer: New York, NY, USA, 2018.
145. Mosiño, A.; Salomón-Núñez, L.A.; Moreno-Okuno, A.T. Estudio empírico sobre el tipo de cambio. *Econoquantum* **2019**, *16*, 33–56. [CrossRef]
146. Pindyck, R.S. The dynamics of commodity spot and futures markets: A primer. *Energy J.* **2001**, *22*. [CrossRef]
147. Vveinhardt, J.; Streimikiene, D.; Rizwan, A.R.; Nawaz, A.; Rehman, A. Mean reversion: An investigation from Karachi stock exchange sectors. *Technol. Econ. Dev. Econ.* **2016**, *22*, 493–511. [CrossRef]

148. Gujarati, D.N.; Porter, D.C. *Econometria Básica*, 4th ed.; Editora Campus: São Paulo, Brasil, 2006.
149. Góngora, D.A.; Azumendi, D. *Predicción de la Velocidad del Viento para la Estimación de la Energía Generada por un Aerogenerador*; Uniandes: Bogotá, Colombia, 2018.
150. Quevedo, F. Distribución normal. *Medwave* **2011**, 1–5. [CrossRef]
151. Latiff, A. La "Curva de Aprendizaje". Qué es y cómo se mide. *Rev. Urol. Colomb.* **2005**, *XIV*, 15–17.
152. Silva, P.P.B.; Riveros, D.P.B.; Naranjo, C.J. Aplicación de la lúdica en la curva de aprendizaje. *Sci. Et Tech.* **2005**, *1*, 185–190.
153. Wolsink, M. Wind power and the NIMBY-myth: Institutional capacity and the limited significance of public support. *Renew. Energy* **2000**, *21*, 49–64. [CrossRef]
154. Wolsink, M. Wind power implementation: The nature of public attitudes: Equity and fairness instead of "backyard motives". *Renew. Sustain. Energy Rev.* **2007**, *11*, 1188–1207. [CrossRef]
155. Wüstenhagen, R.; Wolsink, M.; Bürer, M.J. Social acceptance of renewable energy innovation: An introduction to the concept. *Energy Policy* **2007**, *35*, 2683–2691. [CrossRef]
156. Buijs, P.; Bekaert, D.; Cole, S.; Van Hertem, D.; Belmans, R. Transmission investment problems in Europe: Going beyond standard solutions. *Energy Policy* **2011**, *39*, 1794–1801. [CrossRef]
157. Khademi, E.; Timmermans, H.; Borgers, A. Temporal adaptation to reward schemes: Results of the spitsscoren project. *Transp. Res. Procedia* **2014**, *3*, 60–69. [CrossRef]
158. Jung, N.; Moula, M.E.; Fang, T.; Hamdy, M.; Lahdelma, R. Social acceptance of renewable energy technologies for buildings in the Helsinki Metropolitan Area of Finland. *Renew. Energy* **2016**, *99*, 813–824. [CrossRef]

applied sciences

MDPI

Article

Potential of Onshore Wind Turbine Inertia in Decarbonising the Future Irish Energy System

Henning Thiesen * and Clemens Jauch

Wind Energy Technology Institute (WETI), Flensburg University of Applied Sciences, 24943 Flensburg, Germany; clemens.jauch@hs-flensburg.de

* Correspondence: henning.thiesen@hs-flensburg.de

Abstract: Power system inertia is an essential part for grid frequency stability and decreases due to the replacement of fossil fuel fired power plants with variable renewable energy sources. This development is not represented sufficiently in unit commitment and economic dispatch models. If considered at all, only synchronous inertia from fossil fuel driven power plants is modelled. This results in increased CO_2 emissions, curtailed renewable energy and high system costs. While wind turbines are a source for synthetic inertia and an important renewable energy source, their capability to provide inertia is not incorporated into energy system models. The work at hand closes this research gap and applies a methodology to depict synthetic inertia provided by wind turbines as part of the optimisation dispatch model. A unit commitment and economic inertia dispatch model of the the all-Island Irish power system is created. The potential of wind inertia is analysed and quantified by assessing CO_2 emissions, curtailed renewable energy and system costs. Results show that synthetic inertia provided by wind turbines can save up 30.99% of the CO_2 emissions, reduce curtailment by up to 39.90% and reduce system costs by 32.72%.

Keywords: economic dispatch modelling; energy system modelling; power system inertia; renewable energy; synthetic inertia; wind energy

Citation: Thiesen, H.; Jauch, C. Potential of Onshore Wind Turbine Inertia in Decarbonising the Future Irish Energy System. *Appl. Sci.* **2022**, *12*, 2984. https://doi.org/10.3390/app12062984

Academic Editor: Maria Vicidomini

Received: 14 February 2022
Accepted: 12 March 2022
Published: 15 March 2022

1. Introduction

To limit global warming, energy systems are decarbonised by replacing fossil fuel driven power plants with renewable energy sources (RESs). Disconnecting synchronously connected generators reduces the overall power system inertia [1]. Inertia is an essential part for power system stability since it limits the speed with which the grid frequency changes, commonly referred to as the rate of change of frequency (ROCOF), and provides time for generators supplying grid frequency control reserves [2]. Frequency converter connected generation units like wind turbines (WTs) and photovoltaic systems (PVs) as well as storage units with the same type of interface to the power system do not provide an inherent inertial response [1]. However, they are able to mimic the behaviour of a synchronously connected rotating mass known as synthetic inertia (SI) [3,4]. Variable speed WTs are a source for SI and as such are already in application [3]. To sustain the controllability of the grid frequency, future power systems will consist of an inherent and a non-inherent inertia part [1].

Some power systems already experience periods with low system inertia during high non-synchronous penetration and/or low power demand [1]. The grid operator of the Hydro-Québec electricity transmission system demanded WTs to support the power system in the same manner as synchronous generators do in terms of an inertial response [5,6]. The all Island transmission system operator (TSO) from Ireland defines a grid supporting frequency sensitive mode for WTs [7] as well as system operational constraints like an upper limit for non-synchronous penetration, a maximum ROCOF and an operational limit for inertia [8]. The European Network of Transmission System Operators for Electricity

(ENTSO-E) addresses the need for actions in future power systems to compensate the loss of inherent inertia from synchronous generators [9]. However, the requirement for system inertia can be reduced by introducing energy sources with response times faster than traditional mechanical response from conventional generators [10].

Energy system modelling is an important method in aiding energy system analysis [11]. Its main purpose is to assist in designing, planning and implementing future energy systems [12]. Modelling system inertia in energy system models however, is a topic of minor interest. Table 1 sums up the results of a review of relevant literature in which inertia has been considered as part of energy system models. All papers used unit commitment and (economic) dispatch models applying different constraints to account for minimum system inertia, e.g., ROCOF thresholds or grid frequency nadir thresholds. Actual grid islands like Great Britain or the ERCOT power system were analysed or the interconnected Continental European power systems, which as a whole is a synchronous grid island. In conclusion, the review shows that minimum inertia constraints result in increasing system costs, carbon dioxide emissions and curtailed variable RES.

Table 1. Overview of the research highlights, the applied methodology and scenarios, as well as the key results of the reviewed literature.

Author(s)	Research Highlight(s)	Methodology & Scenario	Result(s)
Teng et al. [13]	Introduction of a novel mixed linear integer programming formulation considering frequency constraint formulations, quantification of wind generation uncertainty on inertia and application of a 2030 model of the Great Britain power system	Unit commitment and economic dispatch formulation applying a multi-stage scenario tree. Analysis of increasing ROCOF threshold from 0.5 Hz/ up to 0.2 Hz/s.	A ROCOF threshold of 0.2 Hz/s would lead to extremely high operation cost (120% increase) and wind curtailment (35% increase)
Collins et al. [14]	Enhancement and robustness check of electricity generation of the PRIMES-REF scenario for 2030 EU by quantifying variable RES curtailment, levels of interconnector congestion and electricity prices.	Creating unit commitment and dispatch model using PLEXOS software, integrating a ROCOF limit of 0.75 Hz/s.	Variable RES congestion increase of 11% in Ireland, overall transmission congestion during 24% of the researched time.
Johnson et al. [15]	Filling the knowledge gap of missing inertia modelling in unit commitment and dispatch models	Mixed integer linear programming model created by PLEXOS optimising for least costs. A system inertia constraint is added to the model for stable grid operations.	Increased dispatch costs by $85 million
Mehigan et al. [16]	Analysing the effect of two different levels of the maximum ROCOF threshold (0.5 Hz/s and 1 Hz/s) on curtailment of variable RES, carbon dioxide emissions and production costs in Europe.	Mixed integer linear programme of a European electricity dispatch system implementing minimum inertia constraints based on ROCOF thresholds.	Increase of generation costs of up to 53.1%, variable RES curtailment increase of 2.15% and CO_2 emission increase of 48.9% for researched system.

None of the reviewed papers and unit commitment and economic dispatch models include SI provision of WTs. Only synchronous inertia is considered. However, it is concluded that "[...] synthetic inertia from wind turbines is believed to play an important role in supporting the frequency performance in the future low carbon power system" [13] as well as that "As grids integrate more renewable energy, relying on the current method of delivering inertia will see escalating costs, which demonstrates the value of integrating new sources of inertia, like synthetic inertia from wind turbines [...]" [15].

The purpose of this work is to close this research gap of unit commitment and economic dispatch models not considering SI provided by WTs and thereby overestimating system costs, RES curtailment and CO_2 emissions. Therefore, a methodology of a unit commitment and economic inertia dispatch model to represent the all-Island Irish power system is applied. A parameter, which determines the SI provision by WTs is analysed as well as two ROCOF thresholds. The potential of wind inertia is quantified by assessing the total CO_2 emissions, RES curtailment and resulting system costs. Other problems related to the increasing penetration of inverter connected units like reactive support, harmonics or black start capability are not researched as these topics are not primarily related to grid frequency stability. The future Irish power systems is analysed in this work, because the present all-Island Irish power system is already characterised by times of a high shares of frequency inverter connected penetration [17]. Such high penetration times occur due to the fact of Ireland being a island with HVDV grid connection to Great Britain as well as a large number of WTs installed. Thus, feed-in from WTs is already curtailed in times of high frequency converter penetration and low system inertia [17]. The Irish grid operator already has reduced the needed system inertia to a minimum [8], requires generation units to provided a fast reacting frequency response [7] and defines an upper limit for non-synchronous penetration and a maximum ROCOF [8].

The following Section 2 provides an overview of the applied methodology. Section 3 describes scenarios and data of the Irish dispatch model. In Section 4 results are summed up and Section 5 concludes the work at hand.

2. Methodology

The introduced methods to depict inertia as part of the energy system dispatch, to provide SI with WTs and incorporated inertia constraints are elaborated in this section. The methodology is incorporated into an open source modelling tool which is presented too. Prior, fundamental basics of inertia provision are introduced.

2.1. Synchronous and Synthetic Inertia in Power Systems

In AC-power systems, power generation and power consumption have to be balanced at all times [2]. The indicator for power balance is the grid frequency, f_{grid} [2]. A power imbalance leads to a deviation of the grid frequency from its nominal value [2]. The speed with which the grid frequency changes is dominated by power system inertia [1]. Equation (1) illustrates this relationship where $\delta f/\delta t$ is the ROCOF, P_{gen} the accumulated power generation, P_{load} the accumulated power consumption and J_{sys} is the overall power system inertia, i.e., the sum of the moment of inertia of all synchronously connected machines [1].

$$\frac{\delta f}{\delta t} = \frac{P_{gen} - P_{load}}{4 \cdot \pi^2 \cdot f_{grid} \cdot J_{sys}} \tag{1}$$

It is state-of-the-art to describe the inertia of a synchronously connected machine with the inertia constant, H_g. The inertia constant represents the proportional expression of the units stored kinetic energy, $E_{g,kin}$, with respect to its apparent power rating, S_g, as depicted in Equation (2) [1,3]. The inertia constant represents the theoretical time, e.g., a synchronous generator, is able to provide its rated power solely using its stored kinetic energy [1]. Likewise, the resilience of a power system against power imbalances can be described with the power system inertia constant [1].

$$H_g = \frac{E_{g,kin}}{S_g} \tag{2}$$

Frequency converter connected generation sources or energy storage units are able to emulate the behaviour of a synchronous generator in the event of a power imbalance known as synthetic inertia (SI) [3,4]. A common strategy to substitute inertia with WTs is to change the electric power output based on the ROCOF and the frequency [3]. Due to the modified power output, the rotational speed of the WT changes [18]. That might cause the WT to operate at non-optimal rotational speed [18]. In the worst case, the WT disconnects from the system due to over- or underspeed protection while providing SI [19]. Thus, the SI support as well as the non-inertia related power feed-in of the WT is lost [19]. Hence, from the TSO's point of view, the situation changes for the worse. To counteract such problems, a novel control methodology to emulate inertia with WTs is introduced by Gloe et al. [19]. The authors propose to scale the inertia constant with the actual operating point of the WT, considering the cut-in speed and the rated speed of the WT as supporting points [19]. Thereby, the risk of disconnecting the WT while providing inertia is diminished. The variable inertia constant is normalised such that the inertia constant demanded by the TSO, H_{dem}, is provided at rated speed. H_{dem} is determined by the TSO and based on power system requirements. Equation (3) depicts how the provided SI of a WT is calculated were $\omega_{actual,WT}$ is the actual rotational speed of the WT, $\omega_{rated,WT}$ is the rated speed of the WT and $\omega_{cut\_in,WT}$ the cut-in wind speed of the WT.

$$\frac{H_{var}}{H_{dem}} = \frac{0.5 \cdot J_{WT} \cdot (\omega^2_{actual,WT} - \omega^2_{cut-in,WT})}{0.5 \cdot J_{WT} \cdot (\omega^2_{rated,WT} - \omega^2_{cut-in,WT})} \tag{3}$$

SI is not a direct substitution of synchronous inertia since it is not an inherent inertial response [3]. Hence, in future power systems, the overall system inertia constant comprises an inherent and and non-inherent term as depicted in Equation (4).

$$H_{sys} = \overbrace{\frac{\sum_g H_{g,sync} \cdot S_{g,sync}}{\sum_g S_g}}^{\text{system synchronous inertia}} + \overbrace{\frac{\sum_g H_{g,synt} \cdot S_{g,synt}}{\sum_g S_g}}^{\text{system synthetic inertia}} \tag{4}$$

2.2. Dispatch Model Formulation

The hereafter presented methods are created applying the novel modelling tool Open Inertia Modelling (OpInMod) [20]. OpInMod is an open source modelling tool extending the basic functionalites of the open energy modeling framework (oemof) package solph [21,22]. oemof is developed considering a strict open source and non-proprietary philosophy and is implemented in python [21]. Thus, scientific principles like reproducibility are transparency fulfilled [23–26]. Additionally, comparing oemof to the proprietary software solutions which have been applied in the reviewed literature [13–16], its free of charge availability makes it easy to access [21].

In general, oemof, and therefore OpInMod as well, allows for an easy representation of energy systems applying a graph structure consisting of buses and components [21]. Components are used to represent power plants, variable RES units or energy storage units in an energy system model [21]. Buses and components are connected via edges representing an energy flow [21]. Components hold basic attributes of the represented energy system component. For example, a representation of a power plant would hold attributes such as the variable costs, the installed capacity or the power plants efficiency. OpInMod extends oemof.solph functionality to depict different sources of inertia as well as creating additional constraints when modelling system inertia. The before example of a power plant representation would in OpInMod be extended by attributes such as the type of inertia source (synchronous inertia or SI), the generators inertia constant or costs associated with the provision of inertia. OpInMod is published under MIT License and available via Github [27] and zenodo [28]

Figure 1 provides an schematic overview of the modelling process of a unit commitment and economic inertia dispatch model created with OpInMod. The process can be devided in three sub processes. In the first sub process, named preperation and initialisation, all packages needed are imported as well as the input files. Next, the energy system is created, using components and buses. In the modelling sub process the energy system representation is translated into an optmisation problem, including all sets of constraints, based on python's optmisation package pyomo [29]. The optimisation problem is solved by a freely selectable solver. In Figure 1 the COIN OR (CBC) solver, GNU Linear Programming Kit (GLPK) solver or gurobi solver are listed as examples. The post processing step of the solver results is the last in the modelling sub process. In the third and last sub process, the results are exported and can be further analysed.

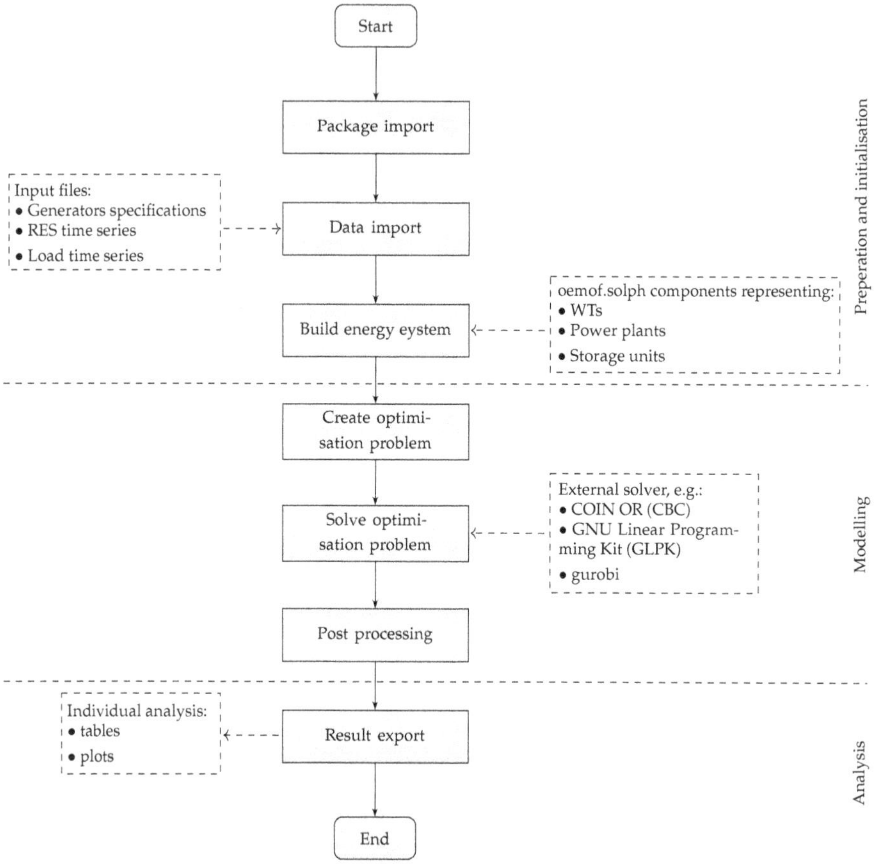

Figure 1. Schematic overview of the full modelling process applying OpInMod. In the preperation and initialisation part of the process, necessary data is imported and the energy system is created using oemof.solph's base components. The base components are extended by attributes and functions to depict synchronous and SI. In the modelling part of the process the optimisation problem, including OpInMod's inertia constraints, is created, solved by an external solver and the results are processed and prepared for the output. In the last section of the process, results are exported and can be further analysed as needed.

The optimisation function depicted in Equation (5) and created by OpInMod is set to minimise the overall system costs subjected to supply power demand for each time step. The first term describes the flow costs by the variable costs of the generator, c_g^{vc} and the flow variable, x_g^{flow} [22]. The second term of the function describes the inertia related costs. The generators moment of inertia is represented by $x_g^{inertia}$, the costs of the generator providing inertia by c_g^{ic}. Traditionally, a synchronously connected rotating mass provides its full inertia to the power system as long as it is connected. This is reflected by the binary variable $x_g^{source\_inertia}$.

$$\min : \sum_t \sum_g \overbrace{c_g^{vc} \cdot x_g^{flow}(t)}^{\text{costs flow}} + \overbrace{x_g^{source\_inertia} \cdot c_g^{ic} \cdot x_g^{inertia}(t)}^{\text{costs inertia}} \tag{5}$$

2.3. System Inertia Constraints

The optimisation function is bound to two constraints introduced by OpInMod representing the need for minimum system synchronous inertia and minimum overall system inertia. Further constraints inherited from oemof.solph are not presented hereafter but described in [21,22,30]. The minimum system synchronous inertia constraint is depicted in Inequation (6), where $x^{min\_sys\_sync\_inertia}$ is the minimum system synchronous inertia and the right side of the equation is the accumulated moment of inertia of all synchronously connected generators connected to the power system.

$$x^{min\_sys\_sync\_inertia} \leq \sum_g x_g^{source\_inertia} \cdot x_g^{sync\_inertia} \tag{6}$$

Inequation (7) depicts the minimum overall system inertia constraint formulation. The minimum overall system inertia is represented by $x^{min\_sys\_inertia}$. The right part of Inequation (7) depicts the inertia sum all connected generators providing either synchronous or synthetic inertia.

$$x^{min\_sys\_inertia} \leq \sum_g x_g^{source\_inertia} \cdot (x_g^{sync\_inertia} + x_g^{synt\_inertia}) \tag{7}$$

In this work, inertia is either provided by synchronously connected generators or emulated in the form of SI by WTs. Synchronous inertia is also provided by synchronously connected loads [31,32]. However, load inertia is not incorporated in this work, because research on this topic is still incomplete [31,32] and in future systems more loads will be connected with the electrical grid via grid frequency converters which will further reduce their inertia contribution [31].

2.4. Inertia by Synchronous Generators

Synchronously connected generators are hereafter referred to as synchronous generators. The inertia provision by synchronous generators is bound to two constraints. OpInMod incorporates a minimum stable operation constraint as depicted in Inequation (8). The output of a synchronous generator, x_g^{flow}, has to be larger than the product of the units rated power, $x_g^{rated\_power}$ and its minimum stable operation value, $x_g^{min\_stable\_op}$, in per unit. The units connection status is depicted by $x_g^{source\_inertia}$.

$$x_g^{flow} \geq x_g^{source\_inertia} \cdot x_g^{min\_stable\_op} \cdot x_g^{rated\_power} \tag{8}$$

The second constraint depicted by Inequation (9) determines the connection status of the synchronous generator. Since $x_g^{source\_inertia}$ is binary, the Inequation's right part is a proportional expression of the generator output and its rated power. Hence, if the generators output flow is larger then zero, the generator is connected to the power system, provides its inertia and $x_g^{source\_inertia}$ is set to the numeric value of 1.

$$x_g^{source\_inertia} \geq \frac{x_g^{flow}}{x_g^{rated\_power}} \tag{9}$$

2.5. Synthetic Inertia by Wind Turbines

A simplified approach to provide SI with WTs based on the control concept developed by Gloe et al. [19] is implemented in OpInMod. This approach is visualised in Figure 2. Since the concept is based on the actual operating point of a WT, normalised characteristics of the fully open source NREL 5 MW WT are applied [33]. The WT's input is used to determine corresponding normalised rotational speed of the WT as depicted in the top subplot of Figure 2. Thereafter, this value is used to determine the variable inertia constant, H_{var}, with respect to the demanded inertia constant, H_{dem}, as illustrated by the bottom subplot in Figure 2.

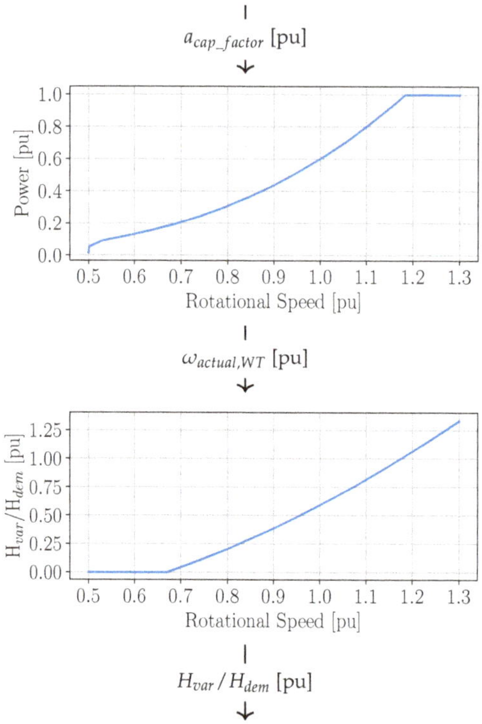

Figure 2. Two step methodology to determine H_{var} / H_{dem} applying the normalised power vs. normalised rotational speed characteristic of the NREL 5 MW WT [34] and the normalised characteristics of the variable H controller [19].

3. Scenario and Data of the Irish Energy Dispatch Model

To assess the potential of wind inertia in the all-Island Irish power system, the before introduced methods are applied to build a unit commitment and economic inertia dispatch model, implementing minimum inertia constraints and SI provided by WTs. Great Britain and France are modelled as well to account for cross border flows with the all-Island Irish system. Since these countries are connected to the Irish system via HVDC interconnectors, inertia is not modelled and these countries are not presented in further detail. However, scenario and data sources for France and Great Britain presented hereafter are the same as for the Irish system.

The scenario applied in this work is based on the 2020 Ten-Year Network Development Plan (TYNDP) of the European Network of Transmission System Operators for Electricity (ENTSO-E) [35,36]. The National Trends (NT) scenario reflects the individual commitments and policies of each single member state to meet CO_2 emission reduction targets. The influence of the EU is reduced to a minimum in favor of national sovereignty. The power sector in particular experiences a high growth in PV and WT generation. Large scale battery energy storage solutions are limited. Coal fired generation units are replaced by gas-fired and nuclear solutions.

To a large extend, data is provided by the 2020 TYNDP of the ENTSO-E [35,37,38]. Installed capacities per generation type, commodity and emission costs are depicted in Table 2. Characteristics of conventional thermal generators like Operation and Maintenance (O&M) costs, the efficiency and minimum stable operation values are listed in Table 3 [38]. Each generator is assigned with an inertia constant based on literature review [16,39–44]. Feed-in from WTs and PVs is modelled via fixed power factor generation profiles in hourly resolution derived from the Pan-European Climate Database [45]. The WT power factor is also used to determine the variable inertia provided as depicted in Figure 2. Hydro inflow for run-of-river is available in a daily time resolution and on a weekly base for hydro reservoir and open loop pump storage [45]. Such inflow profiles are transformed to hourly profiles.

Table 2. Installed Capacities, commodity and emission costs as well as total energy demands [37].

Type	Inst. Cap. [MW]
Wind onshore [MW]	8200
Wind offhore [MW]	5270
Solar PV [MW]	2390
Other RES (e.g., Waste) [MW]	525
Run of river [MW]	238
Natural Gas [MW]	5966
Lignite [MW]	96
Oil [MW]	518
Type	**Costs [EUR/MWh]**
Lignite	3.96
Natural gas	26.316
Light oil	79.92
Heavy oil	61.92
Shale oil	8.28
Biomass	34.89
Other RES (e.g., Waste)	30
CO_2 price [EUR/t]	75.00
Annual demand [TWh]	55.1

Two modelling parameters are varied per optimisation run to assess the influence and potential of SI provided by WTs. First, the demanded inertia constant, H_{dem}, which determines the actual provided SI of WTs [19]. Second, the ROCOF threshold which determines the minimum synchronous system inertia applied to create the constraint depicted in Equation (6) and remaining demand for system SI.

In former grid codes, the TSO of Hydro-Québec requested WTs to emulate a synchronous generator with an inertia constant of 3.5 s and 5 s respectively [5,6]. These two values are applied in this work as well. The Irish grid operator already defines a ROCOF threshold of 1 Hz/s [8]. A 2 Hz/s threshold is considered as well based on recommendations of the ENTSO-E regarding ROCOF withstand capabilities of power generating modules [46]. The resulting minimum synchronous system inertia defined in Equation (6) is calculated by solving Equation (1) for J_{sys}. Therefore, the loss of the largest infeed of 700 MW is used [16]. The overall minimum system inertia applied to create the constraint depicted in Equation (7) is based on system security specifications of Irish TSO. The TSO defines an operational limit for inertia of 23,000 MWs which can be reformulated as a operational limit for the moment of inertia of 466,077 kg·m^2 [8].

Table 3. Generator characteristics.

Prime Energy	Generator Type	O&M [EUR/MWh]	Efficiency [%]	Min. Stable Op. [%]	CO_2 Emissions [t_{CO_2}/MWh_{el}]	Inertia Constant[s]
Nuclear	-	9.00	33.0	50.0	0.0	5
Hard coal	-	28.68	40.0	43.0	0.8495	3.125
Lignite	-	30.57	40.0	43.0	0.9127	3.5
Natural gas	Conventional old 1	16.49	36.0	35.0	0.5700	4.25
Natural gas	Conventional old 2	16.49	41.0	35.0	0.5005	4.25
Natural gas	CCGT [a] old 1	16.99	40.0	35.0	0.513	4
Natural gas	CCGT old 2	16.99	48.0	35.0	0.4275	4
Natural gas	CCGT present 1	16.99	56.0	30.0	0.3664	4
Natural gas	CCGT present 2	16.99	58.0	30.0	0.3537	4
Natural gas	CCGT new	16.99	58.0	35.0	0.3537	4
Natural gas	OCGT [b] old	16.99	35.0	30.0	0.586	5.25
Natural gas	OCGT new	16.99	42.0	30.0	0.4885	5.25
Light oil	-	22.16	35.0	35.0	0.8022	3.25
Heavy oil	-	24.36	37.5	35.0	0.7524	3.25
Shale oil	-	30.30	29.0	40.0	1.0822	3.5
Biomass	-	10.00	40.0	43.0	0.0	3
Other RES	-	10.00	37.5	35.0	0.0	2

[a] Combined cycle gas turbine (CCGT); [b] Open cycle gas turbine (OCGT).

4. Results

OpInMod provides a large set of results of which system costs, RES penetration, storage units state of charge or cross border flows can be analysed. Since it is the objective of the work at hand to assess the potential of SI provided by WTs, results like the SI provided by WTs itself, RES curtailment, carbon dioxide emissions and system costs will be focused on.

The NT scenario is once optimised without the provision of SI by WTs, only considering inertia provision from synchronously connected generators, representing current state of research and acting as the base scenario in this work. The combination of the two ROCOF threshold parameters (1 Hz/s and 2 Hz/s) and two H_{dem} parameters (3.5 s and 5 s) result in four additional scenarios optimised. All optimisation problems designed with respect to the parameter combinations of the future Irish power system are solvable. Hence, the inertia and power demand constraints are satisfied.

Figures 3 and 4 depict results of a 24 h time series, which is an extract from the whole modelled year. The time series is selected randomly with no specific purpose other then illustrating model functionality, i.e., the influence of inertia constraints on the behaviour and output of dispatchable units. Figure 3 shows results of the base scenario and Figure 4 shows results of the same time period and a ROCOF threshold of 2 Hz/s and a demanded inertia constant of 3.5 s.

The top subplot of Figure 3 illustrates the electric power flow of dispatchable and non-dispatchable units and the bottom subplot the thereof resulting inertia provision. The x-axis shows the time of the day and the y-axis of the top subplot shows the electric power flows, imports and exports in MW and the y-axis of the bottom subplot depicts the provided moment of inertia in kg·m^2. Previous publications depicted the provided inertia represented by the accumulated stored kinetic energy of the connected synchronously rotating masses [15,16,39]. However, since this work incorporates SI provided by WTs and applies the inertia provision methodology introduced by Gloe et al. [19], the stored kinetic energy in the WT's rotor is not a valid representation of the actual provided SI.

Additionally, in future systems inertia should be a tradeable commodity [4]. Therefore, the actual and emulated moment of inertia in kg·m^2 is illustrated.

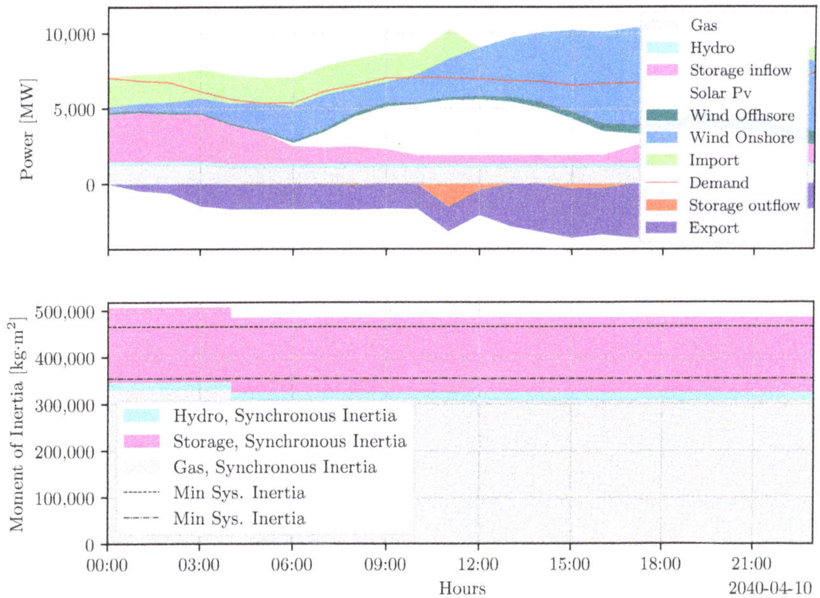

Figure 3. Illustration of a 24 h time series of the base scenario. The top subplot depicts electric power flow results and the bottom subplot flow provided inertia

Figure 3 depicts results of the base scenario. In the top subplot the first six hours of the day are characterised by minor RES feed-in. During these hours, demand is mainly covered by pump storage units, imports, natural gas fired power plants as well as an increasing share of onshore wind feed-in. Electric power flow imports and exports at the same time indicate that the Irish system in this case acts as a transfer system. In other words, electrical energy is transferred from France to Great Britain (or in opposite direction) using the Irish transmission capacities. With increasing onshore wind feed-in, a natural gas fired power plants gets disconnected from the system at a time of 04:00. This results in a downward step in the accumulated provided synchronous inertia by natural gas fired power plants area in the bottom subplot. While feed-in from solar PV and onshore WTs increases during the day and surpasses power demand, feed-in from pump storage units, hydro plants and gas fired power plants remain constant at minimum stable operation due to the minimum stable operation constraint described by Inequation (8). Inertia by such units is needed to satisfy the minimum system inertia constraint described by Inequation (7). Thus, to balance power generation and consumption, power is exported and energy is stored by pump storage units. With decreasing feed-in from solar PV, feed-in from storage units increases. However, since no additional synchronous generator is connected to the power system, provided inertia remains constant throughout the remaining time period. Since feed-in from WTs is high throughout the remaining day and power demand is surpassed by the accumulated feed-in, power exports remain high too.

In conclusion, the before described base scenario demonstrates the influence of the minimum system inertia constraints (see Inequations (6) and (7)). While feed-in from WTs and PV is high throughout the second half of the example time series and surpassing power demand, feed-in from synchronously connected generators such as from gas fired power plants, hydro plants and pump storage units remain at their minimum stable operation points to provide sufficient inertia. This results in exports and electrical energy being stored in energy storage units.

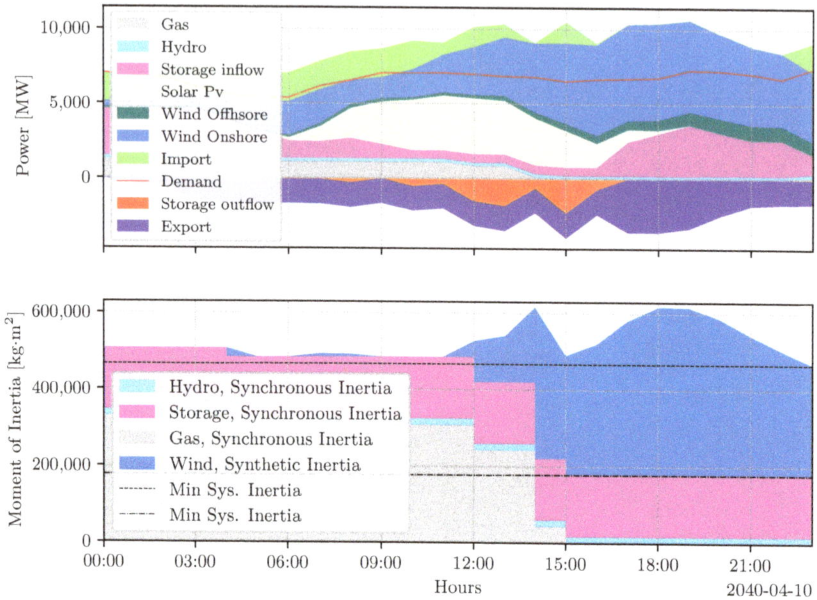

Figure 4. Illustration of a 24 h time series of the parameter combination: ROCOF threshold = 2 Hz/s and demanded inertia constant = 3.5 s. The top subplot depicts electric power flow results and the bottom subplot flow provided inertia.

Figure 4 depicts results of the exact same time period as depicted in Figure 3, but for the parameter combination H_{dem} of 3.5 s and a ROCOF threshold of 2 Hz/s. Hence in this case, WTs provide SI. First 12 h of the time period are almost identical. Minor differences, compared to Figure 3, are visible in the bottom subplot since SI is provided by WTs in the time between 4:00 to 5:00 and 6:00 to 9:00. Moreover, SI provision increases from 11:00 onwards. To satisfy ther overall minimum system inertia constraint (see Inequation (7), natural gas fired power plants, hydro units and pump storage plants are connected to the power system and operate at minimum stable operation. At 12:00 wind SI provision increases significantly. Thus, SI provision from WTs increases. This results in the disconnection of gas fired power plants in the time between 12:00 to 15:00. The system synchronous inertia constraint (see Inequation (6)) is met by the pump storage plants and hydro units provided inertia as well as the provided inertia from gas fired power plants which are disconnected stepwise while feed-in from WTs increases. From 15:00 onwards, the minimum synchronous inertia constraint (see Inequation (6)) can solely be satisfied by the accumulated synchronous inertia from hydro and pump storage plants. Accompanied by sufficient SI from WTs, the overall system inertia constraint (see Inequation (7) is satisfied.

Concluding the model outcome depicted in Figure 4, the beneficial influence of the SI provision by WTs is visible. By the second half of the depicted day, gas fired power plants are completely disconnected from the system, since feed-in from onshore WTs, pump storage units, hydro and offshore WTs is high enough to meet power demand. At the same time, the disconnection of the gas fired power plant is only realisable, due to the fact that overall system inertia constraint is satisfied by the provided wind SI.

Figure 5 illustrates the provided SI (here represented by the synthetic moment of inertia in kg·m^2) by dispatched WTs for the demanded inertia constant (H_{dem}) of 3.5 s and 5 s. The results are illustrated via a histogram plot, where the bars height depicts the frequency of occurrence and the x-axis depicts the provided SI. The frequency of occurrence is equal to the number of hours of the moddeled year. The green bars depict the synthetic moment of inertia provided by WTs considering the demanded inertia constant (H_{dem}) of 3.5 s and the red bars depict the synthetic moment of inertia provided by WTs considering the demanded inertia constant of 5 s. Obviously, for more then 4300 h of the modelled times series no or very little SI is provided. Thus, for the rest of the modelled time series (the accumulated number of the remaining bars), SI is provided by WT. A higher demanded inertia constant results in higher values for the provided synthetic moment of inertia (compare red bars (H_{dem} = 5 s) to green bars (H_{dem} = 3.5 s)).

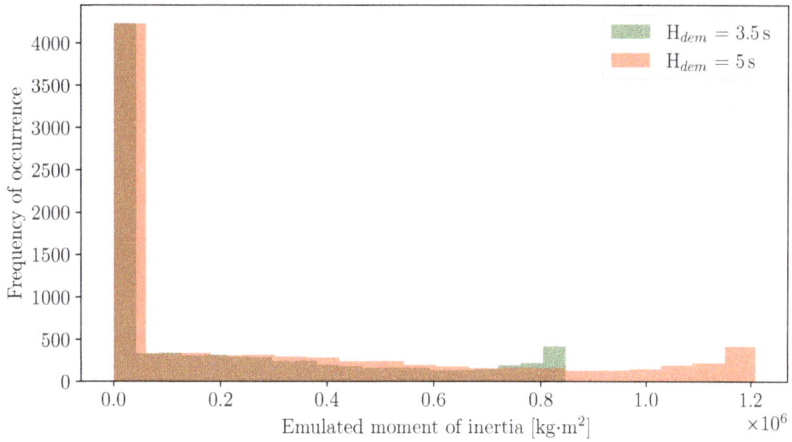

Figure 5. Depiction of the synthetic moment of inertia provided by WTs. The green bars depict the synthetic moment of inertia provided by WTs considering the demanded inertia constant (H_{dem}) of 3.5 s and the red bars depict the synthetic moment of inertia provided by WTs considering the demanded inertia constant of 5 s.

Table 4 shows the accumulated CO_2 emissions and curtailment results per parameter combination. The second column, depicting the CO_2 emissions, indicates that with increasing ROCOF thresholds and higher demanded inertia constants accumulated CO_2 emissions decrease significantly. Equation (10) explains how emission savings in column three of Table 4 are calculated. The respective scenario (e.g., ROCOF = 1 Hz/s, H_{dem} = 3.5 s) results are applied in the numerator of Equation (10). The denominator input is always the result of the base scenario. Increasing the ROCOF threshold up to 2 Hz/s and combining this parameter with an higher H_{dem}, e.g., 5 s, results in CO_2 emission savings of 30.99% with respect to the base scenario. Column four in Table 4 shows accumulated WT and PV curtailment. Similar to the accumulated CO_2 emission results, with increasing demand for WTs to provide SI by higher ROCOF thresholds, the curtailed energy decreases significantly. Equation (11) explains how curtailed energy reduction in column five of Table 4 is calculated. Curtailment reduction up to 39.72% for a demanded inertia constant of 3.5 s and 39.90% for a demanded inertia constant of 39.00% is achievable.

$$reduction\_CO_2\_emissions = \frac{overall\_CO_2\_emissions_{scenario}}{overall\_CO_2\_emissions_{base\_scenario}} \cdot 100 \tag{10}$$

$$reduction\_curtailment = \frac{overall\_curtailed\_energy_{scenario}}{overall\_curtailed\_energy_{base\_scenario}} \cdot 100 \tag{11}$$

Table 4. CO_2 emissions and curtailment results of the analysed scenarios.

Scenario	CO_2 Emissions [t]	Reduction [a] [%]	Curtailment [MWh]	Reduction [a] [%]
Base	4,789,225	0.00	2,613,969	0.00
ROCOF = 1 Hz/s, H_{dem} = 3.5 s	4,085,047	14.70	2,341,290	10.43
ROCOF = 1 Hz/s, H_{dem} = 5 s	4,066,117	15.10	2,341,290	10.43
ROCOF = 2 Hz/s, H_{dem} = 3.5 s	3,474,372	27.45	1,575,412	39.73
ROCOF = 2 Hz/s, H_{dem} = 5 s	3,305,005	30.99	1,570,872	39.90

[a] with respect to the base scenario.

Table 5 shows results of the overall system costs and cost savings with respect to the provided SI. With increasing demand for WTs to provide SI, accumulated system costs decrease significantly, since less feed-in from fossil power plants running at minimum stable operation (must-run capacities) is needed with the purpose to satisfy minimum system inertia constraints. Equation (12) illustrates how cost savings in column two of Table 4 are calculated. Based on the analysed parameter combinations, cost savings up to 32.72% are achievable due to the potential wind inertia provision. The last column of Table 5 depicts the cost savings with respect to the overall provided SI. Increasing the ROCOF threshold parameter and thereby the potential demand for SI results in significantly higher cost savings with respect to the provided SI. Increasing the ROCOF threshold from 1 Hz/s to 2 Hz/s while maintaining the demanded inertia constant at 3.5 s leads to cost savings of 0.09 EURO/kg·m^2 and 0.17 EURO/kg·m^2. Increasing the demanded inertia constant from 3.5 s to 5 s while keeping the ROCOF threshold constant decreases cost savings with respect to the provided SI. While increasing the demanded inertia constant increases system cost saving potential due to SI provision by WTs, the value of the provided SI decreases.

$$reduction\_system\_costs = \frac{overall\_system\_costs_{scenario}}{overall\_system\_costs_{base\_scenario}} \cdot 100 \quad (12)$$

Table 5. Overall system costs and cost savings with respect to the provided SI of the analysed scenarios.

Scenario	System Costs [EURO]	Reduction [a] [%]	Cost Savings [a] for Provided SI [EURO/kg·m^2]
Base	737,265,818	0.00	0.00
ROCOF = 1 Hz/s, H_{dem} = 3.5 s	625,765,315	15.12	0.09
ROCOF = 1 Hz/s, H_{dem} = 5 s	622,733,455	15.53	0.06
ROCOF = 2 Hz/s, H_{dem} = 3.5 s	525,560,884	28.71	0.17
ROCOF = 2 Hz/s, H_{dem} = 5 s	496,034,635	32.72	0.14

[a] with respect to the base scenario.

5. Conclusions

Increasing installed capacities of grid frequency converter connected generation units like WTs and PVs decreases synchronous inertia in energy systems. System inertia is an essential part of grid frequency stability. It limits the instantaneous ROCOF and limits the grid frequency nadir. The trend of decreasing system inertia is not reflected adequately in current unit commitment and economic dispatch models. If inertia in such models is considered at all, only synchronous inertia from fossil power plants is represented. This leads to increased CO_2 emissions, RES curtailment and system costs, since power systems require a minimum system inertia to maintain controllability of the grid frequency. While WTs are a key technology to decarbonise energy systems and a source for an synthetic inertial response, SI is not incorporated into energy system models.

This work closes this research gap. The influence and potential of SI provision by WT on system indicators like CO_2 emissions, variable RES curtailment and system costs is analysed. Therefore, a methodology to depict SI provision in dispatch optimisation problems is presented. The methodology is incorporated into an open source unit commitment and economic inertia dispatch modelling tool and applied to the all-Island Irish power system. Ireland today is already a power system, in which wind feed-in is curtailed due to low power system inertia and the grid operator has to react by defining an operational level for power system inertia. The model framework OpInMod creates minimum inertia constraints which have to be met by synchronous inertia sources and SI provided by WT. Data and the scenario researched are based on the 2020 TYNDP. Two parameters, the ROCOF threshold determining the needed minimum synchronous system inertia and the demanded inertia constant which determines the provided wind inertia are varied for analysis purpose.

The result analysis shows that the introduction of SI provided by WTs results in the disconnection of fossil fuel fired power plants. The thereof missing inertia is replaced by the provided SI by WTs. Thus, less CO_2 is emitted and costs decrease since WTs are a zero marginal cost source. Based on the analysed parameter combinations, up to 30.99% of the CO_2 emissions can be saved due to the provision of SI by WTs. Curtailment reduction for the parameter combinations show similar results. Curtailment reduction of 39.9% is possible for the parameter combination of a 2 Hz/s ROCOF threshold and a demanded inertia constant of 5 s. Cost savings are in the range of 15.12% and 32.72% with respect to the respective base scenario. System costs saving potential can also be expressed as total cost savings with respect to the overall provided SI by WTs. Here, cost savings are in the range of 0.06 EURO/kg·m^2 and 0.17 EURO/kg·m^2. Increasing the demanded inertia constant and thereby the overall provided inertia reduces the utility for the power system expressed by reduced cost saving per provided inertia.

In total, the analysis shows that incorporating inertia constraints in energy system modelling has a significant influence on the results. Further, SI provided by WTs has a very high potential to reduce CO_2 emissions, RES curtailment and system costs.

Author Contributions: H.T.: Conceptualization, Data curation, Formal analysis, Investigation, Methodology, Software, Validation, Visualization, Writing—original draft & editing. C.J.: Project administration, Resources, Supervision, Writing—review. All authors have read and agreed to the published version of the manuscript.

Funding: This research received no external funding.

Institutional Review Board Statement: Not applicable.

Informed Consent Statement: Not applicable.

Data Availability Statement: The data presented in this study is available in the cited references.

Conflicts of Interest: The authors declare no conflict of interest.

Abbreviations

The following abbreviations are used in this manuscript:

CCGT	Closed cycle gas turbine
ENTSO-E	European Network of Transmission System Operators for Electricity
EU	European Union
MILP	mixed integer linear programming
NT	National Trends
oemof	open energy modeling framework
OCGT	Open cycle gas turbine
OpInMod	Open Inertia Modelling
O&M	Operation and Maintenance
PV	photovoltaic system
RES	renewable energy source
ROCOF	rate of change of frequency

SI	synthetic inertia
TSO	transmission system operator
TYNDP	Ten-Year Network Development Plan
WT	wind turbine

References

1. Tielens, P.; Hertem, D.V. The relevance of inertia in power systems. *Renew. Sustain. Energy Rev.* **2016**, *55*, 999–1009. [CrossRef]
2. Kundur, P.; Balu, N.; Lauby, M. *Power System Stability and Control*; EPRI Power System Engineering Series; McGraw-Hill: New York, NY, USA, 1994; ISBN 978-0-07-035958-1.
3. Fernández-Guillamón, A.; Gómez-Lázaro, E.; Muljadi, E.; Molina-García, A. Power systems with high renewable energy sources: A review of inertia and frequency control strategies over time. *Renew. Sustain. Energy Rev.* **2008**, *115*, 119369. [CrossRef]
4. Thiesen, H.; Jauch, C.; Gloe, A. Design of a System Substituting Today's Inherent Inertia in the European Continental Synchronous Area. *Energies* **2016**, *9*, 582. [CrossRef]
5. Hydro Québec TransÉnergie. Technical Requirements for the Connection of Generation Facilities to the Hydro-Québec Transmission System. Supplementary Requirements for Wind Generation. 2005. Available online: https://www.aeeolica.org/uploads/documents/4535-separata-del-borrador-de-po122.pdf (accessed on 25 May 2020).
6. Hydro Québec TransÉnergie. Transmission Provider Technical Requirements for the Connection of Power Plants to the Hydro–QuéBec Transmission System. 2009. Available online: http://www.hydroquebec.com/transenergie/fr/commerce/pdf/exigence_raccordement_fev_09_en.pdf (accessed on 25 May 2020).
7. EirGrid. EirGrid Grid Code–Version 9. Technical Report. 2020. Available online: https://www.eirgridgroup.com/site-files/library/EirGrid/GridCodeVersion9.pdf (accessed on 21 January 2021).
8. EirGrid and SONI. Operational Constraints Update 21 December 2020. 2020. Available online: https://www.eirgridgroup.com/site-files/library/EirGrid/Operational-Constraints-Update-December-2020.pdf (accessed on 23 March 2021).
9. ENTSO-E. *Need for Synthetic Inertia (SI) for Frequency Regulation*; Report; ENTSO-E: Brussels, Belgium, 2017. Available online: https://consultations.entsoe.eu/system-development/entso-e-connection-codes-implementation-guidance-d-3/user_uploads/igd-need-for-synthetic-inertia.pdf (accessed on 3 May 2019).
10. Denholm, P.; Mai, T.; Wallace Kenyon, R; Kroposki, B.; O'Malley, M. *Inertia and the Power Grid: A Guide without the Spin*; Technical Report NREL/TP-6120-73856; National Renewable Energy Laboratory: Golden, CO, USA, 2020. Available online: https://www.nrel.gov/docs/fy20osti/73856.pdf (accessed on 3 March 2022).
11. Lopion, P.; Markewitz, P.; Robinius, M.; Stolten, D. A review of current challenges and trends in energy systems modeling. *Renew. Sustain. Energy Rev.* **2018**, *9*, 156–166. [CrossRef]
12. Lund, H.; Arler, F.; Østergaard, P.A.; Hvelplund, F.; Connolly, D.; Mathiesen, B.V.; Karnøe, P. Simulation versus Optimisation: Theoretical Positions in Energy System Modelling. *Energies* **2018**, *10*, 840. [CrossRef]
13. Teng, F.; Trovato, V.; Strbac, G. Stochastic Scheduling With Inertia-Dependent Fast Frequency Response Requirements. *IEEE Trans. Power Syst.* **2016**, *31*, 1557–1566. [CrossRef]
14. Collins, S.; Deane, J.; Gallachóir, B.Ó. Adding value to EU energy policy analysis using a multi-model approach with an EU-28 electricity dispatch model. *Energy* **2017**, *130*, 433–447. [CrossRef]
15. Johnson, S.C.; Papageorgiou, D.J.; Mallapragada, D.S.; Deetjen, T.A.; Rhodes, J.D.; Webber, M.E. Evaluating rotational in- ertia as a component of grid reliability with high penetrations of variable renewable energy. *Energy* **2019**, *180*, 258–271. [CrossRef]
16. Mehigan, L.; Al Kez, D.; Collins, S.; Foley, A.; Gallachóir, B.O.; Deane, P. Renewables in the European power system and the impact on system rotational inertia. *Energy* **2020**, *203*, 117776. [CrossRef]
17. EirGrid. All Island Quarterly Wind Dispatch Down Report. 2020. Available online: http://www.eirgridgroup.com/site-files/library/EirGrid/Grid-Code.pdf (accessed on 25 May 2020).
18. Riquelme, E.; Fuentes, C.; Chavez, H.A. A Review of Limitations of Wind Synthetic Inertia Methods. In Proceedings of the 2020 IEEE PES Transmission Distribution Conference and Exhibition—Latin America (T D LA), Montevideo, Uruguay, 28 September–2 October 2020; pp. 1–6. [CrossRef]
19. Gloe, A.; Jauch, C.; Craciun, B.; Winkelmann, J. Continuous provision of synthetic inertia with wind turbines: Implications for the wind turbine and for the grid. *IET Renew. Power Gener.* **2019**, *13*, 668–675. [CrossRef]
20. Thiesen, H. Open Inertia Modelling (OpInMod)—An Open Source Approach to Model Economic Inertia Dispatch in Power Systems. *Preprints* **2022**, 2022010419. [CrossRef]
21. Hilpert, S.; Kaldemeyer, C.; Krien, U.; Günther, S.; Wingenbach, C.; Pleßmann, G. The Open Energy Modelling Framework (oemof)—A new approach to facilitate open science in energy system modelling. *Energy Strategy Rev.* **2018**, *22*, 16–25. [CrossRef]
22. Krien, U.; Schñfeldt, P.; Launer, J.; Hilpert, S.; Kaldemeyer, C.; Pleßmann, G. oemof.solph–A model generator for linear and mixed-integer linear optimisation of energy systems. *Softw. Impacts* **2020**, *6*, 100028. [CrossRef]
23. Bazilian, M.; Rice, A.; Rotich, J.; Howells, M.; DeCarolis, J.; Macmillan, S.; Brooks, C.; Bauer, F.; Liebreich, M. Open source software and crowdsourcing for energy analysis. *Energy Policy* **2017**, *49*, 149–153. [CrossRef]
24. Pfenninger, S.; Hawkes, A.; Keirstead, J. Energy systems modeling for twenty-first century energy challenges. *Renew. Sustain. Energy Rev.* **2014**, *33*, 74–86. [CrossRef]

25. Pfenninger, S.; Hirth, L.; Schlecht, I.; Schmid, E.; Wiese, F.; Brown, T.; Davis, C.; Gidden, M.; Heinrichs, H.; Heuberger, C.; et al. Opening the black box of energy modelling: Strategies and lessons learned. *Energy Strategy Rev.* **2018**, *19*, 63–71. [CrossRef]
26. Wiese, F.; Hilpert, S.; Kaldemeyer, C.; PleßPleßmannmann, G. A qualitative evaluation approach for energy system modelling frameworks. *Energy Sustain. Soc.* **2018**, *8*, 13. [CrossRef]
27. Thiesen, H. Open Inertia Modelling (OpInMod). 2021. Available online: https://github.com/hnnngt/OpInMod (accessed on 19 October 2021).
28. Thiesen, H. Open Inertia Modelling (OpInMod) (0.1). Online. 2021. Available online: http://dx.doi.org/10.5281/zenodo.5582502 (accessed on 19 October 2021). [CrossRef]
29. Hart, W.E.; Watson, J.P.; Woodruff, D.L. Pyomo: Modeling and solving mathematical programs in Python. *Math. Program. Comput.* **2011**, *3*, 219. [CrossRef]
30. Oemof-Developer Group. Open Energy Modelling Framework (oemof), v0.4 GitHub. 2021. Available online: https://github.com/oemof/oemof (accessed on 7 June 2021).
31. Bian, Y.; Wyman-Pain, H.; Li, F.; Bhakar, R.; Mishra, S.; Padhy, N.P. Demand Side Contributions for System Inertia in the GB Power System. *IEEE Trans. Power Syst.* **2018**, *33*, 3521–3530. [CrossRef]
32. Thiesen, H.; Jauch, C. Determining the load inertia contribution from different power consumer groups. *Energies* **2020**, *13*, 1588. [CrossRef]
33. Jonkman, J.; Butterfield, S.; Musial, W.; Scott, G. *Definition of a 5MW Reference Wind Turbine for Offshore System Development*; National Renewable Energy Lab. (NREL): Golden, CO, USA, 2009. [CrossRef]
34. Jauch, C. *First Eigenmodes Simulation Model of a Wind Turbine–for Control Algorithm Design*; TechnicaL Report; WETI: Flensburg, Germany, 2020. [CrossRef]
35. ENTSO-E. TYNDP 2020–Scenario Report. 2020. Available online: https://eepublicdownloads.azureedge.net/tyndp-documents/TYNDP_2020_Joint_Scenario_Report_ENTSOG_ENTSOE_200629_Final.pdf (accessed on 15 March 2021).
36. ENTSO-E. TYNDP 2020–Scenario Building Guidelines. 2020. Available online: https://2020.entsos-tyndp-scenarios.eu/wp-content/uploads/2020/06/TYNDP_2020_Scenario_Building_Guidelines_Final_Report.pdf (accessed on 15 March 2021).
37. ENTSO-E. Data Download. 2020. Available online: https://www.entsos-tyndp2020-scenarios.eu/download-data/#download (accessed on 15 March 2021).
38. ENTSO-E. Unit Data. 2020. Available online: https://www.entsoe.eu/Documents/TYNDP%20documents/TYNDP2018/Scenarios%20Data%20Sets/Input%20Data.xlsx (accessed on 15 March 2021).
39. Trovato, V.; Bialecki, A.; Dallagi, A.A. Unit Commitment With Inertia-Dependent and Multispeed Allocation of Frequency Response Services. *IEEE Trans. Power Syst.* **2019**, *34*, 1537–1548. [CrossRef]
40. Machowski, J.; Lubosny, Z.; Bialek, J.; Bumby, J. *Power System Dynamics: Stability and Control*, 2nd ed.; Wiley: West Sussex, UK, 2020; ISBN 978-1-119-52638-4.
41. Chown, G.; Wright, J.G.; van Heerden, R.; Coker, M. System inertia and Rate of Change of Frequency (RoCoF) with increasing non-synchronous renewable energy penetration. In Proceedings of the 8th Southern Africa Regional Conference 2017, Cape Town, South Africa, 14–17 November 2017.
42. Rawn, B.; Gibescu, M. Effects Future Renewable Installations Will Have on System Synchronous and Synthetic Inertia. Master's Thesis, TU Delft, Delft, The Netherlands, 2012. Available online: https://repository.tudelft.nl/islandora/object/uuid:1c94be1d-3619-40a1-86fa-4a93941ccce50/datastream/OBJ/download (accessed on 15 December 2021).
43. Seneviratne, C.; Ozansoy, C. Frequency response due to a large generator loss with the increasing penetration of wind/PV generation-A literature review. *Renew. Sustain. Energy Rev.* **2016**, *57*, 659–668. [CrossRef]
44. Tielens, P.; Van Hertem, D. Grid Inertia and Frequency Control in Power Systems with High Penetration of Renewables. 2012. Available online: https://lirias.kuleuven.be/retrieve/182648 (accessed on 28 October 2021).
45. ENTSO-E. Pan-European Climate Database. 2020. Available online: https://eepublicdownloads.entsoe.eu/clean-documents/sdc-documents/MAF/2020/Pan-European%20Climate%20Database.7z (accessed on 15 March 2021).
46. ENTSO-E. *Rate of Change of Frequency (RoCoF) Withstand Capability*; Technical Report; ENTSO-E: Brussels, Belgium, 2018. Available online: https://eepublicdownloads.entsoe.eu/clean-documents/Network%20codes%20documents/NC%20RfG/IGD_RoCoF_withstand_capability_final.pdf (accessed on 20 August 2021).

applied sciences

MDPI

Article

Development of Technology for Hydromechanical Breakdown of Mud Plugs and Improvement of Well Cleaning by Controlled Buckling of the Drill String

Viacheslav G. Kadochnikov * and Mikhail V. Dvoynikov

Arctic Competence Center, Saint Petersburg Mining University, 2, 21st Line, 199106 Saint Petersburg, Russia; dvoynikov_mv@pers.spmi.ru
* Correspondence: kadochnikov_vg@pers.spmi.ru

Abstract: The article provides brief information about a non-standard experimental setup developed in the laboratory of the St. Petersburg Mining University Well Drilling Department. The developed technique presented makes it possible to simulate well cleaning process of cuttings by incorporating the variation of the parameters (the zenith angle of the well, the volume flow and rheological properties of the cleaning agent, the rotation frequency, the number and length of the drill string half-waves) that cause buckling. For the first time, the positive side of the drill string (DS) buckling phenomenon is considered. A positive hydro-mechanical effect on mud plugs and improved well cleaning were revealed. The results of the experimental study confirm an intense difficulty in transporting cuttings to the surface at a critical zenith angle of the well of 55°. Regularities have been established making it possible to determine the effect of DS buckling on the cutting-carrying capacity when drilling deviated and extended reach wells. It is proposed to use hydromechanical impact on the accumulated cuttings by artificially controlling the resulting DS buckling in order to destroy the mud plug and increase the efficiency of well cleaning without the use of specialized devices. A conceptual solution aimed at implementing a method for hydromechanical destruction of mud plugs—the use of drill pipes equipped with a quasi-distributed differential measuring system of strain gauges based on a fiber-optic Bragg grating—is presented.

Keywords: breakdown of mud plugs; hydromechanical destruction; buckling; drill string; loss of stability; inclined well; cleaning; drilling parameters; fiber Bragg grating (FBG); FBG deformation sensor

Citation: Kadochnikov, V.G.; Dvoynikov, M.V. Development of Technology for Hydromechanical Breakdown of Mud Plugs and Improvement of Well Cleaning by Controlled Buckling of the Drill String. *Appl. Sci.* **2022**, *12*, 6460. https://doi.org/10.3390/app12136460

Academic Editor: Abílio M.P. De Jesus

Received: 28 February 2022
Accepted: 23 June 2022
Published: 25 June 2022

1. Introduction

Every year, the share of directional drilling around the world is increasing and today it is about 85% of the total volume of oil and gas wells [1,2]. This is due to the development of hard-to-reach and deep-lying reservoirs, access to which is possible only through long wells of complex trajectories, which have a large number of areas of change in zenith and azimuth angles. Consequently, the likelihood of complications and accidents increases, which in most cases are associated with poor-quality well cleaning of drill cuttings (mud) [3,4]. Poor cleaning at the well of cuttings leads to their accumulation at the bottom wall and the formation of the so-called "mud dunes" or "mud pads", which affect the drilling tool and hydrodynamic pressure, increasing the equivalent circulation density (ECD) of the drilling fluid (DF) [5–7]. With a local increase in the concentration of cuttings, the annular space between the drill string (DS) and the walls of the well can become completely clogged and form a "mud plug", which leads to the risk of complications and accidents associated with the operation of the tool in the well, rock stability [8–10] and reservoir properties.

Improving the quality of well cleaning from drill cuttings is an important task when drilling directional and, especially, horizontal wells, which can be achieved through the use of special technical means and technological methods. One of these technological methods is the operational control and management of drilling parameters, the essence of which is the timely regulation of key parameters: DS rotation speed, flow rate and axial load on bit. An increase in the flow rate leads to an increase in the upward flow velocity in the annular space and an improvement in well cleaning [11]. The dynamic component of the ECD strongly depends on the consumption of drilling mud. However, it is impossible to improve cleaning by increasing the consumption of mud alone, especially under conditions of a narrow range of fluid loss and kick pressures [12]. Increasing the rotation speed of the DS has a hydrodynamic effect on the cuttings, creating a circumferential flow that does not carry particles out of the well, but has a "turbulent" effect, which contributes to the removal of particles from stagnant zones [13,14]. It has been empirically established that an increase in the rotational speed of the DS to 140 rpm has a positive effect on the quality of cleaning, and a subsequent increase in the rotational speed has practically no effect on the removal of mud. A significant increase in the rotation frequency of the bottom hole assembly (BHA) can have a negative impact on the elements of the BHA, such as reactive torque increase—twists-off, increase vibrations and beats—irreversible deformations and breaks-off. The use of special technical means that are part of the DS and improve well cleaning [15–17], such as turbulators, circulating subs, etc., do not obtain the desired effect, due to the complexity of the trajectories, limitations in the number of installed devices, vugular porosity and rock stability. In oil production, there were situations when the presence of turbulators had a negative effect—the reactive twisting moment increased, and the area of the annular space decreased, which led to an increase in hydraulic shocks and an increase in ECD [16]. The rheology of the mud is one of the factors that have a significant impact on the cleaning of deviated wells, ECD and the state of the bottomhole zone [18,19]. Undoubtedly, mud chemical treatment makes it possible to increase the efficiency of well cleaning and prevent technological complications in the drilling process. However, there are risks of contamination of the oil and gas saturated formation [20], and the selection of complex rheological profiles for drilling with a complex trajectory in specific geological and technical conditions requires significant financial costs.

The increase in drilling speed is affected by the axial load on the bit, as well as the quantity and quality of mud, which must promptly clean the bottomhole from cuttings during drilling. The axial load applied to the entire area of contact with the drilling tool must exceed the strength of the rock and break it down. However, it is known that in directional wells, and especially in wells with complex profiles and large deviations from the vertical, the axial load applied to the bit does not reach the bottom hole in full [14,21]. This is due to the complex stress–strain state of the drilling tool in cramped downhole conditions. The initial increase in the axial load that does not reach the bottom leads to the loss of the longitudinal stability of the DS and the acquisition of the shape of a sinusoid—a bend of the first kind (sinusoidal buckling) is formed. With a subsequent increase in the load, the DS takes the form of a spiral—a bend of the second kind (spiral buckling) is formed [22]. Longitudinal loss of stability of the drilling tool contributes to the occurrence of lapels and fractures of the drilling tool. In drilling practice, engineers try to avoid situations in which the DS loses its stability and takes any form of buckling [14,21–23]. However, the positive effects of artificially created and controlled buckling of the DS on the destruction of mud plugs and cleaning the well from cuttings are proposed in this paper.

Medium- and large-scale full-scale modeling of drilling and well cleaning processes in the laboratory is being carried out by many researchers, both in Russia and abroad. Experiments carried out at M-I SWACO (USA) are considered to be particularly significant studies that have found wide application in the practice of drilling wells. The results of the research were presented in 2001 as a guide for drilling fluid engineers. Scientists have determined the critical zenith angles (range 45–60°) at which cuttings particles are insufficiently carried out of the well. Many other factors influencing well cleaning were analyzed, and recommendations were made for the construction of wells of various trajectories. In 2012, Malaysian researchers conducted full-scale experiments to study the effect of mud viscosity and velocity, as well as the zenith angle, on the transport of drilled particles in deviated and horizontal wells [24]. In 2016, Iranian researchers developed an experimental stand and published the results of studies [25] on the influence of the zenith angle, the speed of the DS, the consumption of DF and the cuttings fraction on the cleaning of an inclined well. In 2019, an employee of RN-Vankor published the results of a study of the effect of DS rotation on cuttings transport in a horizontal well, obtained on a small-scale experimental bench. The influence of DS rotation on mud transport was confirmed, as well as a decrease in the "turbulent effect" with an increase in DF viscosity [13]. In 2018–2019 a group of international researchers published the results of the effect of rotation of the DF on cleaning the well with the use of polypropylene particles in DF at different zenith angles [26–28]. In 2020–2021 a group of Norwegian scientists published the results [29,30] of hydraulic studies and mud transport with oil-based solutions in deviated wells, using a large-scale laboratory facility developed with the support of Aker BP and M-I SWACO. The above studies confirmed the presence of critical zenith angles of the well trajectory, described the effect of the rotation speed of the drilling rig, volumetric flow, technological and mud rheological properties on the quality of wellbore cleaning of cuttings. However, the effect of DS buckling on the destruction of mud plugs and mud carryover has not been addressed in any of the above works, due to the fact that engineers are trying to predict and exclude the buckling phenomenon [21,31–33], and the positive effect of controlled DS buckling on well cleaning has hardly been considered.

The purpose of the study is to determine the effect of DS buckling on the breakdown of mud plugs and the cleaning of deviated wells from cuttings, taking into account drilling parameters, zenith angle and mud properties. To achieve this goal, two series of applied field studies were carried out using muds with different rheological models. The development of a technology to improve the quality of well cleaning, which is based on the operational control and management of the number and lengths of half-waves of DS and their hydromechanical effect on mud plugs, taking into account the drilling parameters, zenith angle and mud properties, will improve the efficiency of the construction of directional wells.

The present study includes the determination of DS spatial shape influence on the sludge plugs' destruction and deviated wells cleaning from cuttings. The main idea of the study is not to fight against DS buckling, as a negative effect that one way or another occurs when drilling wells with long or complex trajectories, but to use this phenomenon to advantage—to control and manage buckling, artificially creating DS bends in places where cuttings accumulate.

2. Materials and Methods

The study of the effect of DS buckling on the breakdown of mud plugs and well cleaning was carried out on a specially designed experimental stand in the laboratory of St. Petersburg Mining University Well Drilling Department [34]. The stand has a closed circulation system and allows modeling of the process of destruction of the mud plug and transport of mud from the well, taking into account the zenith angle, rotational speed, number and length of half-waves string with longitudinal buckling, as well as the mud flow rate, its rheological properties and the mud fraction. The main characteristics and composition of the experimental stand are shown in Figure 1 and in Table 1.

The experimental studies are carried out according to the developed algorithm. A drilling fluid with required composition and quantity is prepared in a tank 8, a sieve cleaning unit 9 is installed on top, in which, depending on the selected cuttings fraction, a sieve with a particular mesh size is installed. The wellhead pipe 3 is connected to the upper part of the "wellbore" pipe 4 installed on the base 15, and the plug 5 is installed in the lower part of the pipe 4. The pipe DS 6 with the selected number of half-waves is installed at the lower end (with holes and a rubber seal) into the pipe 4 on the support cone 7, through the "mouth head" 3, and is connected by a threaded connection to the transit swivel 2, which is connected to the drive 1. The end of the discharge line 10, connected to the swivel, is made of a metal pipe, on which a laser tachometer is installed 12 and ultrasonic flow meter 11, the principle of which is based on the Doppler effect. A reflective element is attached to the lower part of the swivel shaft 2, which is necessary for measuring the DS speed with a laser tachometer 12. The discharge line 10 is connected to the flow controller 13, which is fixed in the tank 8. The pump 14 is installed at the bottom of the tank 8 and connected to the flow controller 13. The slurry of the required amount is poured into the annular space between pipes 4 and 6 through the mouth head 3.

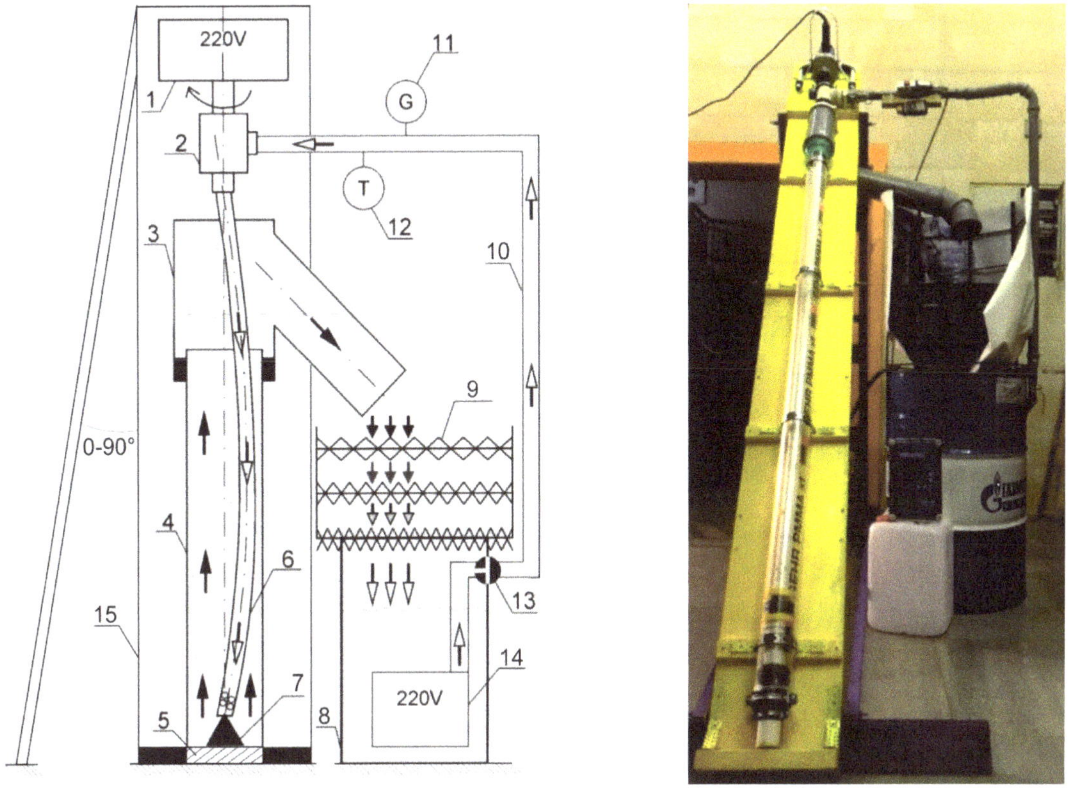

(a) (b)

Figure 1. Schematic (**a**) and photo (**b**) of the experimental stand: 1—drive, 2—swivel, 3—wellhead equipment, 4—wellbore, 5—plug, 6—drill pipes, 7—support cone, 8–mud tank, 9—sieve cleaning unit, 10—discharge line, 11—ultrasonic flowmeter, 12—laser tachometer, 13—mudflow controller, 14—pump.

The experimental stand is set at the required zenith angle. The pipe 6 with the selected number of half-waves (straight pipe L_0 without half-waves, pipe with the 1st half-wave L_1 = 1.8 m pipe with 3 half-waves L_3 = 0.6 m) is installed on the support cone 7 in the pipe 4 "wellbore". Mud weighing m = 1.5 kg fraction 1.0 ÷ 2.0 mm is poured into the annular space between pipes 4 and 6. Pump 14 pumps DF at a flow rate of Q (L/min), regulated by the flow controller 13. The value of the mud flow is displayed on display of the control unit of the acoustic flowmeter 11. Next, the drive is turned on, which rotates the pipe with a frequency of n (rpm), the value of which is displayed on the display of the laser tachometer 12. The time allotted for 1 experiment, within the framework of a series of experiments, t = 120 s. During this time, the "borehole" is cleaned of accumulated mud on the sieves of the corresponding fraction in the cleaning unit 9. After the time has elapsed, the pump and drive are turned off, the sieves are removed and cleaned of mud, which is weighed on electronic scales. Change of pipe with a different number and lengths of half-waves is carried out by disconnecting the swivel from the drive and pipe, with its subsequent extraction from the "borehole". After a series of experiments (No. 1), the mud is disposed of, the circulation system is completely flushed for 10 min, another type of mud is prepared in the tank, and a series of experiments (No. 2) is carried out according to the above algorithm.

Table 1. Main technical characteristics of the experimental stand.

Parameter		Unit	Value
Geometric Characteristics			
Stand base size		mm	2700 × 300
Borehole diameter		mm	74
Pipe diameter		mm	40
The length of the "well" (equal to the length of the pipe)		mm	2000
Zenith angle of the "well"		degree	0 ÷ 90
Sieve cleaning unit:	sieve 1—cell	mm	0.5 × 0.5
	sieve 2—cell	mm	1.0 × 1.0
	sieve 3—cell	mm	2.0 × 2.0
Characteristics of Equipment and Measuring Devices			
Rotation per minute (drive)		rpm	0–600
Maximum pump performance		L/min	258
Tachometer laser (range)		rpm	2.5–99,999
Ultrasonic Doppler flow meter (range)		m/s	0.03–12.2
Scales electronic (range)		g	5–40,000

The input parameters of the experiments are presented in Table 2.

Table 2. The input parameters of the experiments.

Parameters	Unit	Value	
Experiment series No.	-	1	2
Q—consumption of BR	$\times 10^{-3}$ m^3/s (L/min)	1.25 (75); 1.45 (87); 1.67 (100)	0.83 (50); 1.05 (63); 1.25 (75)
n—rotation frequency	s^{-1} (rpm)	0.83 (50); 1.67 (100); 2.50 (150)	0.83 (50); 1.67 (100); 2.50 (150)
Li—number of half-waves p	pcs	0; 1; 3	0; 1; 3
α—zenith angle	degree	35; 55; 70	35; 55; 70
Rheological model of mud	-	DF No. 1—Newtonian model	DF No. 2—Pseudo-plastic model
ρ—density	kg/m^3	1001	1035
PV (η)—viscosity	mPa·s	0.941 (η)	40.283 (PV)
YP—yield point	Pa	-	16.92
T—temperature	°C	23	23

The design of the experimental stand was based on the principle of geometric similarity (1):

$$\frac{D_{w1}}{D_{w2}} = \frac{d_{DS1}}{d_{DS2}} = C, \quad (1)$$

where D_{w1} is the diameter of a real wellbore, mm; d_{DS1} is the diameter of the real string, mm; D_{w2}—diameter of the experimental "wellbore", mm; d_{DS2} is the diameter of the experimental string, mm; and C is the coefficient of proportionality or "constant of geometric similarity".

$$\frac{D_{w1}}{D_{w2}} = \frac{215.9}{74} = 2.92; \frac{d_{DS1}}{d_{DS2}} = \frac{114.3}{40} = 2.86; C \cong 2.89 \approx 3$$

DS deflection (wave amplitude) in the experimental stand (2) is represented as:

$$f = \frac{(D_{w2} - d_{DS2})}{2} \quad (2)$$

$$f = \frac{(74 - 40)}{2} = 17 \text{ mm}$$

In order to determine the effect of buckling on the destruction of mud plugs and well cleaning of cuttings particles, it was necessary to obtain an artificial buckling of the DS. At this stage of research, it was decided to abandon the creation of a system for smooth artificial regulation of the number of half-waves strings in favor of a simplified discrete system, due to the lack of knowledge of this drilling area and the need to test the hypothesis put forward. The half-waves strings were obtained by heating the pipes and subsequent irreversible bending deformation, while maintaining the diameter along the entire length. Thus, the pipes with 1st and 3rd half-waves were obtained, equal to the length of the well.

3. Results

3.1. General Research Results

In addition to comparing the experimental data on the effect of buckling on the removal of cuttings, it was possible to evaluate the process of breakdown of the mud plugs in directional sections of the well visually (Figure 2) due to the transparency of the "wellbore" pipe.

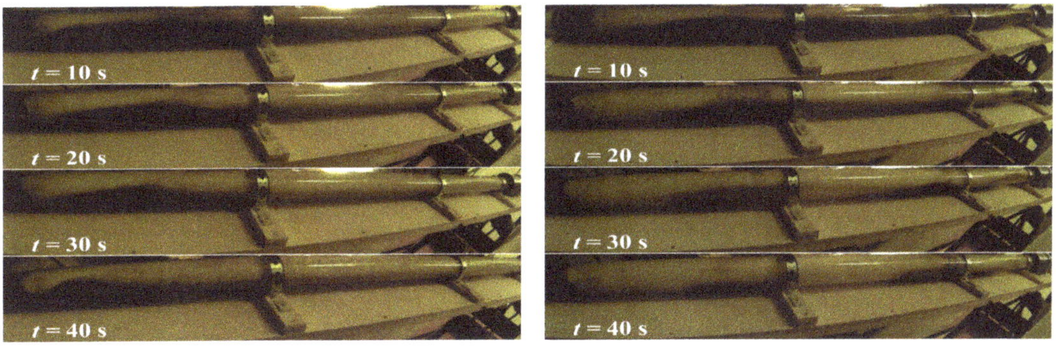

(**a**) (**b**)

Figure 2. Temporal collage of photographs of experiments with mud No. 2 at α = 70°, $Q = 1.25 \times 10^{-3}$ m^3/s, rotation with a frequency of n = 1.67 s^{-1}, direct pipe L_0 (**a**) and pipe with 3 half-waves L_3 (**b**).

To assess the effectiveness of the hydromechanical process of breakdown of mud plugs, which is in a state of buckling of the first kind, as well as to assess the quality of well cleaning from cuttings, it is necessary to fix the mass of fractional sand poured into the experimental stand. Through test measurements, it was found that under the given geometric and temporal conditions, the mass of sand m = 1.5 kg can be taken as a unit. However, in the course of experimental studies, it was found that, in some cases, it is impossible to assess the quality of well cleaning, due to the complete removal of cuttings ahead of the allocated time, and t = 120 s due to certain combinations of values of variable parameters. For this reason, the number of cuttings cleaned out was recalculated in time and the well cleaning intensity was obtained: for series of experiments No. 1, the time $t_1 = 30$ s, and for series of experiments No. 2, the time $t_2 = 60$ s.

The initial and time-recalculated results of the mud removal intensity, using the example of a Newtonian fluid at a zenith angle $\alpha = 35°$, are shown in Figure 3. It can be seen from the graphs that when $L3$ rotates a flat-curved pipe with three half-waves (blue lines) at a frequency of $n = 2.5$ s^{-1}, the quality of wellbore cleaning increases significantly compared to L_0 of a straight pipe (yellow lines), with a changing mud consumption, Q within the range of $1.25 \div 1.67 \times 10^{-3}$ m^3/s.

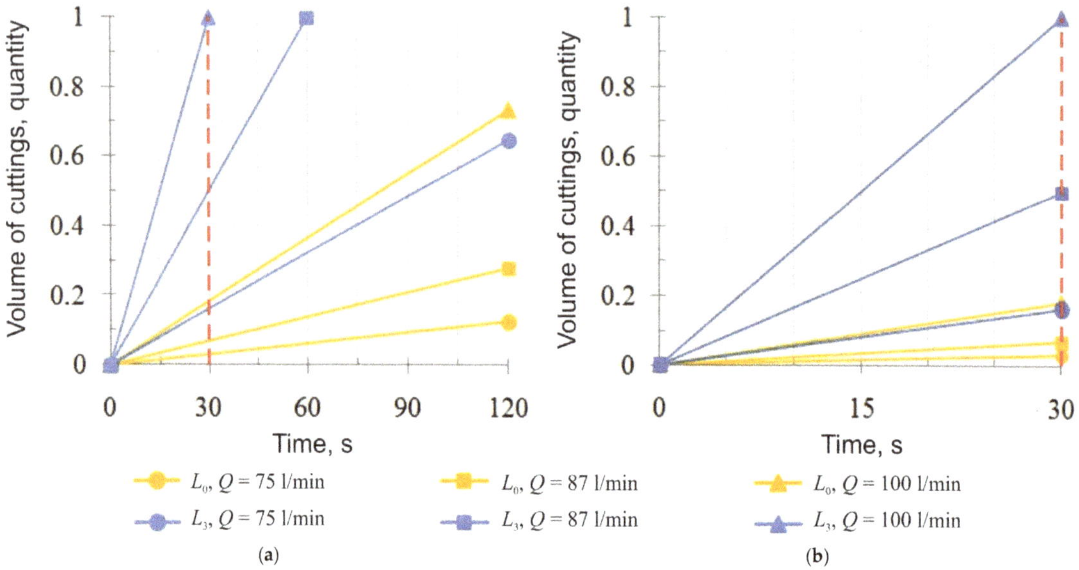

Figure 3. Initial (**a**) and recalculated (**b**) rates of mud removal at $\alpha = 35°$, DF No. 1, $Q = 75 \div 100$ L/min and rotation at a frequency of $n = 2.5$ s^{-1} direct pipe L_0 and DS with 3 half-waves L_3.

The recalculation of the removal of mud cuttings over time made it possible to obtain values for the subsequent assessment of the process of hydromechanical breakdown of the drilling mud plugs, which is in a state of buckling of the first kind, and the quality of well cleaning. The data obtained as a result of experimental studies can be interpreted both on a plane, in two-dimensional space, and in three-dimensional space, in the form of surfaces, depending on the number of fixed variables. For each case of interpretation, mathematical models were obtained—regression equations with correlation coefficients (Figure 4, Table 3). A total of 54 mathematical models have been obtained that describe the change in the quality of well cleaning from cuttings particles depending on variable parameters. For processing the results of experiments, correlation-regression and variance analyzes, "Statistica 13" and "MS Office Excel 2016" software were used.

Analyzing the obtained mathematical models presented in Table 3, the coefficient b_{12} characterizing the strength of the influence of factors on each other can be seen: with a direct pipe $b_{12} = 0.0112$; at the 1st half-wave $b_{12} = 0.0449$; with 3 half-waves $b_{12} = 0.8348$. This allows us to conclude that the mutual influence of the factors of the speed of rotation of the pipe and the mud consumption on each other increases, as a result of the loss of stability of the pipe and the subsequent acquisition of a sinusoidal shape.

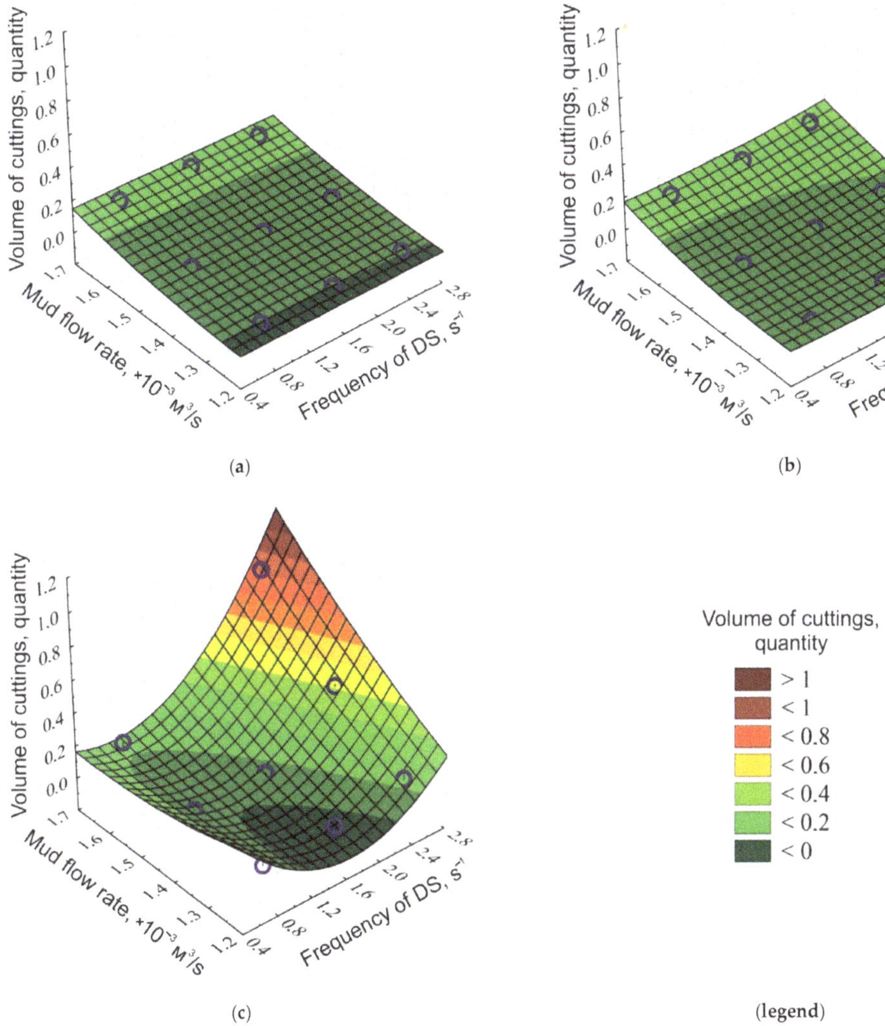

Figure 4. Dependence of the amount of mud carried out (fraction of units) during the destruction of the mud plug on the DF No. 1 consumption, the rotational speed and the number of half-waves pipes: without half-waves (**a**); with the 1st half-wave (**b**) and with the 3rd half-waves (**c**), at the zenith angle α = const = 70°.

At the maximum flow rate of mud No. 1 $Q = 1.67 \times 10^{-3}$ m^3/s, a decrease in the mud carried out on the sieves is observed, in comparison with washing at the average flow rate $Q = 1.45 \times \cdot 10^{-3}$ m^3/s, at a maximum constant frequency $n = 2.50$ s^{-1}. This observation allows us to conclude that an increase in the drill speed does not always improve the removal of cuttings.

Experimental data reflecting the effect of pipe buckling at a rotation frequency of $n = 2.5$ s^{-1}, as well as the zenith angle, flow rate, technological and rheological properties of DF on the destruction of mud plugs and well cleaning, are shown in Figure 5 in the form of histograms, convenient for comparison.

Table 3. Mathematical models of dependencies presented in Figure 4.

№	Mathematical Model	Number of Half-Waves Pipes	X	Y	Constant
(a)	$Z = 0.678 - 0.148x - 1.1825y + 0.0005x^2 + 0.0112xy + 0.5119y^2$	L_0	Frequency of DS, s^{-1}	Mud flow rate, $\times 10^{-3}$ m^3/s	Zenith angle $\alpha = 70°$
(b)	$Z = 1.3531 - 0.0889x - 2.0947y + 0.014x^2 + 0.0449xy + 0.8247y^2$	L_1			
(c)	$Z = 3.0149 - 1.8193x - 3.0061y + 0.2478x^2 + 0.8348xy + 0.824y^2$	L_3			

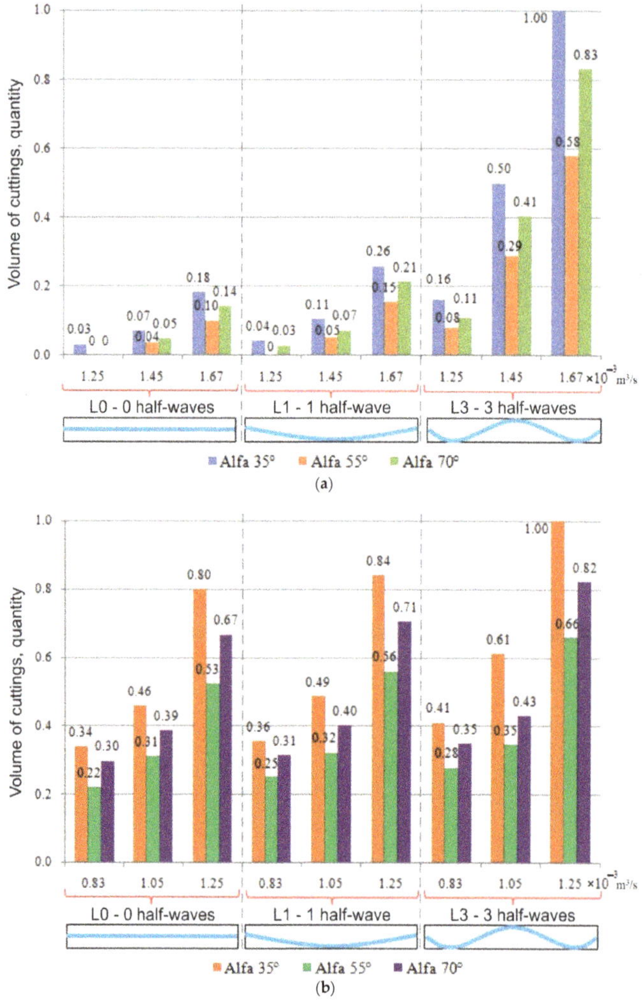

Figure 5. Removed cuttings from mud #1 (**a**) and mud #2 (**b**) depending on the sinusoidal pipe buckling, zenith angle, flow rate, and mud properties at a constant rotational speed $n = 2.5$ s^{-1}.

3.2. Influence of the Zenith Angle on the Destruction of Mud Plugs and Well Cleaning

Analyzing Figure 5a it can be seen that, with an increase in the zenith angle α from 35° to 55°, when cleaning with a Newtonian liquid, the mud removal decreases by 40–56%, depending on the flow rate of mud No. 1 $Q = 1.25 \div 1.67 \times 10^{-3}$ m^3/s and drill speed $n = 0.83 \div 2.50$ s^{-1}. At a flow rate of mud No. 1 $Q = 1.25 \times 10^{-3}$ m^3/s, a sufficient upward flow rate is not provided and cleaning deteriorates by 51–60% in the presence of the three half-waves pipe, up to the complete absence of 100% mud removal during rotation of the straight pipe and the pipe with the first half-wave.

The subsequent increase in the zenith angle α from 55° to 70°, ceteris paribus, has a positive effect on the quality of well cleaning, and cuttings removal is improved by 30–48%.

When comparing the results of experiments, with an increase in the zenith angle α from 35° to 70°, the mud transportation deteriorates by 17–42%. In Pipe L_1, deterioration is by 34–60%, but is still insufficient for pipe L_0.

From Figure 5b it can be seen that when using a pseudo-plastic liquid as a mud, with an increase in the zenith angle α from 35° to 55°, with a flow rate of mud No. 2 $Q = 0.83 \div 1.25 \times 10^{-3}$ m^3/s, the mud removal decreases by 29–46%, when rotating at a frequency of $n = 0.83 \div 2.50$ s^{-1} pipe with the first half-wave, and by 32–44% when rotating the direct pipe L_0 and DS with three half-waves.

A further increase in the zenith angle α from 55° to 70°, ceteris paribus, provides an increase in the removal capacity of the cuttings by 25–44%, 24–48% and 25–43%, with pipe L_0, pipe L_1 and pipe L_3, respectively.

When comparing the results of experiments with an increase in the zenith angleα from 35° to 70°, by passing the angle $\alpha = 55°$, the mud transportation deteriorates by only 13–30%, with a flow rate of pseudo-plastic liquid $Q = 0.83 \div 1.25 \times 10^{-3}$ m^3/s and rotation of the straight pipe L_0, pipe L_1 and pipe L_3.

Experimental data on the change in the quality of well cleaning, depending on the zenith angle α, confirm the presence of a critical angle $\alpha = 55°$ in the zone with a range of 45–60°.

3.3. Influence of Pipe Buckling on the Destruction of Mud Plugs and Well Cleaning

The results of the experimental study confirm the influence of the spatial shape of the pipe, namely the sinusoidal bend, on the destruction of mud plugs and cleaning the well from cuttings.

From Figure 5a it can be seen that, when conducting experiments using Newtonian liquid as a cleaning agent, depending on the flow rate $Q = 1.25 \div 1.67 \times 10^{-3}$ m^3/s and pipe speed $n = 2.50$ s^{-1}, there is a relative increase in the carrying capacity of the mud.

So, when drilling a well with a zenith angle $\alpha = 35°$, the presence of the first half-wave pipe leads to an improvement in cutting removal by 12–50%, relative to the rotation of a straight pipe without half-waves. In the presence of the three half-waves pipe, the mud is better carried out by 3.5 times, relative to the rotation of the pipe with the first half-wave, and by six times relative to the rotation of the straight pipe, with a speed $n = 2.50$ s^{-1}.

In the study of hydromechanical destruction of mud plugs in the zone of critical zenith angles, with $\alpha = 55°$, the presence of the first half-wave of the string leads to an improvement in mud removal by 9–54%, relative to the rotation of the pipe without half-waves. It is especially noteworthy that in the presence of the three half-waves pipe, the mud is cleared out 4.5 times better, relative to the rotation of the pipe with the first half-wave, and more than 6.5 times better, relative to the rotation of a straight pipe without half-waves, at a frequency of rotation of the pipe $n = 2.50$ s^{-1}.

After passing through the critical zone, during the rotation of the pipe in the well with a zenith angle $\alpha = 70°$, the presence of the first half-wave of the pipe leads to an improvement in cuttings removal by 13–49%, relative to the rotation of a straight pipe without half-waves. When rotating with a frequency of $n = 2.50$ s^{-1} and the presence of the three half-waves pipe, the mud is carried out 4.5 times better, relative to the rotation of the string with the first half-wave, and more than 7 times better, relative to the rotation of a straight string without half wave.

From Figure 5b it can be seen that, when a pseudo-plastic liquid is used as a mud, depending on the flow rate $Q = 0.83 \div 1.25 \times 10^{-3}$ m^3/s, the sinusoidal bending of the pipe increases the carrying capacity of the cuttings.

When drilling a well with a zenith angle $\alpha = 35°$, the presence of the first half-wave of string leads to an improvement in cuttings removal by 4–8%, relative to the rotation of a straight string without half-waves. In the presence of the three half-waves string, the mud is better carried out by 6–26%, relative to the rotation of the pipe with the first half-wave, and by 10–33% relative to the rotation of the straight pipe without half-waves.

In the study of hydromechanical destruction of mud plugs in the zone of critical zenith angles, with $\alpha = 55°$, the presence of the first half-wave string leads to an improvement in the removal of mud by 3–15%, relative to the rotation of a straight string without half-waves. It is noteworthy that in the presence of three half-waves pipes, the mud is better carried out by 7–28%, relative to the rotation of the pipe with the first half-wave, and by 10–36% relative to the rotation of a straight pipe without half-waves.

It has been established that, during the rotation of pipe in a well with a zenith angle $\alpha = 70°$, the presence of the first half-wave string leads to an improvement in cuttings removal by 2–7%, relative to the rotation of a straight string without half-waves. In the presence of three half-waves strings, the mud is carried out better by 4–28%, relative to the rotation of the pipe with the first half-wave, and by 8–37% relative to the rotation of a straight pipe without half-waves.

3.4. Influence of Mud Properties on the Destruction of Mud Plugs and Well Cleaning

An analysis of the experimental data reflecting the influence of the properties of the mud on the quality of cleaning the well of cuttings showed that the physical and rheological properties have a significant impact on the carrying capacity of the mud. It has been established that, with an increase in density from 1001 to 1035 kg/m^3, viscosity from 0.941 to 40.283 mPa·s and muds from 0 to 16.92 Pa, when cleaning with a flow rate $Q = 1.25 \times 10^{-3}$ m^3/s, varying drill speed $n = 0.83 \div 2.50$ s^{-1} and the presence of the first and third half-waves strings, the mud removal increases by 4–13 times. In some cases, where the Newtonian fluid did not allow the cuttings to be carried to the surface, and there was no carryover of the cuttings, the pseudo-plastic fluid showed excellent results (Figure 6).

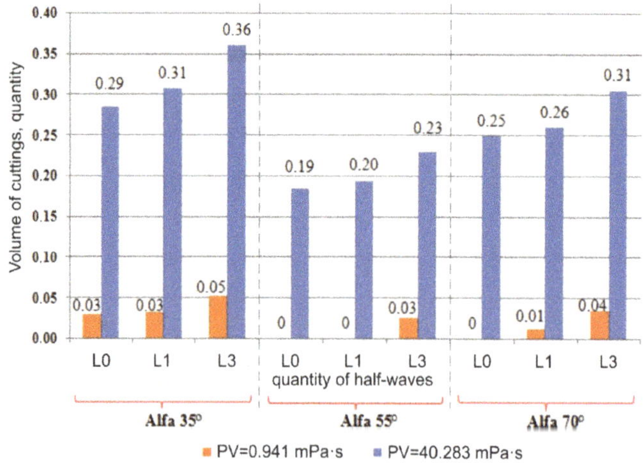

(a)

Figure 6. *Cont.*

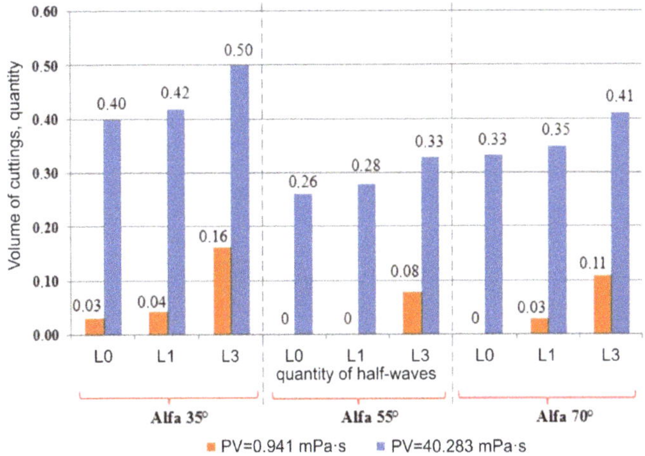

(b)

Figure 6. Comparison of the quality of cleaning from mud with Newtonian and pseudo-plastic liquids with a constant flow of mud, at different zenith angles, the presence of buckling and the speed $n = 1.67\ s^{-1}$ (**a**) and $n = 2.50\ s^{-1}$ (**b**).

4. Discussion

An increase in the number of cuttings on the screens confirmed the improvement in cleaning; in addition, it was clear that the pipe, which is in a state of controlled buckling, has a "double" hydromechanical effect on the cuttings plug and on particles settled on the bottom wall of the well.

The first impact of the string is mechanical—the body of the pipe in the area of maximum deflection had a direct physical impact, "breaking" the mud plug and mud dunes, moving the particles, and involving them in a suspended state.

The second effect of the mud is hydrodynamic—the pipe body, which has lost its longitudinal stability, changes its spatial shape, and, consequently, changes the profile of the mud flow moving along the upper wall of the well along the path of least resistance. This causes the mud to "wash out" the accumulated mud and entrain the particles into the general flow. As a result, local flow turbulence is created without the use of special devices, such as turbulators, circulation subs, etc.

The implementation of a hydromechanical method for the destruction of mud plugs in deviated and extended reach wells in order to improve cleaning and avoid problems associated with mud build-up is carried out by controlling and regulating the artificially created pipe buckling. At the moment, there are no field or experimental drill pipes in the world that, in addition to their main functions, can play the role of a device that measures deformations.

As a conceptual solution, the authors propose to develop and include in the DS special sections of pipes in which strain gauges based on a fiber Bragg grating (FBG) are mounted. Due to their advantages, FBG sensors are increasingly being used in control systems in the aviation and space industries [35,36], and they can also be used to measure deformation, displacement, temperature, pressure, tilt angle, as well as acceleration and vibration [37–40].

FBG fiber optic sensors must be braided from a material that is resistant to the aggressive environment of drilling and cement slurries. The sensors must be rigidly fixed on the inner wall of the drill pipes (Figure 7a) or in the outer wall, in a specially prepared channel (Figure 7b), reinforced with welded material to prevent the weakening of the DS as a result of possible stress concentration. In pipe couplings, the fiber-optic cable is connected to the FBG sensors by means of a quick-disconnect connection and forms a "garland" of sensors—a quasi-distributed differential measuring system that monitors deformations along the pipe body.

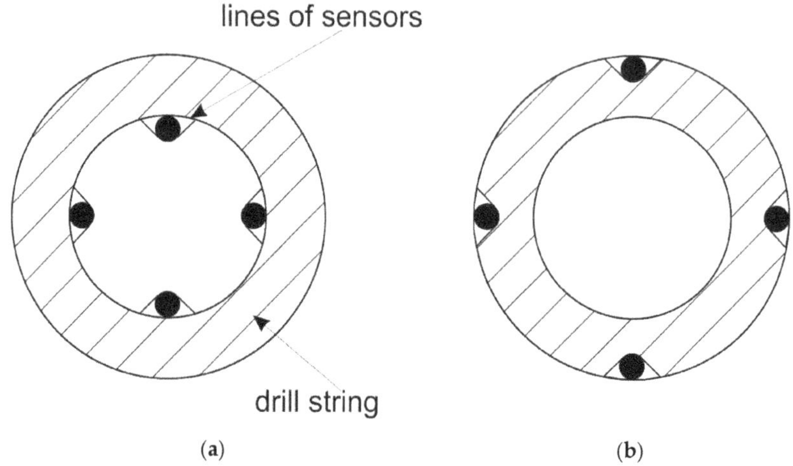

Figure 7. Conceptual diagram of the location of FBG strain sensors in the body pipes on the inner wall (**a**) and in the outer wall (**b**).

The reliability of such a differential measurement system will depend on the mounting methods, materials used and operating conditions. Before including the sensors in the measuring system, it is necessary to calibrate them according to the deformation of the strain gauge and determine the proportionality coefficient between the deformation of the pipe measured by the FBG and the strain gauge.

Thanks to the use of special sections of drill pipe with FBG strain sensors, the operator can control the place of buckling and the spatial shape of the drill bit by adjusting the axial load on the bit, within the established strength limits of the drilling tool. By creating an additional axial load on the bit, the operator forms the buckling of the pipe sections, and the control panel screen displays the spatial shape of the string in the well, taking into account the geometry of the wellbore and drilling parameters.

As a result of controlled buckling, using only mud hydraulics and DS mechanics, without the use of special devices that have a positive effect on well cleaning, hydromechanical destruction of mud plugs and improved removal of broken particles are achieved, and complications associated with the formation of mud pads are prevented.

5. Conclusions

The experimental stand, developed at the St. Petersburg Mining University, is an innovative solution in the field of improving the quality of cleaning directional wells. This study is fundamental, since for the first time the issue of the positive aspect of the loss of stability of the DS was considered, and the positive effect of pipe buckling on the efficiency of cuttings removal was established, taking into account the drilling parameters, technological and rheological properties of the drilling rig, and the angle of inclination of the well.

The analysis of well cleaning quality data, depending on the zenith angle α, confirms the presence of a critical angle $\alpha = 55°$ in the zone with a range of 45–60°, in which the maximum difficulty is noted in the transport of cuttings particles to the day surface—a decrease in the removal of particles to 40–60%.

When conducting a series of experiments No. 1 with a Newtonian liquid, an increase in the efficiency of mud removal by up to eight times was found, at the maximum flow rate of the BR, the rotation frequency $n = 0.83 \div 2.50\ s^{-1}$ and the presence of three half-waves string, in comparison with straight DS and strings with the first half-wave. When conducting a series of experiments No. 2 and using a pseudo-plastic liquid as mud, all other things being equal, an increase in the removal of mud cuttings up to 36% was found.

It has been experimentally established that, when the string rotates at a speed of $n = 0.83 \div 2.50\ s^{-1}$, at a flow rate $Q = 1.25 \times 10^{-3}\ m^3/s$ of a pseudo-plastic liquid with rheological properties, the removal of mud cuttings to the surface increases dozens of times in comparison with the Newtonian fluid.

The authors put forward a hypothesis that the controlled buckling of the string, namely, the number and length of half-waves in the place specified by the operator, allows the breakdown of mud plugs and improves well cleaning, as well as solving the problem of "mud-pads" and the complications and emergencies associated with this phenomenon.

The presented technical and technological solution for the implementation of a hydromechanical method for the destruction of stagnant mud zones is a novel concept and requires further study with the subsequent development of experimental drill pipes based on FBG strain sensors.

The obtained results confirm the need for further study of the effect of pipe buckling on the destruction of mud plugs and cleaning of deviated wells from cuttings, taking into account the multifactorial nature of the drilling process.

Author Contributions: Conceptualization, V.G.K. and M.V.D.; methodology, V.G.K. and M.V.D.; software, V.G.K.; validation, V.G.K. and M.V.D.; formal analysis, V.G.K. and M.V.D.; investigation, V.G.K. and M.V.D.; resources, M.V.D.; data curation, M.V.D.; writing—original draft preparation, V.G.K.; writing—review and editing, M.V.D.; visualization, V.G.K.; supervision, M.V.D.; project administration, M.V.D.; funding acquisition, V.G.K. All authors have read and agreed to the published version of the manuscript.

Funding: This research received no external funding.

Institutional Review Board Statement: Not applicable.

Informed Consent Statement: Not applicable.

Data Availability Statement: All the data used in this study are available in the article.

Acknowledgments: The authors are grateful to Sergey L. Yurtaev, Chief Specialist of Well Construction Laboratory, for his help in designing experimental equipment and support during the research.

Conflicts of Interest: The authors declare no conflict of interest.

References

1. Archegov, V.B.; Nefedov, Y.V. The strategy of oil and gas exploration in the energy potential estimation of Russian Arctic shelf seas. *J. Min. Inst.* **2015**, *212*, 6–13. (In Russian)
2. Vasiltsov, V.S.; Vasiltsova, V.M. Strategic planning of Arctic shelf development with apparatus of fractal theory. *J. Min. Inst.* **2018**, *234*, 663–672. (In Russian) [CrossRef]
3. Gao, D.; Tan, C.; Tang, H. Limit analysis of extended reach drilling in South China Sea. *Pet. Sci.* **2009**, *6*, 166–171. [CrossRef]
4. Menand, S.; Isambourg, P.; Sellami, H.; Simon, C.; Bouguecha, A. Axial force transfer of buckled drill pipe in deviated wells. In Proceedings of the SPE/IADC Drilling Conference and Exhibition, Amsterdam, The Netherlands, 17–19 March 2009; pp. 1–12. [CrossRef]
5. Aldred, W.; Cook, J.; Bern, P.; Carpenter, B.; Hutchinson, M.; Lovell, J.; Rezmer-Cooper, I.; Leder, P.C. Using downhole annular pressure measurements to improving drilling performance. *Oilfield Rev.* **1998**, *10*, 40–55.
6. Dolgopolskiy, A.L. Use of invert-emulsion drilling mud during well construction at Kharyaginsk field. *Eng. Pract.* **2015**, *3*, 92. (In Russian)
7. Zhu, X.; Yi, J.; Liu, Q. Distribution features of cuttings bed and sensitivity analysis of major drilling parameters for cuttings transport in gas drilling horizontal wells. *J. Hydrodyn.* **2015**, *27*, 884–893. [CrossRef]
8. Kiełbasiński, K.; Dobak, P.; Kaczmarek, Ł.; Kowalczyk, S. The influences of local glacitectonic disturbance on overconsolidated clays for upland slope stability conditions: A case study. *Appl. Sci.* **2021**, *11*, 10718. [CrossRef]
9. Blinov, P.A. Determining the stability of the borehole walls at drilling intervals of loosely coupled rocks considering zenith angle. *J. Min. Inst.* **2019**, *236*, 172–179. [CrossRef]
10. Dvoynikov, M.V.; Budovskaya, M.E. Development of a hydrocarbon completion system for wells with low bottomhole temperatures for conditions of oil and gas fields in Eastern Siberia. *J. Min. Inst.* **2022**, *253*, 12–22. [CrossRef]

11. Sandoiu, G.-F. Removing cuttings from deviated and horizontal wells. *J. Min. Inst.* **2008**, *174*, 41–42.
12. Leusheva, E.; Morenov, V.; Liu, T. Dependence of the equivalent circulation density of formate drilling fluids on the molecular mass of the polymer reagent. *Energies.* **2021**, *14*, 7639. [CrossRef]
13. Dobik, Y.A. Effect of drill string rotation on cutting transport in horizontal well. *Constr. Oil Gas Wells Land Sea* **2019**, *3*, 25–29. (In Russian) [CrossRef]
14. Mitchell, R.F. Lateral buckling of pipe with connectors in curved wellbores. In Proceedings of the SPE Drilling Conference, Amsterdam, The Netherlands, 27 February–1 March 2003; pp. 22–32. [CrossRef]
15. Balaba, V.I.; Zinchenko, O.D. Technical devices for efficiency upgrading of cuttings hydrotransport during directional and horizontal drilling. *Equip. Technol. Oil Gas Complex* **2015**, *3*, 23–27. (In Russian)
16. Oganov, A.S.; Rayhert, R.S.; Tsukarenko, M.S. Problems of cleaning directional and horizontal wells. *Neftegaz. Ru* **2015**, *6*, 32–39. (In Russian)
17. Rayhert, R.S.; Tsukarenko, M.S.; Oganov, A.S. Technological solutions of directional and horizontal well's cleaning from cuttings. *Oil Gas Novations* **2016**, *3*, 28–35. (In Russian)
18. Nutskova, M.V.; Sidorov, D.A.; Tsikplornu, D.E.; Sergeev, G.M.; Vasilev, N.I. Investigations of oil based muds primary opening of productive formations. *Perm. J. Pet. Min. Eng.* **2019**, *19*, 138–149. (In Russian) [CrossRef]
19. Ulyasheva, N.M.; Leusheva, E.L.; Galishin, R.N. Development of drilling fluid composition with rheological parameters for directional well. *J. Min. Inst.* **2020**, *244*, 454–461. (In Russian) [CrossRef]
20. Sharafutdinov, Z.Z.; Sharafutdinova, R.Z. Water based drilling fluids and management its rheological parameters. *Neftegazov. Delo* **2004**, *1*, 1–21. (In Russian)
21. Jeong, J.; Lim, C.; Park, B.-C.; Bae, J.; Shin, S.-C. Multi-objective optimization of drilling trajectory considering buckling risk. *Appl. Sci.* **2022**, *12*, 1829. [CrossRef]
22. Basovich, V.S.; Buyanovskiy, I.N.; Sapunzhi, V.V. Perspectives of application of light alloy drilling pipes with external helical finning for drilling horizontal wells and branch holes. *Drill. Oil* **2014**, *5*, 42–46. (In Russian)
23. Liu, X.; Vlajic, N.; Long, X.; Meng, G.; Balachandran, B. Nonlinear motions of a flexible rotor with a drill bit: Stick-slip and delay effects. *Nonlinear Dyn.* **2013**, *72*, 61–77. [CrossRef]
24. Piroozian, A.; Ismail, I.; Yaacob, Z.; Babakhani, P.; Ismail, A.S.I. Impact of drilling fluid viscosity, velocity and hole inclination on cuttings transport in horizontal and highly deviated wells. *J. Pet. Explor. Prod. Technol.* **2012**, *2*, 149–156. [CrossRef]
25. Amanna, B.; Movaghar, M.R.K. Cuttings transport behavior in directional drilling using computational fluid dynamics (CFD). *J. Nat. Gas Sci. Eng.* **2016**, *34*, 670–679. [CrossRef]
26. Katende, A.; Segar, B.; Ismai, I.; Sagala, F.; Saadiah, H.H.A.R.; Samsuri, A. The effect of drill-pipe rotation on improving hole cleaning using polypropylene beads in water-based mud at different hole angle. *J. Pet. Explor. Prod. Technol.* **2019**, *10*, 1253–1262. [CrossRef]
27. Hakim, H.; Katende, A.; Sagala, F.; Ismail, I.; Nsamba, H. Performance of polyethylene and polypropylene beads towards drill cuttings transportation in horizontal wellbore. *J. Pet. Sci. Eng.* **2018**, *165*, 962–969. [CrossRef]
28. Heshamudin, N.S.; Katende, A.; Rashid, H.A.; Ismail, I.; Sagala, F.; Samsuri, A. Experimental investigation of the effect of drill pipe rotation on improving hole cleaning using water-based mud enriched with polypropylene beads in vertical and horizontal wellbores. *J. Pet. Sci. Eng.* **2019**, *179*, 1173–1185. [CrossRef]
29. Ytrehus, J.D.; Lund, B.; Taghipour, A.; Carazza, L.; Gyland, K.R.; Saasen, A. Oil-based drilling fluid's cuttings bed removal properties for deviated wellbores. *ASME J. Energy Resour. Technol.* **2021**, *143*, 103003. [CrossRef]
30. Ytrehus, J.D.; Lund, B.; Taghipour, A.; Kosberg, B.R.; Carazza, L.; Gyland, K.R.; Saasen, A. Hydraulic behavior in cased and open-hole sections in highly deviated wellbores. *ASME J. Energy Resour. Technol.* **2021**, *143*, 033008. [CrossRef]
31. Gao, D.; Huang, W. A review of down-hole tubular string buckling in well engineering. *Pet. Sci.* **2015**, *12*, 443–457. [CrossRef]
32. Gulyayev, V.; Shlyun, N. Influence of friction on buckling of a drill string in the circular channel of a bore hole. *Pet. Sci.* **2016**, *13*, 698–711. [CrossRef]
33. Pei, H.-F.; Yin, J.-H.; Zhu, H.-H.; Hong, C.-Y.; Jin, W.; Xu, D.-S. Monitoring of lateral displacements of a slope using a series of special fibre Bragg grating-based in-place inclinometers. *Meas. Sci. Technol.* **2012**, *23*, 25007. [CrossRef]
34. Kadochnikov, V.G.; Dvoynikov, M.V.; Blinov, P.A. Influence of the drill string spatial form on transport of cuttings in directional wells. *Bull. Assoc. Drill. Contract.* **2020**, *2*, 12–19. (In Russian)
35. Mikhailovskiy, K.V.; Bazanov, M.A. Measuring of residual technological deformations in carbon-plastic with injection optical fiber Bragg sensors. *Constr. Compos. Mater.* **2016**, *2*, 54–58. (In Russian)
36. Leduc, D.; Lecieux, Y.; Morvan, P.-A.; Lupi, C. Architecture of optical fiber sensor for the simultaneous measurement of axial and radial strains. *Smart Mater. Struct.* **2013**, *22*, 075002. [CrossRef]
37. Liu, H.-L.; Zhu, Z.-W.; Zheng, Y.; Liu, B.; Xiao, F. Experimental study on an FBG strain sensor. *Opt. Fiber Technol.* **2018**, *40*, 144–151. [CrossRef]
38. Takeda, N.; Tajima, N.; Sakurai, T.; Kishi, T. Recent advances in composite fuselage demonstration program for damage and health monitoring in Japan. *Struct. Control Health Monit.* **2005**, *12*, 245–255. [CrossRef]

39. Ma, T.; Chen, P.; Zhao, J. Overview on vertical and directional drilling technologies for the exploration and exploitation of deep petroleum resources. *Geomech. Geophys. Geo-Energy Geo-Resour.* **2016**, *2*, 365–395. [CrossRef]
40. Makhsidov, V.V.; Yakovlev, N.O.; Ilichev, A.V.; Shienyuk, A.M. Measuring of carbon-plastic deformation with inside-integrated optical fiber Bragg sensors. *Mech. Compos. Mater. Constr.* **2015**, *21*, 360–369. (In Russian)

applied sciences

MDPI

Article

Comparative Analysis of Multi-Criteria Decision-Making Techniques for Outdoor Heat Stress Mitigation

Aiman Mazhar Qureshi and Ahmed Rachid *

Laboratory of Innovative Technologies, University of Picardie Jules Verne, 80000 Amiens, France
* Correspondence: ahmed.rachid@u-picardie.fr

Abstract: Decision making is the process of making choices by organizing relevant information and evaluating alternatives. MCDMs (Multi-Criteria Decision Methods) help to select and prioritize alternatives step by step. These tools can help in many engineering fields where the problem is complex and advanced. However, there are some limitations of the different MCDMs that reduce the reliability of the decision that needs to be improved and highlighted. In this study, Elimination and Choice Expressing Reality (ELECTRE) NI (Net Inferior), NS (Net Superior), Technique for Order Preference by Similarity to Ideal Solutions (TOPSIS), Preference Ranking Organization Method for Enrichment Evaluation (PROMETHEE), VIekriterijumsko KOmpromisno Rangiranje (VIKOR), Multi-Objective Optimization Ratio Analysis (MOORA), Weight Sum Method (WSM) and Weighted Product Method (WPM) are applied for the selection of urban heat mitigation measurements under certain criteria. The models were applied using weighting criteria determined by two ways, (i) the direct weighting method and (ii) the Analytic Hierarchy Process (AHP), for precise weighting factoring through pairwise comparison. This numerical research evaluated the reliability of MCDMs using the same decision matrix under different normalization techniques and shows the impact of AHP on the decision. The results show that WSM and PROMETHEE provided reliable and consistent results for all normalization techniques. The combination of AHP with applied MCDMs improved the frequency of consistent ranking, except with ELECTRE-NS.

Keywords: heat stress; multi criteria decision making; analytic hierarchy process; priority; interventions

Citation: Qureshi, A.M.; Rachid, A. Comparative Analysis of Multi-Criteria Decision-Making Techniques for Outdoor Heat Stress Mitigation. *Appl. Sci.* **2022**, *12*, 12308. https://doi.org/10.3390/app122312308

Academic Editor: Maria Vicidomini

Received: 8 August 2022
Accepted: 29 November 2022
Published: 1 December 2022

1. Introduction

Multi-criteria Decision Methods (MCDMs) are valuable tools to handle the selection problem. They are based on five components, which are: goals, thoughts of the specialist, alternatives, criteria, and results. The MCDM requires human recognitions as sources of information where uncertainty and subjective aspects exist. The decision maker's assessments can be expressed by using linguistic terms such as "low importance" or "brilliant performance". The idea of these assessments is often subjective because some criteria that do not have an objective measure, which forces the decision makers to express their thoughts using numeric scales. There are many MCDM methods which are used in different fields of study; for example, the Previously Fuzzy logic method [1] has been applied in the soil sciences [2], supplier's performance [3], for imprecise information related to distribution problems [4], in the field of accounting and finance to develop guidelines for investment decisions [5,6], and in the selection of the appropriate process performance [7]. However, the fuzzy logic method has no potential to measure the level of consistency in the judgments provided by a decision maker.

AHP is a one of the oldest and most trusted decision-making methods [8]. It is a comprehensive technique that has ability to solve the complex decision-making problems by assembling, quantifying, and evaluating the alternative solution through hierarchies [9,10]. Furthermore, it is easy to implement by experts of other fields and overcomes the resulting risk of inconsistency. Consistency plays a vital role in AHP. When the consistency ratio of the pairwise comparison matrix is greater than 10%, it requires a review of the inputs to make the results consistent with the provided judgments [11]. The modified version of Fuzzy-AHP is aimed at removing the vagueness and uncertainty in decision-making, but due to heavy calculations and a high risk of errors, it is difficult to adapt. In contrast, the conventional AHP is quite easy to update, completely reliable, and cost effective, and its analysis can easily be performed by software [12,13]. Previously, the AHP was coupled with other MCDMs, such as MOORA for public transport service quality [14], ELECTRE for personnel selection [15], TOPSIS for the evaluation of knowledge sharing capabilities of supply chain partners [16] and suitable technology transfer strategy for wind turbines [17], PROMETHEE for the selection of policy scenarios for vehicle fleet [18], VIKOR for the assessment of school teachers [19], WSM to evaluate the knowledge in supply chain patterns [20], and WPM for the selection of open-source electronical medical records [21]. In one research, the DEA model was used to assess the performance of small- to medium-sized manufacturing enterprises, and it showed that the MCDM model combined with the AHP was more consistent than stand-alone models where the decision was entirely based on quantitative inputs [22]. The main reason why many companies do not rely on MCDM methods can be due to the fact that decision makers intuitively notice ranking errors [23]. However, there is a need to determine the comparative study on the reliability of the MCDMs. So far, the application of several MCDMs and their comparative study in the field of Urban Heat Stress (UHS) has not been carried out due to long reasoning, which is difficult to quantify and scale. This research is based on the selection of interventions to mitigate outdoor heat stress by using different multicriteria decision applications, such as PROMETHEE, VIKOR, MOORA, ELECTRE (NS, NI), TOPSIS, WPM, and WSM. The applied methods are also combined with AHP, which identifies the effectiveness of standalone application using direct criteria weightage and the impact of AHP on the decision process in field of UHS. The impact of different normalization methods and coherent frequency in ranking results obtained from the application of different MCDMs are also investigated.

The article is organized as follows. Section 2 presents the research background and related work. The underlying concepts and mathematical formulas are given in Section 3. The research methodology is explained in Section 4. Sections 5 and 6 provides the simulation results and discussion. The article ends with a conclusion and some perspectives.

2. Research Background

UHS is the current crucial concern for scientists and residents of medium- and large-sized cities, which are at more risk for heat events. Prolonged exposure can cause heat exhaustion, cramps, stroke, as well as exacerbate pre-existing chronic conditions, such as various respiratory, cerebral, and cardiovascular diseases, especially for vulnerable people. There are many heat mitigation strategies for improving thermal comfort in urban areas. As human beings, it is our sacred duty to save the environment in which we live, but decision maker responsibility is much bigger because many remedial measures need to be implemented on a large scale. The application of MCDMs could be first approach that assists decision makers to quantitively assess the importance of criteria and the performance of alternatives in selection processes.

In previous studies, AHP-SWOT, multi-criteria outranking approach, EFDM, FDEMATEL, multi-criteria method by linear regression, TOPSIS, SMCE, Fuzzy-AHP TOPSIS have been used in the field of UHS [24]. In addition, the AHP method was specifically used to select the urban heat resilience intervention under certain criteria [25].

From the literature, it was investigated that most popular interventions considered to deal with thermal stress are white roofs, extensive green roofs, intensive green roofs [26], planting trees in cities [27], green parking lots, cool roofs, watering methods (sprays in public areas) [28], green walls, and shades. Figure 1 shows the interventions chosen for this study that can mitigate thermal discomfort in an urban setting.

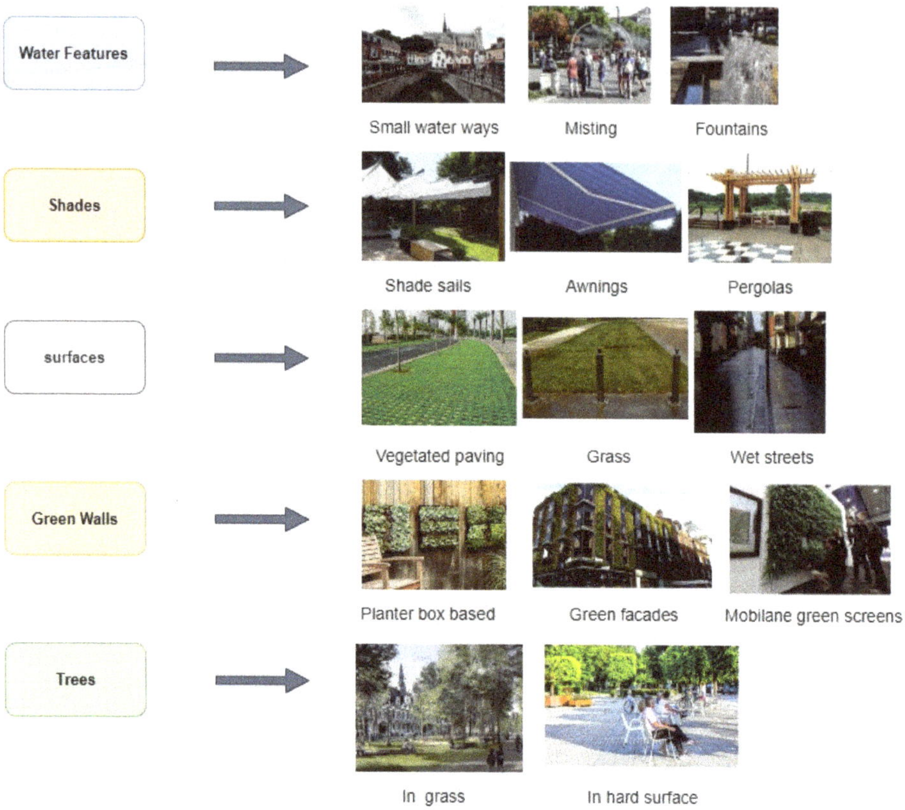

Figure 1. Type of heat resilience interventions.

For our studies, we considered four general criteria to address for the selection process of interventions (alternatives). These criteria are the following:

- *Cost*: capital and running cost of the intervention, which is often taken as a non-beneficial (NB) criterion.
- *Environment*: impact of the intervention on the level of air, land, and water. For example, it might be necessary to know if a recently introduced intervention significantly improves the previous mean level of air quality.
- *Efficiency*: cooling effect of intervention in open spaces.
- *Durability*: intervention capability to withstand the level of heat and remain useful without requiring additional maintenance after extreme weather events throughout the service life.

3. Mathematical Models of MCDMs

Eight different MCDMs and seven normalization methods were computed for deciding the UHS mitigation intervention. Additionally, the AHP was also applied for calculating the weightage of criteria. The procedure and key equations of methods are given in Table 1. Mathematical formulas of applied normalized methods are presented in Table 2.

Table 1. Key equations of MCDMs used in this study.

MCDM	Steps	Reference
TOPSIS: Technique for Order Preference by Similarity to Ideal Solutions	Step 1: make decision matrix Step 2: normalize decision matrix Step 3: weighted normalized decision matrix r_{ij}: normalized decision matrix $\mathcal{V}_{ij} = w_j r_{ij}$ w_j: weight of the Jth criteria (attribute) Condition $\sum_{j=1}^{n} w_j = 1$ Step 4: ideal best $\mathcal{V}_j^+$ and ideal worst $\mathcal{V}_j^-$ values If beneficial criteria: $\mathcal{V}_j^+ = \max_i \left(v_{ij}\right) = \max\left\{v_{ij}, i = 1, \ldots, m\right\}$ If cost criteria: $\mathcal{V}_j^+ = \min_i \left(v_{ij}\right) = \min\left\{v_{ij}, i = 1, \ldots, m\right\}$ Step 5: Calculate the distances of each alternative from the positive ideal solution and the negative ideal solution $S_i^{\pm} = \sqrt{\sum_{j=1}^{n} \left(\mathcal{V}_{ij} - \mathcal{V}_j^{\pm}\right)^2}$ Step 6: Calculate the relative closeness to the ideal solution (performance score) $R_i = \frac{S_i^-}{S_i^+ + S_i^-}$ Step 7: Ranking the best alternative	[29]
ELECTRE(NI-NS): Elimination and Choice Expressing Reality	Step 1: make decision matrix Step 2: normalize decision matrix Step 3: weighted normalized decision matrix $v_{ij} = R \times W = \begin{vmatrix} r_{11} \cdot \mathcal{W}_1 & r_{12} \cdot \mathcal{W}_2 & \mathrm{K} & r_{1n} \cdot \mathcal{W}_n \\ r_{21} \cdot \mathcal{W}_1 & r_{22} \cdot \mathcal{W}_2 & \mathrm{K} & r_{2n} \cdot \mathcal{W}_n \\ \mathrm{K} & \mathrm{K} & \mathrm{K} & \mathrm{K} \\ r_{m1} \cdot \mathcal{W}_1 & r_{m2} \cdot \mathcal{W}_2 & \mathrm{K} & r_{mn} \cdot \mathcal{W}_n \end{vmatrix}$ Step 4: concordance and discordance interval sets $C_{ab} = \left\{j \mid v_{aj} \geq v_{bj}\right\}$ $D_{ab} = \left\{j \mid v_{aj} < v_{bj}\right\} = J - C_{ab}$ Step 5: calculation of the concordance interval matrix $C_{ab} = \sum_{j \in C_{ab}} \mathcal{W}_j$ $C = \begin{vmatrix} - & c(1,2) & \ldots & c(1,m) \\ c(2,1) & - & \ldots & c(2,m) \\ \vdots & \vdots & \vdots & \vdots \\ c(m,1) & c(m,2) & \ldots & - \end{vmatrix}$ Step 6: Determine the concordance index matrix $\bar{c} = \frac{\sum_{a=1}^{m} \sum_{b}^{m} c(a,b)}{m(m-1)}$ $e(a,b) = 1 \text{ if } c(a,b) \geq \bar{c}$ $e(a,b) = 0 \text{ if } c(a,b) < \bar{c}$ Step 7: Calculation of the discordance interval matrix $d(a,b) = \frac{\max_{j \in D_{ab}} \left\| \mathcal{V}_{aj} - \mathcal{V}_{bj} \right\|}{\max_{j \in Jmn} \left\| \mathcal{V}_{aj} - \mathcal{V}_{bj} \right\|}$ $D = \begin{vmatrix} - & d(1,2) & \ldots & d(1,m) \\ d(2,1) & - & \ldots & d(2,m) \\ \vdots & \vdots & \vdots & \vdots \\ d(m,1) & d(m,2) & \ldots & - \end{vmatrix}$	[30]

Table 1. *Cont.*

MCDM	Steps	Reference
	Step 8: determine the discordance index matrix $\bar{d} = \frac{\sum_{a=1}^{m}\sum_{b}^{m} d(a,b)}{m(m-1)}$ $f(a,b) = 1 \text{ if } d(a,b) \leq \bar{d}$ $f(a,b) = 0 \text{ if } d(a,b) > \bar{d}$ Step 9: calculate the net superior and inferior value C_a: net superior $C_a = \sum_{b=1}^{n} C(ab) - \sum_{b=1}^{n} C(ba)$ d_a: net inferior $d_a = \sum_{b=1}^{n} d(ab) - \sum_{b=1}^{n} d(ba)$ Step 10: select the best alternative choose highest value of net superior (C_a) and lowest value of net inferior (d_a)	
PROMETHEE: This method utilizes a preferential function to drive the preference difference between alternative pairs.	Step 1: decision matrix Step 2: normalized the decision matrix Step 3: deviation by pairwise comparison $d_j(a,b) = R_j(a) - R_j(b)$ Step 4: preference function $p_j(a,b) = 0 \text{ if } d_j(a,b) \leq 0$ $p_j(a,b) = d_j(a,b) \text{ if } d_j(a,b) > 0$ Step 5: multi-criteria preference index $pi_j(a,b) = \sum_{j=1}^{n} p(a,b) W_j$ Step 6: positive and negative outranking flows ($a \neq b$) $\phi^+(a) = \frac{1}{m-1}\sum_{b=1}^{n} pi(a,b)$ Step 7: net flow $\phi = \phi^+ - \phi^-$ Step 8: Ranking the best alternative by using highest value of net flow	[31]
VIKOR: multi-criteria optimization and compromise solution which focuses on ranking and selecting from a set of alternatives in the presence of conflicting criteria.	Step 1: Determine the objective and identify the pertinent evaluation attributes. Step 2: normalized decision matrix f Step 3: Find best and worst Best: $f_j^+ = \left(f_{ij}\right)_{max} = \max_i\left(f_{ij}\right)$ Beneficial attribute $f_j^+ = \left(f_{ij}\right)_{min} = \min_i\left(f_{ij}\right)$ Non beneficial attribute Worst: $f_j^- = \left(f_{ij}\right)_{min}$ Beneficial attribute $f_j^- = \left(f_{ij}\right)_{max}$ Non beneficial attribute Step 4: utility measure S_i and regret measure R_i $S_i = \sum_{j=1}^{n} w_j\left[\frac{f_j^+ - (f_{ij})}{f_j^+ - f_j^-}\right]$ $R_i = \max_j\left\{w_j\left[\frac{f_j^+ - (f_{ij})}{f_j^+ - f_j^-}\right]\right\}$ Step 5: calculate the value of Q_i $Q_i = v\left[\frac{S_i - (S_i)_{min}}{(S_i)_{max} - (S_i)_{min}}\right] + (1-v)\left[\frac{R_i - (R_i)_{min}}{(R_i)_{max} - (R_i)_{min}}\right]$ $v = 0 \ldots 1$ generally taken as 0.5 Step 6: ranking the best alternative with lowest value of Q_i	[32]

Table 1. *Cont.*

MCDM	Steps	Reference
MOORA: Multi-Objective Optimization on the Basis of Ratio Analysis	Step 1: The alternatives and attributes values in the decision matrix: Step 2: Normalize decision matrix Step 3: positive and negative effects: maximization for beneficial criteria, minimization for non-beneficial (cost) $yb_i = \sum_{j=1}^{g} x_{ij}^{*} w_j$ $ynb_i = \sum_{j}^{n-g} x_{ij}^{*} w_j$ g is the number of criteria to be maximized $n-g$ is the number of criteria to be minimized x_{ij}^{*} is normalized decision matrix Step 4: determine the weighted assessment value $y_i = yb_i - ynb_i$ $y_i = \sum_{j-1}^{g} x_{ij}^{*} w_j - \sum_{j=1}^{n-g} x_{ij}^{*} w_j$ Step 5: ranking the best alternative Where alternative has the 1st rank with highest value of y_i	[33]
WSM: Weighted Sum Method	Step 1: make decision matrix. Step 2: normalized decision matrix Step 3: weighted normalized decision matrix r_{ij}: normalized decision matrix $v(i,j) = r(i,j).w(j)$ Step 4: weighted sum $ws(i) = \sum_{j=1}^{n} v(i,j)$ Step 5: ranking the best alternative	
WPM: Weighted Product Method	Step 1–3: same as WSM Step 4: weighted product $wp(i) = \prod_{j=1}^{n} v(i,j)$ Step 5: ranking the best alternative	
AHP	Step 1: Pair-wise comparison matrix of criteria or alternatives $P(i) = \prod_{j=1}^{n} A(i,j)$ $Pn(i) = (P(i))^{1/N}$ $Pn(i) = \left(\prod_{j=1}^{n} A(i,j)\right)^{1/N}$ $sp = \sum_{i=1}^{m} Pn(i)$ Step 2: Criteria weights or alternatives scores: $w(i) = Pn(i)/sp$ Step 3: Calculate consistency $w(i) = w(j)$, where $n = m$ size of A matrix $v(i,j) = x(i,j).w(j)$ Step 4: calculate weighted sum value: $sw(i) = \sum_{j=1}^{n} v(i,j)$ Step 5: calculate consistency error $R(i) = sw(i)/w(i)$ $\lambda \sum_{i=1}^{m} R(i)/m_{max}$ $CI = \frac{\lambda_{max} - n}{n-1}$ For $n = 4$, $Ri = 0.9$	[34]

Table 2. Mathematical formulas of different normalization methods.

Normalization	Abbreviation	Beneficial	Non-Beneficial
Linear	$L_{N\text{-}i}$	$\overline{X}_{ij} = \frac{X_{ij}}{X_j^{max}}$	$\overline{X}_{ij} = 1 - \frac{X_{ij}}{X_j^{max}}$
	$L_{N\text{-}ii}$	$\overline{X}_{ij} = \frac{X_{ij}}{X_j^{max}}$	$\overline{X}_{ij} = \frac{X_j^{min}}{X_{ij}}$
	$L_{N\text{-}max\text{-}min}$	$\overline{X}_{ij} = \frac{X_{ij} - X_j^{min}}{X_j^{max} - X_j^{min}}$	$\overline{X}_{ij} = \frac{X_j^{max} - X_{ij}}{X_j^{max} - X_j^{min}}$
	$L_{N\text{-}Sum}$	$\overline{X}_{ij} = \frac{X_{ij}}{\sum_{i=1}^{m} X_{ij}}$	$\overline{X}_{ij} = \frac{\frac{1}{X_{ij}}}{\sum_{i=1}^{m}\left(\frac{1}{X_{ij}}\right)}$
Enhanced accuracy	E_{AN}	$\overline{X}_{ij} = 1 - \frac{X_j^{max} - X_{ij}}{\sum_{i=1}^{m}\left(X_j^{max} - X_{ij}\right)}$	$\overline{X}_{ij} = 1 - \frac{X_{ij} - X_j^{min}}{\sum_{i=1}^{m}\Sigma\left(X_{ij} - X_j^{min}\right)}$
Logarithmic	L_{nN}	$\overline{X}_{ij} = \frac{lnX_{ij}}{ln(\Pi_{i=1}^{m} X_{ij})}$	$\overline{X}_{ij} = 1 - \frac{1 - \frac{lnX_{ij}}{ln(\Pi_{i=1}^{m} X_{ij})}}{m-1}$
Vector	V_N	$\overline{X}_{ij} = \frac{X_{ij}}{\sqrt{\sum_{i=1}^{m} X_{ij}^2}}$	$\overline{X}_{ij} = 1 - \frac{X_{ij}}{\sqrt{\sum_{i=1}^{m} X_{ij}^2}}$

4. Research Methodology

The collection of human perspectives is the first step for implementing any MCDM. The survey was distributed with the explanation of the purpose of the study. Participants were asked to participate in quantitative judgment using the linguistic scale, shown in Figure 2, to assess the importance of criteria and the performance of alternatives (interventions) for UHS mitigation. The experts belonged to academics in the field of urban climate.

Figure 2. Linguistic scale for rating the importance of criteria and performance of alternatives.

The collected questionnaires were checked, and assessments that contained inconsistencies were discarded and not used for further analysis. After this quality check, it was observed that 25 ratings could be useful. These judgments were aggregated using geometric means and then MCDMs, such as ELECTRE-NS, ELECTRE-NI, MOORA, PROMETHEE, TOPSIS, VIKOR, WPM, and WSM, were applied to prioritize the UHS mitigation alternatives (A1, A2 ...). These methods were implemented in two ways: stand-alone, where direct criteria weights were used, and coupled with AHP (criteria weights calculated by AHP using the judgment matrix shown in Table 3). Seven different normalization methods were used for the simple application of MCDMs (stand-alone). The research methodology is shown in Figure 3.

Table 3. Criteria weight matrix by AHP.

Criteria	Cost	Efficiency	Durability	Environment Impacts
Cost	1	2	3	2
Efficiency	1/2	1	2	1
Durability	1/3	1/2	1	1/2
Environment	1/2	1	2	1

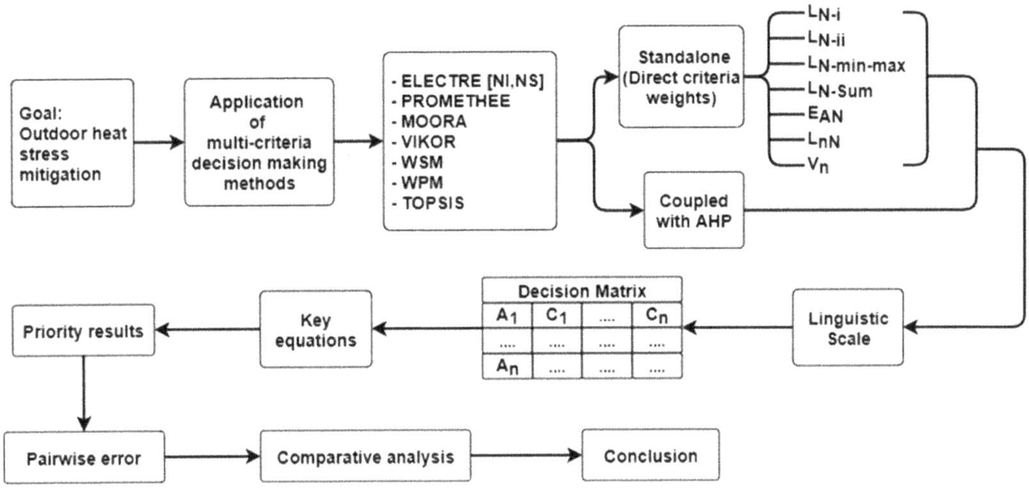

Figure 3. Research methodology.

The inputs for criteria and alternatives using the linguistic scale 1–10 was used, where high value of scale represents the high importance; for example, cost is a non-beneficial criterion, and in the case of shades the score was 8, which means it is very expensive, but the durability score was 5, which means that sometimes shades require maintenance after windstorms. All inputs were formulated in the decision matrix shown in Table 4, where criteria weights are given by direct method for the stand-alone application, and criteria weights are calculated by AHP for the combined approach by using the related formulas in Table 1.

Table 4. Decision matrix.

Direct weightage	0.45	0.15	0.20	0.20
Weightage by AHP	0.42	0.23	0.12	0.23
Interventions/Criteria	Cost	Efficiency	Durability	Environment Impacts
Water features	6	4	4	5
Surfaces	5	4	5	3
Green walls	7	6	6	7
Trees	4	7	8	8
Shades (shelter canopies)	8	4	5	2
	NB	B	B	B

5. Results

5.1. Comparative Analysis of Normalization Methods for Applied MCDM

The results show that the logarithmic normalization method had no impact on the ranking results calculated by the applied MCDMs (except for ELECTRE and VIKOR) when compared to other normalization methods. In contrast, the same ranking was observed in PROMETHEE and WSM results using all normalization techniques. Table 5 shows the consistency of similar ranking results using MCDMs under different normalization techniques.

Table 5. Comparative analysis of normalization methods.

Name	Results Consistency
$L_{N\text{-}i}$	ELE-NS, ELE-NI, PROMETHEE, WSM
$L_{N\text{-}ii}$	ELE-NS, ELE-NI, PROMETHEE, WSM
$L_{N\text{-}max\text{-}min}$	ELE-NS, PROMETHEE, WSM
E_{AN}	ELE-NS, PROM, WSM, WPM
L_{nN}	WSM, WPM, TOPSIS, PROMETHEE, MOORA
V_N	WSM, PROMETHEE, ELE-NS
$L_{N\text{-}Sum}$	WSM, PROMETHEE, ELE-NS, ELE-NI

5.2. Priority Ranking

The priority ranking for the selection of heat resilience interventions were calculated by using direct and AHP criteria weights. The ranking results obtained by stand-alone MCDM using L_{nN} and AHP-MCDM are shown in Tables 6 and 7, where alternative A1 is water feature, A2 is surfaces, A3 is green wall, A4 is trees, and A5 is shades, respectively.

Table 6. Rank calculated using MCDMs with L_{nN} method for normalization.

Methods	Alternatives/Interventions Priority Results				
	A1	A2	A3	A4	A5
1—ELE-NS	3	5	1	2	4
2—ELE-NI	4	5	1	3	2
3—MOORA	2	3	4	1	5
4—PROMETHEE	4	5	1	2	3
5—TOPSIS	3	2	4	1	5
6—VIKOR	2	3	5	1	4
7—WPM	4	5	1	2	3
8—WSM	4	5	1	2	3

Table 7. Rank calculated by AHP-MCDM.

Methods	Alternatives/Interventions Priority Results				
	A1	A2	A3	A4	A5
1—ELE-NS	4	5	1	2	3
2—ELE-NI	3	4	1	2	5
3—MOORA	3	4	2	1	5
4—PROMETHEE	3	4	2	1	5
5—TOPSIS	3	4	2	1	5
6—VIKOR	3	2	4	1	5
7—WPM	3	4	2	1	5
8—WSM	3	4	2	1	5

5.3. Ranking Frequency Error of Stand-Alone MCDMs and AHP-MCDMs

The obtained ranking results were showing frequency errors when comparing each applied method, which could mislead the decision. This problem was solved by evaluating the frequency of the same rankings. The pairwise frequency error was calculated by Equations (1) and (2). The sum of the standard deviation defines the frequency error and assist to observe the variation in decision by applying different MCDMs using the same judgments. The method aims to check the consistency of the results and to evaluate the reliability of outcomes. The ends results are shown in the pairwise matrices in Table 8.

$$E_{ij} = \sqrt{\sum_{k=1}^{5} \left(A_{kM_i} - A_{kM_j} \right)^2} \quad (1)$$

$$E_j = \sum_{i=1}^{8} E_{ij} \quad (2)$$

where: $i =$ no of rows, $j =$ no of columns, $k =$ no of alternatives, $M =$ no of methods, A = ranking result of alternative, and E_j = sum of the variation in ranking results.

Table 8. Pairwise comparison of frequency error matrix.

Ranking Frequency of Standalone MCDM								
	ELE-NS	ELE-NI	MOORA	PROMETHEE	TOPSIS	VIKOR	WPM	WSM
ELE-NS	0	2.45	4	1.41	4.47	4.69	1.41	1.41
ELE-NI	2.45	0	5.48	1.41	5.66	5.66	1.41	1.41
MOORA	4	5.48	0	4.69	1.41	1.41	4.69	4.69
PROMETHEE	1.41	1.41	4.69	0	4.90	5.10	0	0
TOPSIS	4.47	5.66	1.41	4.90	0	2	4.90	4.90
VIKOR	4.69	5.66	1.41	5.10	2	0	5.10	5.10
WPM	1.41	1.41	4.69	0	4.90	5.10	0	0
WSM	1.41	1.41	4.69	0	4.90	5.10	0	0
Sum	19	23.48	26.37	17.5	28.24	29.06	17.5	17.5
Ranking Frequency of AHP-MCDM								
	ELE-NS	ELE-NI	MOORA	PROMETHEE	TOPSIS	VIKOR	WPM	WSM
ELE-NS	0	2.45	2.83	2.83	2.83	4.90	2.83	2.83
ELE-NI	2.45	0	1.41	1.41	1.41	3.74	1.41	1.41
MOORA	2.83	1.41	0	0	0	2.83	0	0
PROMETHEE	2.83	1.41	0	0	0	2.83	0	0
TOPSIS	2.83	1.41	0	0	0	2.83	0	0
VIKOR	4.90	3.74	2.83	2.83	2.83	0	2.83	2.83
WPM	2.83	1.41	0	0	0	2.83	0	0
WSM	2.83	1.41	0	0	0	2.83	0	0
Sum	21.5	13.24	7.07	7.07	7.07	22.79	7.07	7.07

6. Discussion

The applied MCDMs were analyzed by considering three criteria in negative and positive attributes (presented in Table 9) which are:

- *Normalization*: Positive evaluation is performed for MCDMs that gives the same results under different normalization techniques, where variations in results are taken as negative.
- *MCDM Frequency*: similar ranking results obtained by stand-alone MCDMs are assessed as positive, and high variations are considered as negative.
- *AHP-MCDM Frequency*: this criterion is used to investigate the impact of coupling AHP with applied MCDMs, where positive and negative signs show the decrease and increase in frequency variation of final ranking results, respectively.

Table 9. Comparative analysis of MCDM.

Methods	Assessment		
	Normalization	Frequency Error	
		MCDM	AHP-MCDM
TOPSIS	-	-	+
MOORA	-	-	+
PROMETHEE	+	+	+
WPM	-	+	+
WSM	+	+	+
VIKOR	-	-	+
ELE-NS	+	+	-
ELE-NI	-	-	+

The frequency error in the ranking results is illustrated in Figure 4, which shows that AHP helps to reduce the irregularity of the final ranking due to the pairwise subjective judgment for calculating criteria weights, which makes more reliable results in decision-making. It is compatible to combine with all the methods except ELECTRE-NS. The increased frequency error in ranking was noticed when the ELECTRE-NS model was coupled with the AHP. It was observed that WSM and PROMETHEE were not affected using any normalization techniques and gave the same ranking, which proves that they are the most reliable methods. Moreover, TOPSIS, MOORA and VIKOR models provided improved results after coupling, but the different normalization techniques could affect the final outcome.

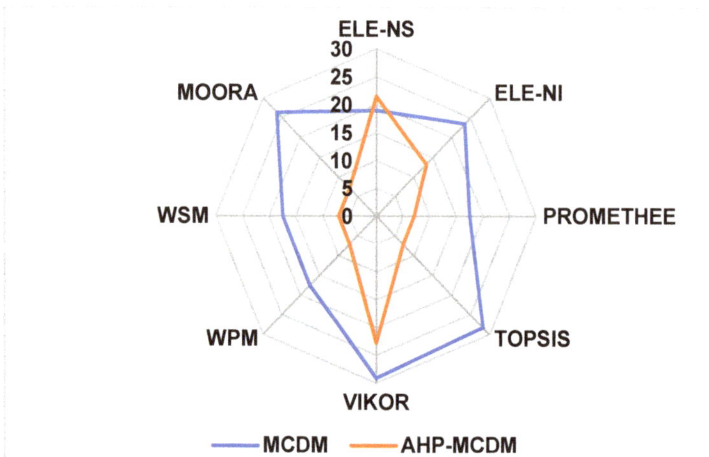

Figure 4. Graph of calculated error in pairs.

Based on provided judgments (Table 4), the priority ranking obtained from the majority of the MCDMs showed that planting trees in the urban area was an effective cooling strategy that provides shade, improves air quality, and gives good cooling in certain areas where green walls improve indoor and outdoor air temperature. Additionally, the green walls enhance the aesthetics of the property. The watering method to wet the streets in the summer, planting grass on the surfaces, and water features such as fountains provides the limited cooling extent and requires extra care. While artificial shadings are expensive to install in hotspots, there was no co-benefit associated with this intervention.

7. Conclusions

This study was performed for the selection of intervention to mitigate outdoor UHS by applying multiple MCDMs. Eight different well known and classical techniques were computed to evaluate the priority ranking of interventions. A major concern with decision-making is that different MCDM methods provide different results for the same problem. For reliability of the outcomes, a comparative study was conducted on the basis of three criteria, which were (i) influence of normalization techniques, frequency of similar ranking results by (ii) stand-alone MCDMs, and (iii) AHP-MCDM application. It was observed that PROMETHEE and WSM were reliable methods in this field among other applied MCDMs, namely, MOORA, WPM, ELECTRE-NS, ELECTRE-NI, TOPSIS, VIKOR, which are sensitive methods and, due to variations, these MCDM models provided different priority results. Additionally, the L_{nN} was a more reasonable normalization technique, and it provided similar rankings in the majority of applied MCDMs. It was noticed that the coupling of AHP helped to minimize the frequency error through the pairwise method for criteria weights, which increased the reliability of the decision.

In this study, the priority of green walls and trees is an arbitrary output of decision makers. The ranking obtained on the parameters was not a general rule, and this procedure was carried out to check the reliability. The results were entirely based on the terrain, the perspectives, characteristics of the pilots, climatic conditions, and inputs of the decision-makers.

The improved frequency of consistent results by AHP-MCDM revealed that the ranking results mainly depended on the nature and the values of the criteria. The reasonable disagreement that was observed among the methods did not affect their reliability. As a result, MCDM models proved generally very effective for dealing with UHS problems before their implementation and selection of the best ones.

However, a possible limitation of this work is that this comparative conclusion is based only on the evaluation of ranking errors. Future works are to extend the experiment with more MCDM models and to perform sensitivity analysis to confirm that the results would not change.

Author Contributions: The paper was a collaborative effort between the authors. A.M.Q. and A.R. contributed collectively to developing the methodology of this study, analysis, development of HVI and the manuscript preparation. All authors have read and agreed to the published version of the manuscript.

Funding: This paper has been produced within the COOL-TOWNS (Spatial Adaptation for Heat Resilience in Small and Medium Sized Cities in the 2 Seas Region) project which receives funding from the Interreg 2 Seas programme 2014–2020 co-funded by the European Regional Development Fund under subsidy contract N° 2S05-040.

Institutional Review Board Statement: Not applicable.

Informed Consent Statement: Not applicable.

Conflicts of Interest: The authors declare no conflict of interest.

Abbreviations

AHP	Analytic hierarchy process
EFDM	Enhanced fuzzy Delphi method
FDEMATEL	Fuzzy decision-making trial and evaluation laboratory
SMCE	Spatial Multi-Criteria Evaluation
NI	Net inferior
NS	Net superior
SWOT	Strength weakness opportunities and threat
DEA	Data envelopment analysis
PROMETHEE	Preference Ranking Organization Method for Enrichment Evaluation

VIKOR	Viekriterijumsko Kompromisno Rangiranje
MOORA	Multi-Objective Optimization Ratio Analysis
MCDA	Multi Criteria Decision Analysis

References

1. Ansari, A.Q. The basics of fuzzy logic: A tutorial review. *Comput. Educ. Staff. Comput. Educ. Group* **1998**, *88*, 5–8.
2. McBratney, A.; Odeh, I.O. Application of fuzzy sets in soil science: Fuzzy logic, fuzzy measurements and fuzzy decisions. *Geoderma* **1997**, *77*, 85–113. [CrossRef]
3. Ordoobadi, S.M. Development of a supplier selection model using fuzzy logic. *Supply Chain Manag. Int. J.* **2009**, *14*, 314–327. [CrossRef]
4. De Korvin, A.; Strawser, J.; Siegel, P.H. An Application of Control System to Cost Variance Analysis. *Manag. Finance* **1995**, *21*, 17–35. [CrossRef]
5. Tanaka, H.; Okuda, T.; Asai, K. A formulation of fuzzy decision problems and its application to an investment problem. *Kybernetes* **1976**, *5*, 25–30. [CrossRef]
6. Gunasekaran, N.; Rathesh, S.; Arunachalam, S.; Koh, S. Optimizing supply chain management using fuzzy approach. *J. Manuf. Technol. Manag.* **2006**, *17*, 737–749. [CrossRef]
7. Chan, D.C.; Yung, K.L.; Lp, A.W.H. An application of fuzzy sets to process performance evaluation. *Integr. Manuf. Syst.* **2002**, *13*, 237–283. [CrossRef]
8. Khaira, A.; Dwivedi, R.K. A State-of-the-Art Review of Analytical Hierarchy Process. *Mater. Today Proc.* **2018**, *5*, 4029–4035. [CrossRef]
9. Kusumawardani, R.P.; Agintiara, M. Application of Fuzzy AHP-TOPSIS Method for Decision Making in Human Resource Manager Selection Process. *Procedia Comput. Sci.* **2015**, *72*, 638–646. [CrossRef]
10. Jurík, L.; Horňáková, N.; Šantavá, E.; Cagáňová, D.; Sablik, J. Application of AHP method for project selection in the context of sustainable development. *Wirel. Netw.* **2020**, *28*, 893–902. [CrossRef]
11. Calabrese, A.; Costa, R.; Levialdi, N.; Menichini, T. Integrating sustainability into strategic decision-making: A fuzzy AHP method for the selection of relevant sustainability issues. *Technol. Forecast. Soc. Chang.* **2018**, *139*, 155–168. [CrossRef]
12. Stojčić, M.; Zavadskas, E.K.; Pamučar, D.; Stević, Ž.; Mardani, A. Application of MCDM Methods in Sustainability Engineering: A Literature Review 2008–2018. *Symmetry* **2019**, *11*, 350. [CrossRef]
13. Waris, M.; Panigrahi, S.; Mengal, A.; Soomro, M.I.; Mirjat, N.H.; Ullah, M.; Azlan, Z.S.; Khan, A. An Application of Analytic Hierarchy Process (AHP) for Sustainable Procurement of Construction Equipment: Multicriteria-Based Decision Framework for Malaysia. *Math. Probl. Eng.* **2019**, *2019*, 6391431. [CrossRef]
14. Moslem, S.; Çelikbilek, Y. An integrated grey AHP-MOORA model for ameliorating public transport service quality. *Eur. Transp. Res. Rev.* **2020**, *12*, 68. [CrossRef]
15. Mojaheed, M.; Marjani, M.E.; Afshari, A.R.; Marjani, S. Using ELECTRE-AHP as a mixed method for personnel selection. In Proceedings of the Traineeational Symposium on the Analytic Hierarchy Process, Kuala Lumpur, Malaysia, 23–26 June 2013. [CrossRef]
16. Maheshwarkar, M.; Sohani, N. Combined AHP-TOPSIS based approach for the evaluation of knowledge sharing capabilities of supply chain partners. *Manag. Sci. Eng.* **2013**, *7*, 27–32.
17. Dinmohammadi, A.; Shafiee, M. Determination of the Most Suitable Technology Transfer Strategy for Wind Turbines Using an Integrated AHP-TOPSIS Decision Model. *Energies* **2017**, *10*, 642. [CrossRef]
18. Turcksin, L.; Bernardini, A.; Macharis, C. A combined AHP-PROMETHEE approach for selecting the most appropriate policy scenario to stimulate a clean vehicle fleet. *Procedia Soc. Behav. Sci.* **2011**, *20*, 954–965. [CrossRef]
19. Waas, D.V.; Suprapto, S. Combination of AHP method and vikor method for assesing sunday school teacher. *IJCCS Indones. J. Comput. Cybern. Syst.* **2020**, *14*, 45–56. [CrossRef]
20. Combined AHP-WSM Based Approach for the Evaluation of Knowledge Sharing Capabilities of Supply Chain Patterns. Available online: https://www.researchgate.net/publication/234032617_Combined_AHP-WSM_based_approach_for_the_evaluation_of_knowledge_sharing_capabilities_of_supply_chain_patterns (accessed on 7 October 2022).
21. Zaidan, A.; Zaidan, B.; Hussain, M.; Haiqi, A.; Kiah, M.M.; Abdulnabi, M. Multi-criteria analysis for OS-EMR software selection problem: A comparative study. *Decis. Support Syst.* **2015**, *78*, 15–27. [CrossRef]
22. Ahmad, N.; Berg, D.; Simons, G.R. The integration of analytical hierarchy process and data envelopment analysis in a multi-criteria decision-making problem. *Int. J. Inf. Technol. Decis. Mak.* **2006**, *5*, 263–276. [CrossRef]
23. Asadabadi, M.R.; Chang, E.; Saberi, M. Are MCDM methods useful? A critical review of Analytic Hierarchy Process (AHP) and Analytic Network Process (ANP). *Cogent Eng.* **2019**, *6*, 1623153. [CrossRef]
24. Qureshi, A.M.; Rachid, A. Review and Comparative Study of Decision Support Tools for the Mitigation of Urban Heat Stress. *Climate* **2021**, *9*, 102. [CrossRef]
25. Qureshi, A.M.; Rachid, A. An Analytic Hierarchy Process for urban heat stress mitigation. In Proceedings of the 2022 2nd International Conference on Digital Futures and Transformative Technologies (ICoDT2), Rawalpindi, Pakistan, 24–26 May 2022; pp. 1–6. [CrossRef]

26. Using Green Roofs to Reduce Heat Islands. Available online: https://www.epa.gov/heat-islands/using-green-roofs-reduce-heat-islands (accessed on 4 April 2022).
27. Using Trees and Vegetation to Reduce Heat Islands. Available online: https://www.epa.gov/heat-islands/using-trees-and-vegetation-reduce-heat-islands (accessed on 4 April 2022).
28. Hendel, M. Pavement-Watering for Cooling the Built Environment: A Review. 2016. Hal-01426167. Available online: https://hal.archives-ouvertes.fr/hal-01426167/ (accessed on 2 July 2022).
29. Uzun, B.; Taiwo, M.; Syidanova, A.; Ozsahin, D.U. The Technique for Order of Preference by Similarity to Ideal Solution (TOPSIS). In *Application of Multi-Criteria Decision Analysis in Environmental and Civil Engineering. Professional Practice in Earth Sciences*; Uzun Ozsahin, D., Gökçekuş, H., Uzun, B., LaMoreaux, J., Eds.; Springer: Cham, Switzerland, 2021; pp. 25–30. [CrossRef]
30. Roy, B.; Skalka, J. *ELECTRE IS: Aspects Méthodologiques et Guide D'utilisation*; Université de Paris Dauphine: Paris, France, 1987. Available online: https://books.google.com.hk/books?id=iAPxHAAACAAJ (accessed on 2 July 2022).
31. Brans, J.P.; Vincke, P.; Mareschal, B. How to select and how to rank projects: The Promethee method. *Eur. J. Oper. Res.* **1986**, *24*, 228–238. [CrossRef]
32. Yazdani, M.; Graeml, F.R. VIKOR and its applications: A state-of-the-art survey. *Int. J. Strateg. Decis. Sci. IJSDS* **2014**, *5*, 56–83. [CrossRef]
33. Chakraborty, S. Applications of the MOORA method for decision making in manufacturing environment. *Int. J. Adv. Manuf. Technol.* **2010**, *54*, 1155–1166. [CrossRef]
34. Chourabi, Z.; Khedher, F.; Babay, A.; Cheikhrouhou, M. Multi-criteria decision making in workforce choice using AHP, WSM and WPM. *J. Text. Inst.* **2018**, *110*, 1092–1101. [CrossRef]

applied sciences

MDPI

Article

Synthesis of the Energy-Saving Dry Dual Clutch Control Mechanism

Nikolay Sergienko [1], Pavel Kalinin [2], Ivan Pavlenko [3], Marek Ochowiak [4,*], Vitalii Ivanov [5,*], Anton Sergienko [6], Natalia Pavlova [1], Yevheniia Basova [1], Oksana Titarenko [2], Aleksandr Nazarov [7], Andżelika Krupińska [4], Magdalena Matuszak [4] and Sylwia Włodarczak [4]

1 Department of Car and Tractor Industry, National Technical University "Kharkiv Polytechnic Institute", 2, Kyrpychova St., 61002 Kharkiv, Ukraine
2 National Academy of the National Guard of Ukraine, 3, Maidan Zakhysnykiv Ukrainy, 61001 Kharkiv, Ukraine
3 Department of Computational Mechanics Named after Volodymyr Martsynkovskyy, Faculty of Technical Systems and Energy Efficient Technologies, Sumy State University, 2, Rymskogo-Korsakova St., 40007 Sumy, Ukraine
4 Department of Chemical Engineering and Equipment, Poznan University of Technology, 5 M. Sklodowskej-Curie Sq., 60-965 Poznan, Poland
5 Department of Manufacturing Engineering, Machines and Tools, Faculty of Technical Systems and Energy Efficient Technologies, Sumy State University, 2, Rymskogo-Korsakova St., 40007 Sumy, Ukraine
6 "Intechresurs" Ltd., 199, Kharkiv Heroes Ave., 61037 Kharkiv, Ukraine
7 Transport Technologies Department, National Automobile and Highway University, 25, Yaroslava Mudroho St., 61002 Kharkiv, Ukraine
* Correspondence: marek.ochowiak@put.poznan.pl (M.O.); ivanov@tmvi.sumdu.edu.ua (V.I.)

Abstract: A promising direction in developing up-to-date transport vehicles is the use of transmissions, an essential element of which is a dual-clutch. Improving functional performance, energy efficiency, and environmental friendliness are relevant and require appropriate research. The object of study is a dry dual-clutch working with a manual transmission with high energy efficiency. In the proposed design scheme, a rotary lever and a movable carriage can change the structural diagram of the force interaction between the pressure spring and the clutch discs. The developed mathematical model of the clutches control mechanism allowed for analyzing the process of controlling the dual-clutch transmissions (DCT) and obtaining functional dependencies between pushing forces in friction pairs and the carriage position. As a result, a method for synthesizing DCT was proposed to ensure the transmission of a given torque. Furthermore, it was determined and numerically proven that the same radial movement of the carriage when switching-on the clutches does not guarantee equal loading for the frictional discs of each clutch. This fact increases the uneven dynamics of torque and disc wear. Overall, the synthesis of the energy-saving dry dual-clutch control mechanism providing the same clutch margin for each clutch was developed. The method can be generalized and applied to designing dry dual-clutch transmissions for machines of various purposes.

Keywords: robotic transmission; transport vehicle; dual-clutch transmission; torque; clutch margin; process innovation; energy efficiency

Citation: Sergienko, N.; Kalinin, P.; Pavlenko, I.; Ochowiak, M.; Ivanov, V.; Sergienko, A.; Pavlova, N.; Basova, Y.; Titarenko, O.; Nazarov, A.; et al. Synthesis of the Energy-Saving Dry Dual Clutch Control Mechanism. *Appl. Sci.* **2023**, *13*, 829. https://doi.org/10.3390/app13020829

Academic Editor: Maria Vicidomini

Received: 9 December 2022
Revised: 28 December 2022
Accepted: 5 January 2023
Published: 6 January 2023

1. Introduction

One of the essential elements of the transmission is a clutch. With the development of transmission designs, it becomes necessary to develop new clutches and control systems. They should increase the vehicle and the transmission elements resource, positively influence dynamics and performance, and control and operation energy efficiency.

DCT has advantages in manufacturing, operating, and maintenance costs. They work without interruption in power flow or can switch clutches rapidly. Unlike traditional design schemes, existing automatic transmissions use closed clutches. They require permanent energy consumption when changing switches and maintaining the frictional pairs of one of the clutches in the closed position.

In DCT, opening one clutch and closing the second one occurred in a reduced or overlapped time. Thus, gear shifting is ensured with no interruption in power flow. This significantly reduces gearshift time, improves acceleration dynamics, and reduces fuel consumption and emissions to the environment.

One of the ways to improve such clutches is to reduce the energy consumption for control while ensuring reliable transmission of the required torque and safety indicators. Studies of the energy-saving dry clutch control mechanism allow recommending rational choice and a unification of the designs while ensuring the proper functioning of the drive and control system.

Recently, significant progress has been reached in the development of transmissions. Particularly, the number of speeds has been increased, the range of gear ratios has been extended, and the gearshift rate has been raised [1].

Electric drives are permanently expanding in response to tightening emission control regulations and urgent fuel economy requirements in automotive transmissions. As a result, integrating electric drives [2] into conventional transmissions occurs. Such an approach simplifies modular designs and the production of transmissions. However, it does not solve the problems of developing and improving the mechanical parts of transmissions.

Improving transmissions involves researching dry clutches related to vehicle dynamics [3–9], reliability [10], clutch architecture [5], durability, noise, switching strategy, and control technology, as well as energy efficiency and comfort [11]. The importance of dry clutch control was previously discussed in refs. [12–14].

Based on the models of workflows in dry clutches, several control algorithms have been proposed: optimal control [15], branched control [16], adaptive control [17], control based on fuzzy logic [8], robust control [18], combined management strategy [19], and predictive control [5]. Notably, effective clutch control requires developing a mathematical model for torque transmission with a minimum deviation from the actual process [20–23]. In ref. [4], a detailed analysis of the dry clutch architecture was carried out, allowing one to understand the main phenomena affecting torque transmission by friction forces. However, the literature does not provide enough information on the justification and analysis of DCT's design schemes with reduced energy costs for control, functional features, and rational selection of types and parameters for drives and actuators.

The clutch design is mainly based on the types of transmission and gearbox [24]. Nowadays, a number of vehicles are equipped with preselective gearboxes. Based on the positive effects of their application, a rational choice of a gearbox is realized [25]. Moreover, preselective transmissions are structurally simpler, have higher energy efficiency, and do not need to be repaired by highly qualified specialists. Using a dry dual-clutch eliminates the need to use expensive components and materials from the hydraulic system. During the operation of DCT, the switched-off multi-plate wet clutch creates additional resistance due to the difference in the angular velocities of the driving and driven discs.

Ref. [26] is devoted to shifting gears with friction clutches in the transmissions of wheeled and tracked vehicles. It analyzes the gear-shifting process with and without interruption of the power flow. However, more attention should be paid to investigating the properties of dry DCTs with reduced energy consumption for control.

The dynamic behavior of automotive dry clutches depends on the friction characteristics of the contact between the material of the friction lining, the flywheel, and the pressure disc when the clutch is engaged [27,28]. When switched on, the friction causes contact heat due to high interphase sliding and relatively high contact pressures. This affects the behavior of the material and the related frictional characteristics [29].

One of the critical problems in designing DCT is clutch durability. When the temperature is exceeded in dry clutches, the friction lining begins to suffer irreversible damage. For a dry DCT, a thermal model should be developed to predict parameter changes. In this regard, it is essential to compare experimental and theoretical results in the contact pressure and wear distribution [30]. However, only a single-clutch transmission was considered.

Overall, parameters and the design scheme of the mechanism that ensure the compression of the friction pairs and the control system predominantly affect the pressure force and the switching-on/off time for each clutch.

Due to the mentioned above, the research aims to analyze the operational process and evaluate the impact of the parameters of the dry DCT's control mechanism with high energy efficiency on changing the force. Simultaneously, the synthesizing parameters of the control mechanism should ensure uniform clutch margins for each clutch.

2. Materials and Methods

2.1. The Design Model

The energy efficiency of drive control for DCTs of different designs was analyzed. A comparison with the original DCT with reduced control costs was also carried out. As a result, it was determined that the proposed transmission [31] requires energy only when starting the vehicle and shifting gears (Figures 1 and 2).

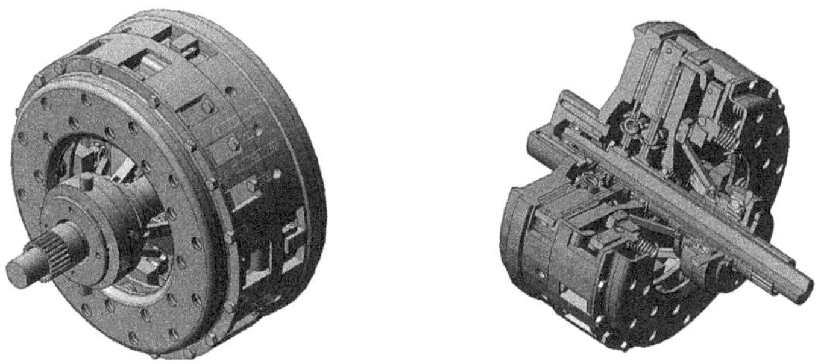

(**a**) (**b**)

Figure 1. 3D model of the DCT: (**a**)—general view; (**b**)—cross-sectional view.

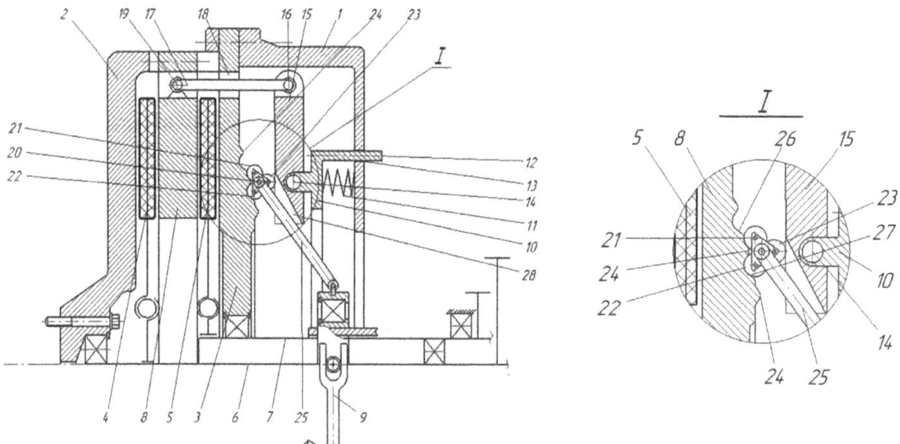

Figure 2. The original DCT design: 1—casing; 2—flywheel; 3—platter; 4,5—driven discs; 6, 7—shafts; 8—pressure plate; 9—drive mechanism; 10—additional disc; 11—pressure springs; 12—elements; 13, 18—holes; 14, 16, 19—hinges; 15—rotary stops; 17, 25—levers; 20—movable supports; 21–23—rollers; 24—clamps; 26, 27—surfaces; 28—inclined grooves.

DCT contains casing 1 (mounted on the engine's crankshaft through the leading flywheel 2), platter 3, driven discs 4 and 5 (mounted on splines on coaxially located main shafts 6 and 7), and shift drive mechanism 9. Flywheel 2 (used as drive disc), platter 3, and casing 1 are rigidly interconnected in the axial direction along the periphery, forming the driving elements of the clutches.

Between platter 3 and casing 1, additional disc 10 is installed in the axial direction. It is loaded with pressure springs 11 and moved through elements 12 with the possibility of interaction with holes 13. On an additional drive 10, from the platter side in the radial direction, stops 15 are fixed with hinges 14. They can rotate in radial planes. The ends of swivel stops 15 are connected with levers 17 from the periphery of the discs (in a horizontal plane) with hinges 16. The last ones (passed through holes 18 in platter 3 through hinges 19) are fixed on pressure plate 8. This plate has a possibility of forced axial movement.

The second free end of the rotary stops 15 interacts with movable supports 20. These supports are made as carriages with rollers 21, 22, and 23, placed between stops 15 and platter 3. The pair of rollers 21 and 22 on carriages can interact with clamps 24. Movable supports 20 are pivotally connected to drive mechanism 9 for switching clutches through levers 25.

The designed DCT operates as follows. In the switched-off (neutral) position, tdrive mechanism 9 is installed in such a way that the supports 20 are fixed in the middle position, excluding the rotation of stops 15 in the radial planes. Driven discs 4 and 5 do not interact with the leading flywheel 2, pressure plate 8, and platter 3. Main shafts 5 and 6 of even and odd gears rotate freely through driven discs 3 and 4 without transmitting torque from the engine.

When switching on the 1st clutch (Figures 2 and 3), drive mechanism 9 is installed in such a way that the supports 20 move from the periphery of the discs to the rotation axis of the shafts through levers 25. In this case, support 20, overcoming the resistance of the clamps 24, moves along surface 27. It is installed in a fixed position, where torque occurs between hinges and rollers 23.

Additional disc 10 moves in the axial direction towards support disc 8 under the action of springs 11. Stops 15 transmit the pushing force to levers 17, turning on hinges 14 through hinges 16. Levers 7 transmit the force to pressure plate 8 through hinges 19. Pressure plate 8 acts on the driven disc 4 and flywheel 2. The friction pairs circuit for the 1st clutch closes. As a result, torque from the engine is transmitted to the input shaft 6 of the gearbox through drive flywheel 2, pressure plate 8, and drive plate 4.

When the 1st clutch is switched off, drive mechanism 9 is set to a position in which stops 20 moves to the periphery of the discs (Figures 2b and 3b). In this case, rollers 21 and 22 move along surface 27, overcoming the resistance of clamps 24. These rollers are installed in a fixed position. However, rollers 23 are mounted on the same horizontal axis as hinges 14 of swivel stops 15 and pressure springs 11. Additional disc 10 moves in the axial direction away from support disc 3, overcoming the resistance of springs 11. Stops 15 transmit the pulling force to levers 17, turning on hinges 14 through hinges 16. Levers 17 transmit the force to pressure plate 8 through hinges 19. Pressure disc 8 opens the power circuit, stopping interaction with driven disc 3. As a result, the torque transmission from the engine to shaft 6 of the even rows of the gearbox stops.

When the 2nd clutch is switched on (Figures 2c and 3c), mechanism 9 is located in such a way that stops 20 moves to the periphery through levers 25. In the rotary stops 15, inclined grooves 28 are made to ensure free movement of the levers 25. As a result, the contact of rollers 23 is ensured along one initial plane of stops 15. In this case, support 20 moves along surface 26, overcoming the resistance of clamps 24. Torque occurs between the axes of hinges 14 and rollers 23. Additional disc 10 moves in the axial direction towards support disc 3 under the action of compression springs 11. Stops 15 transmit the pulling force to levers 17, turning on hinges 14 through hinges 16. Levers 17 transmit the force to pressure plate 8 through hinges 19. Pressure disc 8 closes the power circuit, acting on driven disc 5 and platter 3. As a result, torque from the engine is transmitted to shaft 7 through pressure plate 8, driven disc 5, and support disc 3.

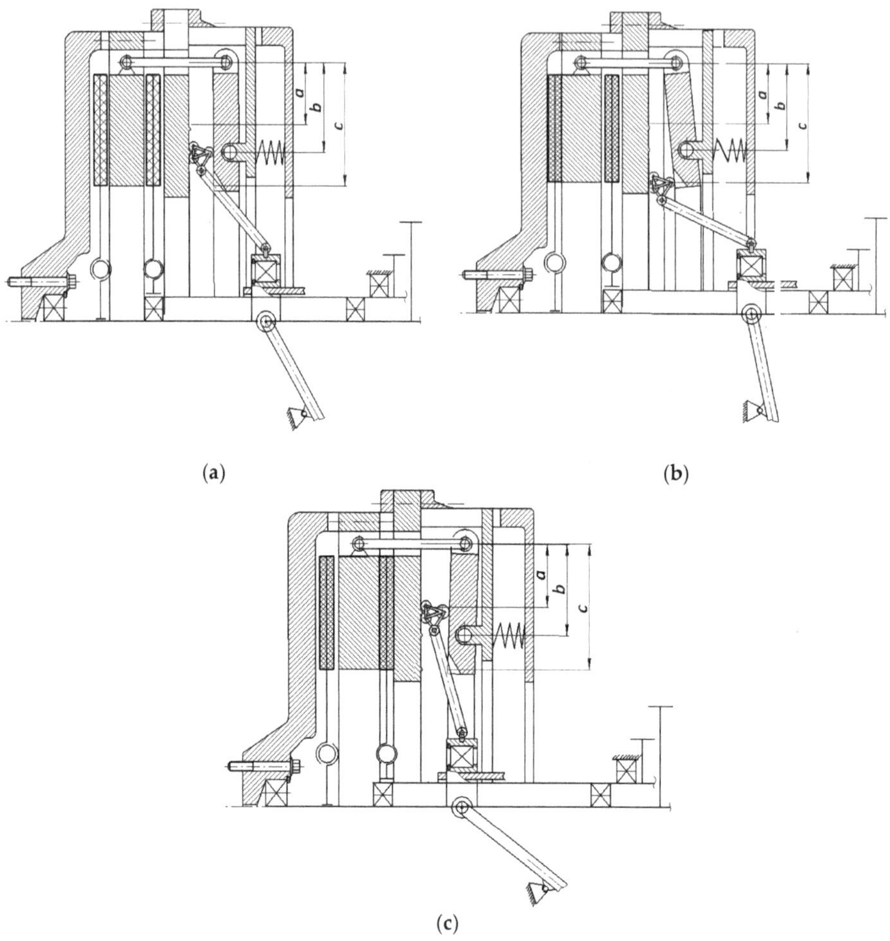

Figure 3. Positions of the movable carriage of the control mechanism: (**a**)—neutral position; (**b**)—1st clutch; (**c**)—2nd clutch; *a*—length determining the position of the carriage under switched-on 2nd clutch, *b*—length determining the position of the carriage under switched-off 1st and 2nd clutches, *c*—length determining the position of the carriage under switched-on 1st clutch.

When the 2nd clutch is switched off, mechanism 9 is set to a position in which stops 20 move from the periphery of the discs to the center through levers 25 (Figure 3a). Rollers 21 and 22 move along inclined conical surface 26, overcoming the resistance of clamps 27. These rollers are installed in a fixed position. Rollers 23 are mounted on the same axis of horizontal hinge 14. The additional disc 10 is pushed into holes 13 of casing 1 through protrusions 12, overcoming the resistance of compression springs 11. They also move axially away from platter 3. In this case, stops 15 transmit the pushing force to levers 17, turning on hinges 14 through hinges 16. Levers 17 transmit the force to pressure plate 8 through hinges 19. Pressure disc 8 stops interacting with driven disc 5 and opens the circuit. As a result, the torque transmission from the engine to shaft 7 stops.

2.2. The Mathematical Model

A design model of the clutch shift mechanism is proposed to analyze the effect of DCT's carriage movement on clutch margins (Figure 4).

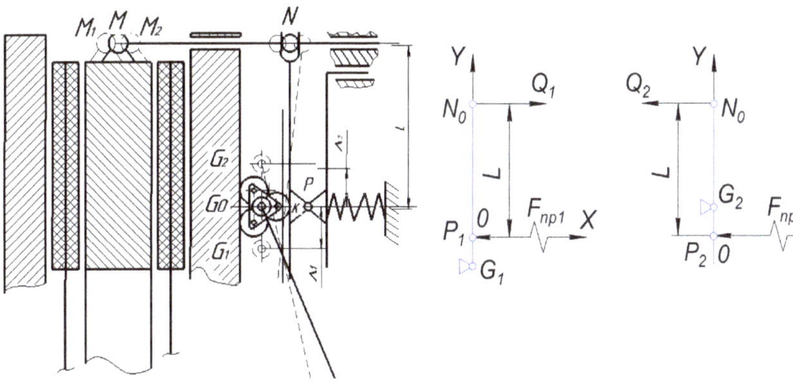

(a) (b) (c)

Figure 4. The original DCT: (**a**)—the design scheme; (**b**)—1st clutch; (**c**)—2nd clutch.

The maximum engine torque, the clutch margin, and the number of friction pairs determine the required compression force Q for the clutch friction pairs. The clutch margin for the light vehicle is β = 1.2, 1.7–2.0 for all-terrain vehicles and 3.0 for heavy machinery. The calculation conditionally considers a single rotary lever and pressure spring. However, the number of levers can reach 3–4, and the number of springs can reach 18–42, depending on the transmitted engine torque.

The total pressing force F_{np} is summarized by all the pre-compressed springs. It is conventionally attached to the clutch swing lever NP, which is connected to the clutch pressure plate via trailing lever NM at point M. When moving the carriage from position G_0 to positions G_1 and G_2, the pressure plate moves from position M to positions M_1 and M_2 (Figure 3a). The action line of the total force Q coincides with the lever NM. It is also located at a distance L from the pressure spring line of action.

The coordinate system XOY is used to study the clutch operation. The axis OX is directed along the compression spring action line. The axis OY is directed along the swivel disc in its neutral position ON_0. The full switch-on of the clutch is considered. Vertical lines M_1 and M_2 determine the extreme positions of the friction pressure plates for the switched-on 1st and 2nd clutch, respectively, with coordinates x_{M1} and x_{M2}.

A pre-compressed pressure spring is used to provide the required force Q. The proposed design scheme allows determining dependencies between the force Q_j (j = 1, 2) on the pressure plate and force F_{npj} of the clutch pressure spring when switching on the 1st (Figure 3b) and the 2nd (Figure 3c) clutch, so during vertical movement of the carriage from the neutral position G_0 to working positions G_j by the distance of Δ_{Kj} (Figure 3a).

The equilibrium condition of the j-th lever relative to the reference point G_j is as follows:

$$\Sigma M_{G_j}(F) = 0, \tag{1}$$

or for the cases of switching on the 1st and the 2nd clutches, respectively:

$$F_{np1}\Delta_{K1} = Q_1(L + \Delta_{K1});\; F_{np2}\Delta_{K2} = Q_2(L - \Delta_{K2}). \tag{2}$$

from which the following expressions can be obtained:

$$Q_1 = \frac{F_{np1}\Delta_{K1}}{L+\Delta_{K1}} = \frac{F_{np1}}{\gamma_1+1};\ Q_2 = \frac{F_{np2}\Delta_{K2}}{L-\Delta_{K2}} = \frac{F_{np2}}{\gamma_2-1};\ (\gamma_2 > 1), \tag{3}$$

where the following parameter is introduced:

$$\gamma_j = \frac{L}{\Delta_{Kj}}. \tag{4}$$

Formula (3) shows that the inequality $F_{np1} > F_{np2}$ ensures equal forces on the pressure disc ($Q_1 = Q_2 = Q$) in the case of similar carriage movements. To fulfill this condition, the design schemes in Figure 5 are considered.

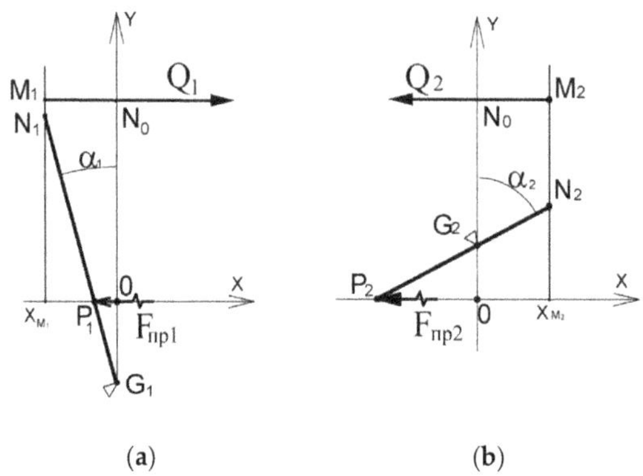

Figure 5. The design schemes of the switched-on DCT: (**a**)—1st clutch; (**b**)—2nd clutch.

When the support carriage G is moved to position G_j, the support lever rotates under the force F_{npj} from the pressure spring at point P_j. This fact provides the required compression force Q_j of the clutch friction pairs. For this case, pressure spring forces on the platter surface will be as follows:

$$F_{npj} = {}_{np}(\Delta_0 - \Delta_j), \tag{5}$$

where c_{np}—compression spring stiffness [kN/mm]; Δ_0—spring pre-deformation [mm] for the neutral position of the carriage (when clutches are switched off); Δ_j—additional deformation of the pressure spring when turned on the j-th clutch (from Figure 4, $\Delta_1 = OP_1$, $\Delta_2 = OP_2$). Particularly, after switching on the 1st clutch (Figure 4), the rotary lever reaches the position N_jG_j, where $x_{N1} = -z_1$. Gap neglecting z_{c1} occurs for the freewheel, and elastic deformation of friction linings equals z_{y1}. Therefore, $z_1 = z_{c1} + z_{y1}$.

Assuming that the lever N_jG_j does not deform, the following geometric expression can be written:

$$\Delta_j = \Delta_{Kj} tg\alpha_j, \tag{6}$$

where α_j—j-th lever angle [deg], determined from the equation:

$$L\ sin\ \alpha_j = z_j^{-}1j + 1\Delta_{Kj} tg\alpha_j. \tag{7}$$

Thus, Equation (5) can be rewritten as follows:

$$F_{npj} = {}_{np}\left[\Delta_0 - \Delta_{Kj} tg\alpha_j\right]. \tag{8}$$

The dependencies (1)–(8) allow studying the switching-on clutches in the DCT.

3. Results

Figure 6 shows the effect of carriage displacement Δ_K on the force F_{npk} in the pressure spring and the force Q_k on the pressure disc when switching on the k-th clutch.

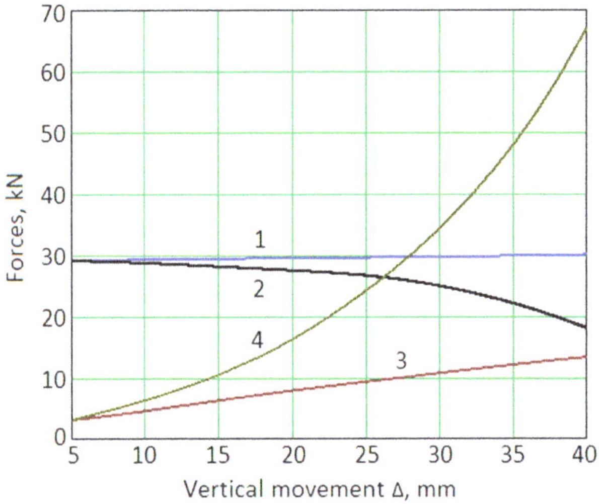

Figure 6. Dependencies of forces Q_k and F_{npk} [kN] on the pressure disc (1,2) and spring (3,4) on the vertical movement Δ [mm] of the carriage: 1—F_{np1}; 2—Q_1; 3—F_{np2}; 4—Q_2.

A dry clutch is considered a test case for simulating the DCT workflow. This clutch is widely used in the automotive and agricultural industries. It also has the following parameters: $L = 0.053$ m, $z_1 = z_2 = z = z_c + z_y$, $z_c = 2.0$ mm, $z_y = 0.8$ mm, $c = 950$ kN/m, $\Delta_0 = 30$ mm.

The analysis results presented in Figure 6 shows that the force on the pressure plate will be different for a uniform vertical movement of the carriage Δ_K (when switching on the 1st and the 2nd clutches). Moreover, $Q_2 > Q_1$ and $F_{np2} > F_{np1}$. In this case, F_{np1} changes insufficiently.

An increase in Δ_K increases the difference $Q_{21} = Q_2$—Q_1 between forces on the pressure disc and decreases the clutch margin. Notably, such a difference in the pressure disc also increases but is lower. Hence, it is not possible to provide the same clutch margin. Moreover, equality $Q_1 = Q_2$ cannot be provided by choosing different carriage displacements ($\Delta_{K1} \neq \Delta_{K2}$).

Therefore, to equalize the clutch margin, it is proposed to abandon the strictly vertical movement of the carriage. It is also suggested to constructively realize an additional horizontal movement of the carriage. In this case, to equalize forces Q_{K1} and Q_{K2}, the force Q_{K1} should increase, and the force Q_{K2} should be decreased. The corresponding change in the position of points P_1 and P_2 of the carriage contact with the pressure spring can achieve the last condition. In other words, the platter profile along which the carriage moves under switching on/off the DCT should be changed.

In the proposed improved DCT design [32], rollers 21 and 22 move on the inclined conical surfaces 26 and 27 of the support disc. The corresponding design schemes are presented in Figure 7.

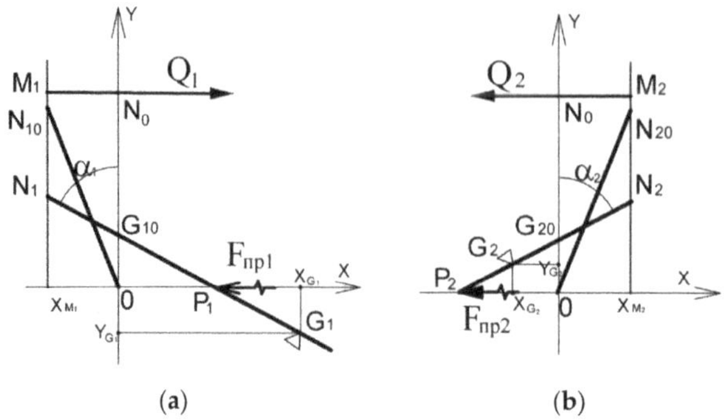

Figure 7. Design schemes of the improved DCT: (**a**)—1st clutch; (**b**)—2nd clutch.

It is convenient to characterize the clutch operation when the k-th clutch is switched on by changing the position of the rotary lever N_kP_k with the inclination angle α_k. The extreme positions of the friction pressure plates are conditionally marked by vertical lines M_k. Point N_k of the rotary lever has the coordinate x_{Mk} = const.

The following dependencies determine the coordinates of characteristic points N_k and P_k for the swivel disc:

$$\begin{array}{lll} N_k: & x_{N_k} = (-1)^k z; & y_{N_k} = L\cos\alpha_k; \\ P_k: & x_{P_k} = -(-1)^j (L\sin\alpha_k - z_k); & y_{P_k} = 0, \end{array} \tag{9}$$

where $z_k = |x_{Mk}|$.

When the clutch is entirely switched on, gap neglect occurs, and the friction linings are elastically deformed. Therefore, the initial position of the rotary lever (at the initial position of the carriage G_0) can be represented by levers ON_{10} and ON_{20}. In this case, $x_{N10} = x_{M1}$, $x_{N20} = x_{M2}$, and the initial inclination of the rotary disc is characterized by angles α_{10}, α_{20}.

At a specific position of the rotary lever, the carriage position is determined from the following expressions:

$$\begin{array}{lll} G_1: & x_{G_1} = x_{P_1} + \Delta_{K1} tg\alpha_1; & y_{G_1} = -\Delta_{K1}; \\ G_2: & y_{G_2} = -\Delta_{K2}; & x_{G_2} = x_{P_2} - \Delta_{K2} tg\alpha_2, \end{array} \tag{10}$$

where Δ_{Kk}—vertical movement of the carriage [mm].

The pressing force Q_k of the friction pairs and the force F_{np1} from the pressure spring (when the j-th clutch is switched on) are determined by Formulas (5)–(8).

The current deformation Δ_k of the pressure spring (the value of OP_k) is determined by the coordinate x_{Pk}. For the investigated clutch design (Figure 1b), under the switching-off of the 1st clutch, the pressure spring is additionally compressed ($\Delta_1 > 0$). For the 2nd clutch, the force in the pressure spring is decreased ($\Delta_2 < 0$). This fact increases the value of Q_1 and decreases the value of Q_2.

Figure 8 shows the impact of specified horizontal carriage displacement (considered by the angle α_k) with a vertical displacement of the carriage by $\Delta_k = \mp 15$ mm for the considered case study.

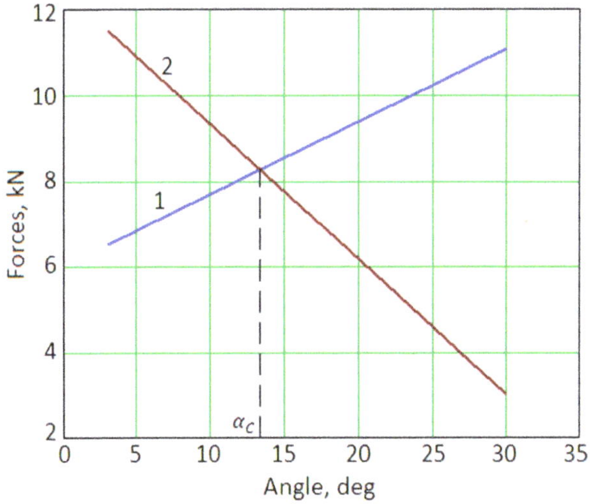

Figure 8. Dependencies of forces Q_k [kN] on the lever rotation angle α_k [deg]: 1—Q_1; 2—Q_2.

As the study results show, after considering the horizontal displacement of the carriage, an increase in the angle α_k (an increase in the value of $|x_{Gk}|$) decreases Q_2 and increases Q_1. However, despite the previous design scheme when $x_{Gk} = 0$, the following equality $Q_2 = Q_1$ takes place under a specific value of the angle α_c. Therefore, the problem of equality for the clutch margins has been solved.

Figure 9 shows results on changes in the forces $Q_1(\alpha_1)$ and $Q_2(\alpha_2)$ on the pressure disc at different values of vertical displacements of the carriage Δ_K = 5–30 mm with a step of 5 mm.

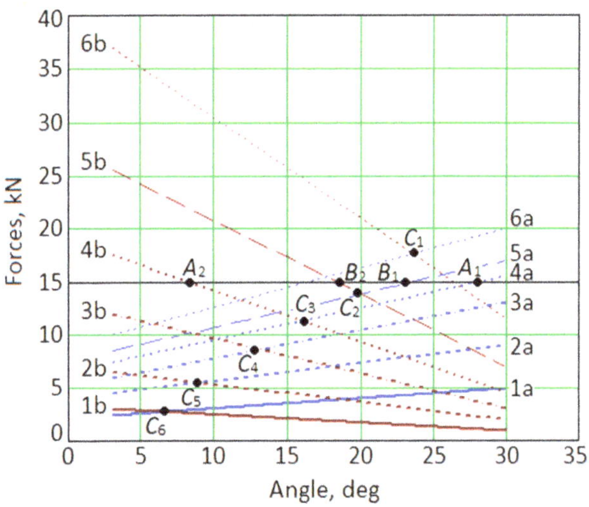

Figure 9. Dependencies of $Q_k(\alpha_k)$ on the angle α [deg] for different displacements Δ [mm]: a—Q_1; b—Q_2; 1—Δ = 5 mm; 2—Δ = 10 mm; 3—Δ = 15 mm; 4—Δ = 20 mm; 5—Δ = 25 mm; 6—Δ = 30 mm.

After applying the obtained results, it is possible to select the *k*-th position of the rotary lever and, accordingly, the position of the support carriage, which provide equal friction coefficients (since $Q_1 = Q_2$) in the case of choosing similar vertical displacements of the carriage ($\Delta_{K1} = \Delta_{K2}$) when the clutches are switched on.

Particularly, ensuring equal forces on the pressure plate ($Q_1 = Q_2 = 15$ kN) under the switched-on clutches 1 and 2 can be realized using the vertical movement of the carriage by $\Delta_{K1} = \Delta_{K2} = 20$ mm (points A_1, A_2) or through vertical displacement by $\Delta_{K1} = \Delta_{K2} = 25$ mm (points B_1, B_2). The levers' positions (angles α_1 and α_2) are different in this case. This complicates surface treatment technology for the platter on which the carriage moves and the corresponding control system.

However, as the results show, this shortcoming can be overcome. Particularly, for the chosen vertical movement Δ_K of the carriage, the lever position (points C_i, $I = 1, 2, \ldots, 6$) can be determined to satisfy the following equalities: $Q_1(C_i) = Q_2(C_i)$, $\alpha_1(C_i) = \alpha_2(C_i) = \alpha(C_i)$. This case corresponds to equal ratios for the 1st and 2nd friction clutches. It also facilitates the platter surface design on which the carriage moves.

Figure 10 shows the results for the dependencies between the forces difference $Q_{21} = Q_2 - Q_1$ (on the pressure plate) and the corner $\alpha = \alpha_1 = \alpha_2$ for some values of vertical movements $\Delta_K = \Delta_{K1} = \Delta_{K2}$ of the carriage.

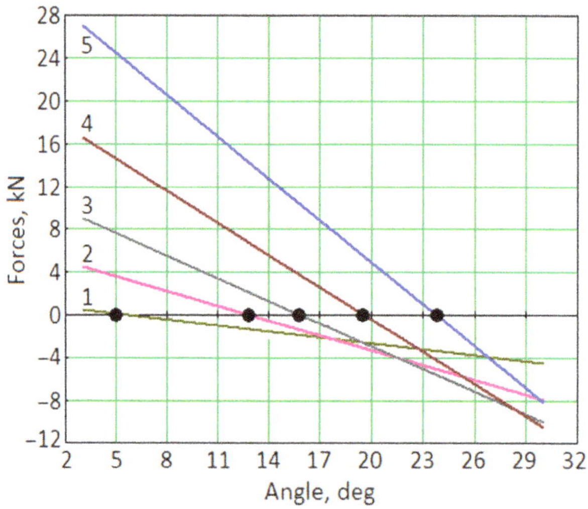

Figure 10. Dependencies of $Q_{21}(\alpha_{21})$ [kN] on the displacement Δ [mm]: 1—Δ = 5 mm; 2—Δ = 10 mm; 3—Δ = 20 mm; 4—Δ = 25 mm; 5—Δ = 30 mm.

When synthesizing the clutch control mechanism, based on the conditions $\Delta_K = \Delta_{K1} = \Delta_{K2}$ for providing the equal position of the rotary lever ($\alpha_1 = \alpha_2 = \alpha_c$), the primary attention should be focused on the results presented in Figure 11.

Based on the required loading Q_P for the clutch friction pair compression, values for the carriage movement Δ_K and the inclination angle α_c should be chosen properly. Such a rational choice allows for designing the clutch disc's surface profile, providing equal ratios for the 1st and 2nd clutches. The following case study illustrates this. Let the required force $Q_P = 11$ kN be given, which corresponds to point A (Figure 11). Furthermore, point B (on the curve Q) should be found, which corresponds to the vertical movement $\Delta_{K1} = \Delta_{K2} = \Delta K = 20$ mm of the carriage (point D). This point D allows determining point C (on the curve α_C) and the angle $\alpha_c = 16°$ (point E).

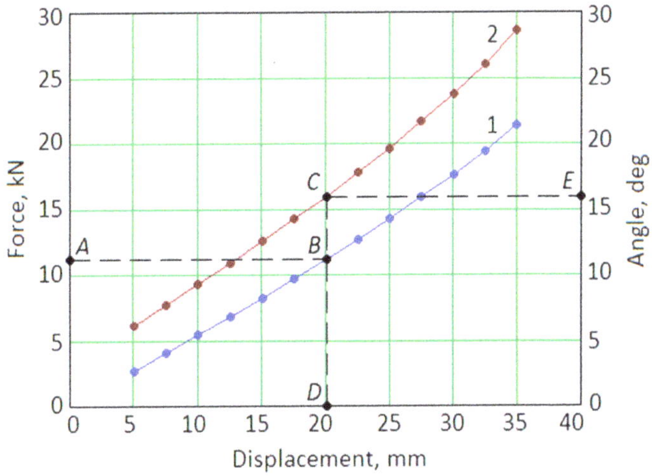

Figure 11. Dependencies of $Q(\Delta_K)$ [kN] (1) and $\alpha_c(\Delta_K)$ [deg] (2) on the displacement Δ [mm].

Suppose that according to the conditions of the task, it is required to change the force Q_P. In that case, new values of Δ_K and α_c can be similarly determined, or the calculated values $\Delta_K = 20$ mm and $\alpha_c = 16°$ can be left (which correspond to the force $Q^* = 11$ kN), but the stiffness c_{np} of the pressure spring should be changed $\chi = Q^*/Q_P$ times.

Remarkably, when choosing the vertical displacement Δ_K of the carriage, the following limits should be considered: $\Delta_{K2} < \Delta_{K2max}$, where $\Delta_{K2max} = \Delta z_c + y_{K2max}$ (Figure 12).

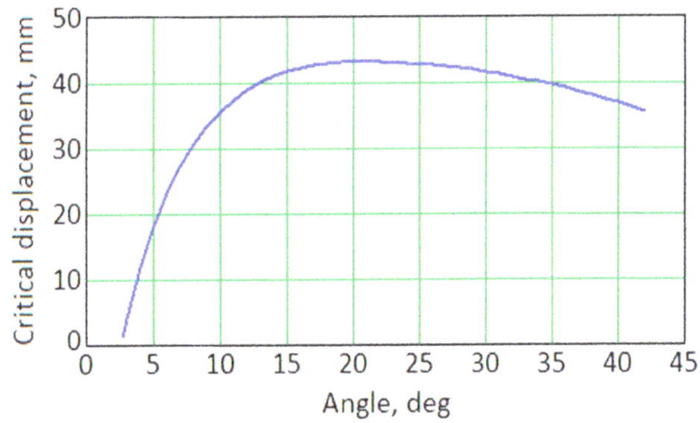

Figure 12. Limitation for the vertical displacement Δ_{k2max} [mm] of the carriage for different values of the angle α [deg] when the 2nd clutch is switched-on ($L = 53$ mm and $z = 2.8$ mm).

The above recommendations are mainly related to ensuring the equal vertical movement of the carriage ($\Delta_K = \Delta_{K1} = \Delta_{K2}$) when switching on the clutch. However, there may be other criteria for synthesizing the clutch control mechanism. Particularly, if there is a need to ensure the equal displacement ($S_1 = S_2$) of the carriage, these displacements should be determined as follows:

$$S_k = OG_k = \sqrt{x_{G_k}^2 + y_{G_k}^2}, \tag{11}$$

where x_{G_k}, y_{G_k}—coordinates of the carriage on the platter surface.

In this case, the equation of a straight line that passes through the points N_K and P_K (Figure 4) should be written as follows:

$$\frac{y - y_{P_K}}{y_{N_K} - y_{P_K}} = \frac{x - x_{P_K}}{x_{N_K} - x_{P_K}}. \tag{12}$$

Using expressions (9) for the coordinates of points N_K and P_K, this dependence can be rewritten as follows:

$$y = (-1)^K ctg\alpha_K \cdot (x - x_{P_K}). \tag{13}$$

The movement of the carriage equals S_K [mm]. Therefore, point $G_K(x_{Gk}, y_{Gk})$ is placed both on the circle S_K and straight line N_kP_k:

$$\begin{cases} x_{G_k}^2 + y_{G_k}^2 = S_k^2; \\ y_{G_k} = (-1)^k ctg\alpha_k \cdot (x_{G_k} - x_{P_k}). \end{cases} \tag{14}$$

The solution of this system allows determining the angle α_K of the rotary lever for carriage located at point $G_K(x_{Gk}, y_{Gk})$ on the distance of S_K, using parameters α_K and $\Delta_K = y(G_k)$.

However, to synthesize the surface profile of the support disc, it is more convenient to use the polar coordinate system (S_k, β_k), where β_k is the angular coordinate (Figure 13):

$$\beta_K = -arctg\left(\frac{x_{G_K}}{y_{G_K}}\right). \tag{15}$$

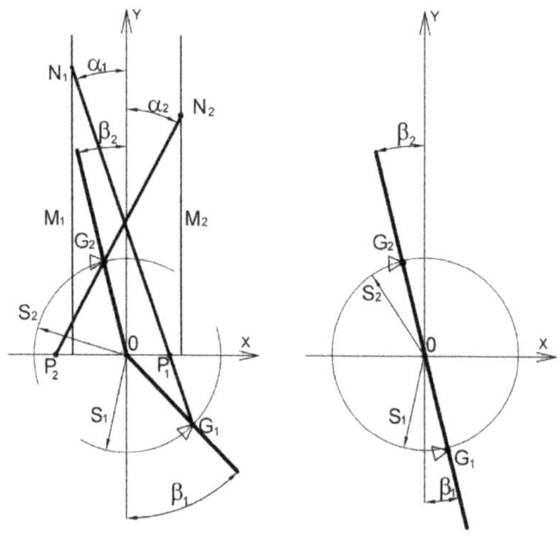

(a) (b)

Figure 13. The conical contact surface of the platter: general (**a**) and simplified (**b**) case studies.

After considering the issues of simplifying the technology of manufacturing a conical surface, the clutch control system can be simplified using the following limitation:

$$\beta_1 = \beta_2, \tag{16}$$

Figure 14 shows the changes in the forces $Q_K = Q(S_k, \beta_k)$ acting on the pressure plate.

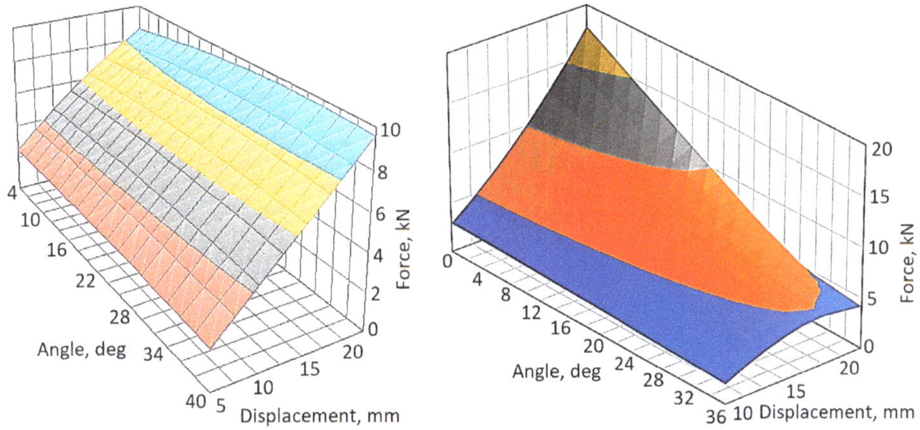

(a) (b)

Figure 14. Surfaces $Q_K = Q(S_1, \beta_1)$ [kN] (**a**) and $Q_K = Q(S_2, \beta_2)$ [kN] (**b**) on the displacement S_k [mm] and the angle β_k [deg].

The results show that condition (16) can be satisfied. Particularly, the locus of these points that meets this condition is presented in Figure 15.

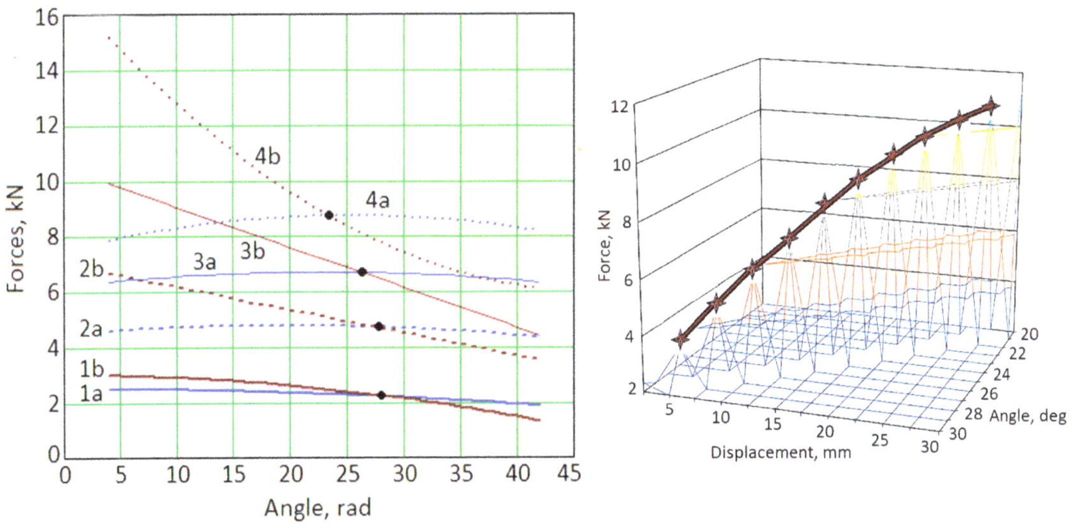

(a) (b)

Figure 15. Dependencies of $Q_k(S_k, \beta_k)$ [kN] (**a**) and $Q(S, \beta)$ [kN] (**b**) on the displacements S, S_k [mm] and the angle β_k and β [deg]: a—Q_1; b—Q_2; 1—S = 5 mm; 2—S = 10 mm; 3—S = 15 mm; 4—S = 20 mm.

The synthesis results presented in Figure 16 are also of interest. Particularly, it demonstrates the dependence of the carriage movement on the force acting on the pressure disc.

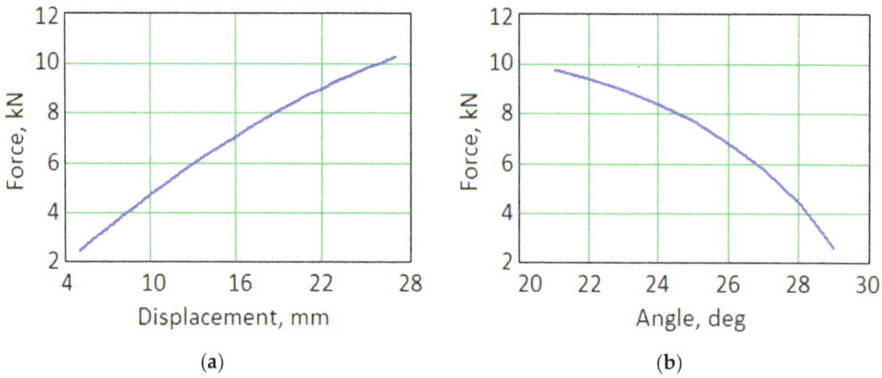

Figure 16. Dependencies of the forces Q [kN] on the displacement S_k (**a**) and the angle β [deg] (**b**).

Notably, the maximum carriage movement depends on the length L of the rotary lever and the parameter z. For the considered case study, $S < 24.5$ mm.

The minimum carriage movement S_{min} is determined by the wear resistance of the carriage movement mechanism. In this case, the inclination angle of the surface profile of the platter is in the range of $\beta = 20–29°$ (Figure 16).

Finally, the carried-out research, the developed methodology, and the corresponding algorithm allow for designing the platter surface profile for the DCT, providing equal clutch margins (Figure 17).

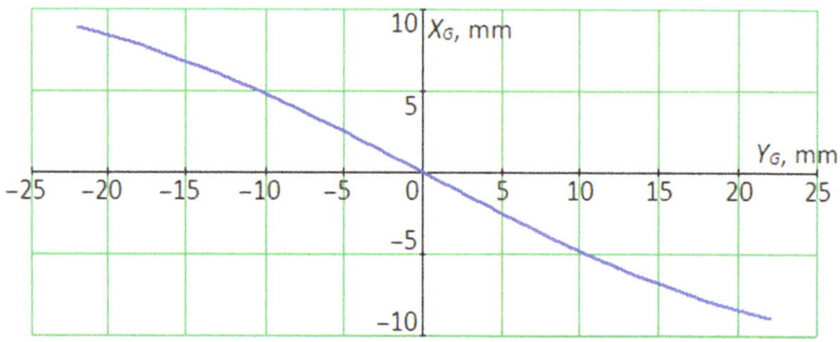

Figure 17. Clutch platter surface profile for equal clutch factors in coordinates X_G, Y_G [mm].

4. Discussion

Comparing the proposed DCT with traditional dual-clutch designs, it can be summarized that it has a reduced energy consumption for clutch control. Particularly, in a fairly common dry dual clutch of the automotive industry, each clutch is switched on by a pressure plate and actuator. The drive of such a clutch uses a separate mechanism and actuator (hydraulic cylinder) connected through a distributor to a pump. This requires additional energy costs. Remarkably, energy is permanently spent on maintaining the clutch in the switched-on position. However, there are no shortcomings in the proposed DCT.

Thus, the considered dry DCT provides fast and practically uninterruptible power flow from the DCT through the clutch. This can improve the vehicle's acceleration dynamics and reduce the energy costs of switching clutches.

Nevertheless, the kinematic and dynamic analyses of the device [33] showed that the loading of the driven discs for each clutch is uneven. Particularly, clutch margins when switching are different. This impacts the service life of each clutch and dynamic processes in the vehicle's transmission.

The novelty of the research is in the substantiation of the mathematical model, the development of the methodology, and the related algorithm for choosing the rational design parameters of the DCT and its drive. This increases the functional performance of the unit while reducing energy costs for control.

The practical significance of the results is to substantiate the possibility of increasing the efficiency of the original DCT. It is also in developing recommendations for determining the rational design parameters and the algorithm's versatility when designing such clutches for automotive, agricultural, and other industries.

5. Conclusions

In the article, the working processes in DCTs have been analyzed. A functional relationship has been established between the drive control parameters and the pressure forces (load capacity) on the clutch friction pairs. The impact of the vertical movement of the control carriage has also been analyzed.

It has been established that only vertical movement of the control carriage does not provide the equality of the clutch margins for the 1st and the 2nd clutch with the equal movement of the carriage.

As a result, a methodology for synthesizing the parameters of the DCT control mechanism has been proposed. The proposed approach is oriented towards the transmission of the required clutch torque. The proposed design solution for the control mechanism substantiates the equalization of the load capacity of the clutch.

The problem of synthesizing the parameters of the surface profile of the disc has also been solved. It allows for providing the equality of clutch margins for the 1st and 2nd clutches. However, the functional limitation regarding the equality of vertical (radial) movements or the general carriage movement for the control mechanism should be satisfied.

The economic effect from the implementation of the obtained results is primarily associated with a decrease in the energy losses of the internal combustion engine and the costs of a vehicle's driver during the operation. The efficiency of the proposed novel clutch is also due to the achievement of the same operating conditions for each clutch. Mainly, equal wear rates of friction pairs and dynamics equalization of the drive wheels (when switching clutches and transmitting the torque of the engine) have been provided. This extends the service life of the clutch and transmission, decreases fuel consumption, and improves the vehicle's environmental safety, traction, and dynamic performance.

Further studies will aim to generalize the proposed algorithm for synthesizing parameters of various types of DCT control mechanisms. Particularly, as part of the strategic problem of the systematic improvement of the vehicle's quality, it is planned to develop a multi-criteria optimization approach considering economic, energy, and environmental quality indicators. This will allow for the optimal design of DCTs with an extended number of parameters.

Author Contributions: Conceptualization, N.S.; methodology, P.K.; software, I.P.; validation, M.O. and I.P.; formal analysis, A.S., A.K., M.M. and S.W.; investigation, N.P. and I.P.; resources, Y.B. and A.K., M.M. and S.W.; data curation, O.T.; writing—original draft preparation, I.P.; writing—review and editing, A.N., M.O. and S.W.; visualization, I.P. and V.I.; supervision, V.I.; project administration, I.P.; funding acquisition, M.O. All authors have read and agreed to the published version of the manuscript.

Funding: The publication was funded by the Polish Ministry of Education and Science (SBAD).

Institutional Review Board Statement: Not applicable.

Informed Consent Statement: Not applicable.

Data Availability Statement: Not applicable.

Acknowledgments: The results were partially obtained at Poznan University of Technology within the project "Application of artificial intelligence system for diagnostics and predictive maintenance of rotary machines" of the Ulam NAWA Programme ordered by the Polish National Agency for Academic Exchange, grant number BPN/ULM/2022/1/00042. The authors also acknowledge the International Association for Technological Development and Innovations for support while conducting the research.

Conflicts of Interest: The authors declare no conflict of interest.

References

1. Xu, X.; Dong, P.; Liu, Y.; Zhang, H. Progress in automotive transmission technology. *Automot. Innov.* **2018**, *1*, 187–210. [CrossRef]
2. Shevchenko, S.; Mukhovaty, A.; Krol, O. Gear clutch with modified tooth profiles. *Procedia Eng.* **2017**, *206*, 979–984. [CrossRef]
3. Sankaranayanan, V.; Emekli, M.E.; Gilvenc, B.A.; Guvenc, L.; Ozturk, E.S.; Ersolmaz, E.S.; Eyol, I.E.; Sinal, M. Semiactive suspension control of a light commercial vehicle. *IEEE/ASME Trans. Mechatron.* **2008**, *13*, 598–604. [CrossRef]
4. Tripathi, K. Some design-objectives and design-guidelines for automotive friction clutch based on clutch engagement dynamics. *J. Inst. Eng. India Ser. C* **2014**, *95*, 51–61. [CrossRef]
5. Pisaturo, M.; Senatore, A. Thermal compensation control strategy in automated dry clutch engagement dynamics and launch manoeuvre. *Int. J. Automot. Technol.* **2019**, *20*, 1089–1101. [CrossRef]
6. Minas, I.; Morris, N.; Theodossiades, S.; O'Mahony, M.; Voveris, J. Automotive dry clutch fully coupled transient tribodynamics. *Nonlinear Dyn.* **2021**, *105*, 1213–1235. [CrossRef]
7. Fan, X.; Walker, P.D.; Wang, Q. Modeling and simulation of longitudinal dynamics coupled with clutch engagement dynamics for ground vehicles. *Multibody Syst. Dyn.* **2018**, *43*, 153–174. [CrossRef]
8. Chen, Y.; Wang, X.; He, K.; Yang, C. Model reference self-learning fuzzy control method for automated mechanical clutch. *Int. J. Adv. Manuf. Technol.* **2018**, *94*, 3163–3172. [CrossRef]
9. Deng, T.; Hu, F.; Lu, R. Research on DCT shifting torque control and a benchmark test. *J. Mech. Sci. Technol.* **2015**, *29*, 3581–3589. [CrossRef]
10. Pourgol-Mohammad, M.; Hejazi, A.; Soleimani, M.; Ghasemi, P.; Ahmadi, A.; Jalali-Vahid, D. Design for reliability of automotive systems; case study of dry friction clutch. *Int. J. Syst. Assur. Eng. Manag.* **2017**, *8*, 572–583. [CrossRef]
11. Park, J.; Choi, S. Optimization method of reference slip speed in clutch slip engagement in vehicle powertrain. *Int. J. Automot. Technol.* **2021**, *22*, 55–67. [CrossRef]
12. Lucente, G.; Montanari, M.; Rossi, C. Modelling of an automated manual transmission system. *Mechatronics* **2007**, *17*, 73–91. [CrossRef]
13. Senatore, A. Advances in the automotive systems: An overview of dual-clutch transmissions. *Recent Pat. Mech. Eng.* **2009**, *2*, 93–101. [CrossRef]
14. Yang, Y.-V.; Zhu, Z.-B.; Wang, X.-Y.; Chen, Z.; Ma, Z.-L. Optimal launching-intention-aware control strategy for automated clutch engagement. *Int. J. Automot. Technol.* **2017**, *18*, 417–428. [CrossRef]
15. Van Der Heijden, A.C.; Serrarens, A.F.; Camlibel, M.K.; Nijmeijer, H. Hybrid optimal control of dry clutch engagement. *Int. J. Control* **2007**, *80*, 1717–1728. [CrossRef]
16. Glielmo, L.; Iannelli, L.; Vacca, V.; Vasca, F. Gearshift control for automated manual transmissions. *IEEE/ASME Trans. Mechatron.* **2006**, *11*, 17–26. [CrossRef]
17. Ma, K.; Sun, D.; Sun, G.; Wang, D. Wet dual clutch launching adaptive control considering service time. *J. Mech. Sci. Technol.* **2022**, *36*, 2759–2773. [CrossRef]
18. Glielmo, L.; Gutman, P.O.; Iannelli, L.; Vasca, F. Robust Smooth Engagement of an Automotive Dry Clutch. *IFAC Proc. Vol.* **2006**, *39*, 632–637. [CrossRef]
19. Deng, T.; Hu, F.B.; He, Z.Y.; Yin, Y.L. Simulation, experimental testing and optimization of starting and shifting control strategies of DCT wet dual clutches with respect to sliding friction. *Iran. J. Sci. Technol. Trans. Mech. Eng.* **2019**, *43*, 693–705. [CrossRef]
20. Cho, J.; Lee, Y.; Kim, W.; Jang, S. Wet single clutch engagement behaviors in the dual-clutch transmission system. *Int. J. Automot. Technol.* **2018**, *19*, 463–472. [CrossRef]
21. Pica, G.; Cervone, C.; Senatore, A.; Lupo, M.; Vasca, F. Dry dual clutch torque model with temperature and slip speed effects. *Intell. Ind. Syst.* **2016**, *2*, 133–147. [CrossRef]
22. Zhu, M.; Yao, P.; Pu, Y.; Liu, T. Comparative study on the temperature rise of a dry dual clutch under different starting conditions. *Automot. Innov.* **2019**, *2*, 35–44. [CrossRef]
23. Abdullah, O.I.; Schlattmann, J. Thermal behavior of friction clutch disc based on uniform pressure and uniform wear assumptions. *Friction* **2016**, *4*, 228–237. [CrossRef]
24. Bukashkin, A.Y.; Dobretsov, R.Y.; Galyshev, Y.V. Split transmission of tractor with automatic gearbox. *Procedia Eng.* **2017**, *206*, 1728–1734. [CrossRef]
25. Darko, S.; Spasojevic, V.; Stevanovic, I. The contemporary automatic gearboxes—Review of the current state and interpretation of advantages and disadvantages of their use with respect to vehicle performance and traffic safety. *Istraz. Proj. Privredu* **2013**, *11*, 89–97. [CrossRef]
26. Gkinis, T.; Rahmani, R.; Rahnejat, H.; O'Mahony, M. Heat generation and transfer in automotive dry clutch engagement. *J. Zhejiang Univ. Sci. A* **2018**, *19*, 175–188. [CrossRef]

27. Hebbale, K.; Samie, F.; Kish, J. Dry dual clutch transmission (DCT) thermal model. In Proceedings of the SAE 2015 World Congress & Exhibition, Detroit, MI, USA, 21–23 April 2015; SAE Technical Paper. SAE International: Warrendale, PA, USA, 2015. [CrossRef]
28. Romanchenko, O.; Sokolov, V.; Krol, O.; Baturin, Y.; Stepanova, O. Automatic control of electrohydraulic drive for technological equipment. In Proceedings of the International Conference on Reliable Systems Engineering (ICoRSE) 2022, Bucharest, Romania, 8–9 September 2022; Lecture Notes in Networks and Systems. Cioboată, D.D., Ed.; Springer: Cham, Switzerland, 2023; Volume 534, pp. 329–337. [CrossRef]
29. Grzelczyk, D.; Awrejcewicz, J. Wear processes in a mechanical friction clutch: Theoretical, numerical, and experimental studies. *Math. Probl. Eng.* **2015**, *2015*, 725685. [CrossRef]
30. Vanyeyev, S.M.; Miroshnichenko, D.V.; Rodymchenko, T.S.; Protsenko, M.; Smolenko, D.V. Data measuring system for torque measurement on running shafts based on a non-contact torsional dynamometer. *J. Eng. Sci.* **2019**, *6*, E15–E23. [CrossRef]
31. Sergienko, M.Y.; Sergienko, A.M.; Khudolii, O.I. Two-Loop Friction Clutch. Patent for an Invention of Ukraine No. 101,711, 25 April 2013.
32. Sergienko, M.Y.; Sergienko, A.M.; Khudolii, O.I.; Sergienko, D.Y.; Pavlova, N.M.; Maidaniuk, V.G.; Zarubina, A.O.; Svidlo, V.S. Two-Loop Friction Clutch. Patent for an Invention of Ukraine No. 125,475, 23 March 2022.
33. Sergienko, N.; Kalinin, P.; Gasanov, M.; Pavlova, N.; Svidlo, V.; Okun, A. Analysis of the parameters of the double clutch drive with reduced control energy consumption. *Bull. Natl. Tech. Univ. "KhPI" Ser. New Solut. Mod. Technol.* **2021**, *4*, 49–60. [CrossRef]

applied sciences

MDPI

Article

Novel Multibus Multivoltage Concept for DC-Microgrids in Buildings: Modeling, Design and Local Control

Heriberto Rodriguez-Estrada [1,*], Elias Rodriguez-Segura [2], Rodolfo Orosco-Guerrero [2], Cecilia Gordillo-Tapia [2] and Juan Martínez-Nolasco [3]

1 Doctorado en Ingeniería Electrónica, Tecnológico Nacional de México/IT de Celaya, Celaya 38010, Mexico
2 Departamento de Ingeniería Electrónica, Tecnológico Nacional de México/IT de Celaya, Celaya 38010, Mexico
3 Departamento de Ingeniería Mecatrónica, Tecnológico Nacional de México/IT de Celaya, Celaya 38010, Mexico
* Correspondence: heriberto.rodriguez@itcelaya.edu.mx

Abstract: In this paper, a novel microgrid (MG) concept suitable for direct current (DC) multibus architectures is depicted. Multibus feature is improved in order to distribute power in DC using a number of buses at different voltage level. A teachers offices building that houses several kinds of loads, including a charging station for electric vehicles (EV), is considered to validate the strategy. Several topologies of power electronics converters (PECs) are included in the system to perform specific tasks and providing isolation between bus and final loads. In order to develop the PECs, first, a switching function is used to obtain average model of each converter. Then, converters design is done by using well known methods that allow to obtain parameter values of all the devices in every version of each kind of converter. A hierarchical control is selected to govern the direct current microgrid (DCMG). At a lower control level, local control stage is implemented and tuned using models and designs obtained, with linear controllers in some PECs and classic strategies in others. In higher control level, there is a supervisory strategy that prioritizes the use of generated power to supply the building's loads. This energy management system (EMS) is based in Petri net theory; it consists of a start-up test, then source condition synchronous algorithm and load condition synchronous algorithm operate the DCMG according to the mentioned priority. Finally, PECs are tested on standalone, performing in closed loop, facing load changes to verify the adequate operation. Some trajectories of a simplified version of the CDMG are tested with local control in order to validate the multibus multivoltage concept. In order to verify coordinated control, some events managed by EMS are presented.

Keywords: direct current microgrid; multibus; multivoltage; space state model; feedback state controller; hierarchical control, electric vehicle charge station; energy management system; Petri net

Citation: Rodriguez-Estrada, H.; Rodriguez-Segura, E.; Orosco- Guerrero, R.; Gordillo-Tapia, C.; Martínez-Nolasco, J. Novel Multibus Multivoltage Concept for DC-Microgrids in Buildings: Modeling, Design and Local Control. *Appl. Sci.* **2023**, *13*, 2405. https://doi.org/10.3390/app13042405

Academic Editor: Maria Vicidomini

Received: 22 November 2022
Revised: 27 January 2023
Accepted: 31 January 2023
Published: 13 February 2023

1. Introduction

More than a decade ago, news about fossil fuel resources reduction in underground layers of the Earth appeared; at the same time, human concern about environmental problems has increased. Taking care on these issues, the electric power systems evolve in two ways. First, the vision of a classic centralized generation turns into distributed generators (DG) development and commissioning. DGs are able to supply $\leq$100 kW nominal power and most of them operate from renewable energy sources (RES) such as eolics, photovoltaics, hydroelectric, geothermal and biofuel. Second, nowadays, distribution aims are: reliability, high efficiency and able to meet demand response in spite of daily aging infrastructure. The main objective is to supply commercial spaces, industrial sites, offices buildings and residential places at any geographical area with no disturbances.

Besides of DGs, a modern power system includes energy storage systems (ESSs) based in energy storage units (ESUs) such as batteries or supercapacitors to support the intermittent nature of some RES such as photovoltaics or eolics. The aim of this set is to supply power to different types of loads; some of them can be called recent in the market, such as electric vehicles (EV). The concept that considers DGs and ESSs operating together to supply power to loads appeared just over a decade ago. In order to make this idea come true, PECs are included to provides control and flexibility to a general structure called MG [1].

For any MG, it is possible to define three general objectives: higher reliability, losses reduction in feeders and improved efficiency in local supply. In order to reach them, three critical components are needed as part of the MG [2]. They are: local control in each PEC, high level control layers or/and EMS and protections array. In this manuscript, we describe the beginning stage of a long-term development; the final goal is the implementation of a MG placed in a teachers offices building for experimental purposes. Specifically in this paper, a novel concept for a MG architecture is presented with in situ linear controllers are proposed and includes a supervisory control level with a EMS based in Petri net theory [3]. Protections arrays are not part of this document.

An MG can be classified according to two criteria. First, the voltage at common connection point (CCP), power distribution could be in alternating current (AC) or in DC [4], also hybrid architectures have been presented in literature. The second criteria is about the system operation. If the MG interacts with mains, it is called interconnected mode; else, when MG is not connected to mains, it is said that the MG is operating in island mode [5]. Some systems are able to change its operation mode eventually. When the MG elements are integrated in a DC bus, parameters such as reactive power flow, frequency variations and amplitude disturbances do not take place. The result in the MG is a less complex system to control [6]; even a higher efficiency is reached in DC distribution [7]. These are the reasons to select a DCMG able to interacts with mains for this development.

PECs are an important part of the DCMG architecture because they add control and flexibility to the system to work in a harmonic way. In order to reach this system condition, the most popular option is to implement a coordinated control structure. Agreed with norm ISA-95, several layers of control can be established in a hierarchical control structure [8]. Operations such as voltage and current regulation, maximum power point tracking (MPPT), battery state of charge (SoC) estimation and some other basic functions take place at zero level [9,10] . At level number one: droop control, virtual impedance loop and power calculation that can be performed with variables measured and processed in the own converter [11–13]. The local control stage includes level zero and level one of the hierarchical control structure. In higher levels, supervisory control brings extended functions to MG such as: secondary control to eliminate voltage deviations, tertiary control to manage connection–disconnection events with mains and the ability to change the operation mode in a specific converter [14,15]. All of these tasks can be combined as algorithms of an EMS, which fulfill one of several specific system purposes. Coordinated control is reached when local control and supervisory control works together.

A popular option to implement zero level of local control is a cascade control configuration with an internal current loop and a external voltage loop. Control cascade schemes are reported using different strategies for controllers; proportional–integral (PI) classical controller is the most popular because is easy to tune, brings zero error in steady state and adds robustness [16]. Some other designs for loops are based in: classical proportional–derivative (PD) as filter [17], fuzzy logic [18] and boundary controller [19]. For level one, there are two possible techniques for droop control design: select power as signal output [5] or choose current as feedback signal [20]. In droop control strategy, droop coefficients influence system stability and current sharing, and there is a trade-off between current sharing and voltage bus deviation [21,22]. Sometimes, a virtual impedance loop, droop controller and power calculator strategy can be included. In this paper, a linear technique is selected to design a state feedback as local controller to reach well regulated buses, to support equally shared current distribution and get over input voltage variations; all of these issues are possible because all the dynamics in the PEC feedbacked.

Once the local control stage has been defined and taking into account the benefits of state feedback control, an EMS is included as a higher coordinated control level. Coordinated control provides an improved power collected manage, and bus stability can be enhanced [23]. An EMS allows to harness power collected from DGs to the loads in a defined way. The DCMG structure and the EMS design can be diverse, and they are made up in accordance with the specific objectives that are desired to be achieved. Some of EMS are reported perfoming basic tasks such as management of the energy between source and load based on optimal scheduling [24] and power flow management depending on ESS batteries Soc [25], using virtual impedance [26], or based in fuzzy logic [27]. In this paper, a novel EMS is focused to harness all the power harvested in PV arrays to supply the loads, to charge the ESS batteries or to inject to mains is presented. The EMS is conceived in a similar way to an industrial production system, under the premise that all of the energy that is produced in the DGs is consumed in its entirety. The EMS is modeled as a Petri net and has three stages: startup test, source condition establishment stage and load condition establishment stage.

In the MG proposed architecture for teachers building, an electric vehicle charge station (EVCS) is included. The EVCSs can be classified in three groups according to the input voltage and the power rating [28–30]. Table 1 shows the information ordered and summarized. Then, a commercial sigle phase AC-EVCS is an adequate selection to be supplied by the highest voltege level bus of the proposed DCMG.

Table 1. Classification of charge stations based on power levels.

EVCS Type	Power Supply	Charger Power	Charging Time
AC L1 residential	120/220 V_{AC}, 12–16 A, single phase	1.44–1.92 kW	Approx. 17 h
AC L2 commercial	208–240 V_{AC}, 15–80 A, single/split phase	3.1–19.2 kW	Approx. 8 h
DC L3 fast charger	300–600 V_{DC}, *Max* 400 A,	120–240 kW	Approx. 30 min

The objective of manuscript is to validate the design and performance of a novel concept for DCMG called multibus multivoltage architecture; it is the starting stage of an design suitable to supply power, voltage and current requirements of loads inside a teacher's offices building. In Section 2, the DCMG particular architecture is presented in a detailed way. In Section 3, the procedure to obtain an average model for all of the PECs in the system using a particular switching function is depicted. In Section 4, design procedures for PECs are mentioned. The aim is to size the passive elements of each converter version using typical design methods; inductor current ripple into a fixed limit criteria is set to avoid high frequency components propagation when a PECs cascade connection occurs. In Section 5, the development of the local control stage for every converter is shown; the innovation of using a linear control technique based on feedback state controllers that allows to keep the output voltage regulated, good response to transients and favors equal current distribution from source converters to loads, then a three stage EMS that harness all the power collected in PV to supply loads in the place is included. Section 6 includes simulation results for standalone converter with load changes; also, some critical power trajectories for the proposed DCMG are enabled to test local control, and some events managed by EMS are presented to validate the concept. Discussion and conclusions are presented in Section 7.

2. Proposed MG Architecture

About the DCMG scheme, multibus architectures, according to the original concept, are able to perform with higher reliability, basically because there is more than one bus in the system to overpass a fail [31]. Each bus gathers at least one DG, one ESS and some loads.

The proposed system must be able to satisfy the power demand in the teachers offices building as well as to supply an EVCS or, eventually, it can be able to inject power to mains. Some months ago, a quality energy logger equipment was installed to quantify power consumption. At the same time, a loads characterization study was performed to identify voltage and current requirements at every load inside the building. The highlighted results are: the bulding peak power consuption is 6 kW; in the other hand, it is necessary to set three DC buses at different voltage level because of the different type of the loads in the building. A primary bus at 380 V to supply electronic loads at every office; another secondary bus at 190 V to power the lighting system of the place (this bus can support the primary bus as an emergency action eventually). From the secondary bus, unidirectional 48 V buses are derivated; every one of them supplies six tandems driver-LED luminary for a specific area lighting. The primary bus and the secondary bus each incorporate an ESS based in batteries. At this point, it can be noticed that the MG has a particular architecture; then, it can be mentioned, according to the aforementioned two criteria, as multibus multivoltage DCMG.

The selected DCMG is a flexible architecture able to satisfy the building power demand. In the place, four photovoltaic (PV) arrays are installed as DGs, each at 2.5 kW power rating, eventually 10 kW total power. Then, three buses distribute the power to every load in the building mainly using primary bus (380 V) and secondary bus (190 V). Harvested power at PV arrays is processed through boost converters—three of them linked to the primary bus and one more setting the secondary bus. This kind of structure is able to transfer power from one bus to another at a power level lower than its rated power. In typical multibus architectures, power is transfered between buses with similar level voltage using static switches, but in the DCMG proposed, primary and secondary buses are at different voltage levels. This feature is the main difference with DC typical multibus systems in [32,33]. This development uses bidirectional power converters to transfer between buses, instead of static switches; then, it is possible to set the secondary bus from the primary bus and vice versa. A couple of PECs with bidirectional feature allows batteries charge/discharge at the ESSs. From secondary bus, 48 V unidiretional buses are derived to supply six tandems driver-LED luminary, four 48 V unidirectional buses are able to supply twenty-four luminaries inside the place. In order to improved DCMG security, all of the eventual power consumers in the system use isolated converters as interface to satisfy current and voltage required by final load. The architecture can be defined as multibus multivoltage; this concept brings flexibility to design a DCMG at any place. In order to validate the concept, a simplified DCMG with all the PECs but not all the loads is shown in Figure 1.

Inside the building, there is a distribution center where most of the PECs are placed. Those are the boost converters, bidirectional PECs to transfer power between buses, bidirectional PECs to charge–discharge ESSs batteries and unidirectional converters to supply lighting system. The remaining PECs are placed at different distances from load center—in this case, line resistive–inductive impedance effects must be taken into account. Line capacitance is not significant and can be neglected. Inductance and resistance values for THW-LS wire are calculated according to expressions in [34]; results for different lengths appear in Table 2.

The solar panel used for DGs, as PV arrays, is a similar version to part number JC250M-24/Bb, series Virtus II, manufactured by ReneSola. PV panel simplified model parameters are shown in Table 3.

This architecture concept allows to add some other kind of DGs. Then, addition of new DGs can increase the system complexity, but the performance and controllers can mainly be kept.

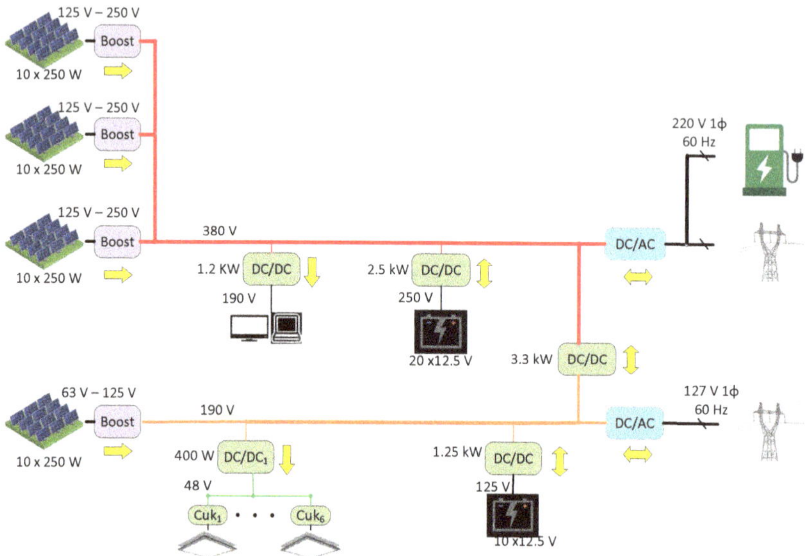

Figure 1. Multibus multivoltage DCMG architecture proposed.

Table 2. Line impedance values for different wire gauges and lenghts.

Lengh (Meter)	AWG 14 (µH; mΩ)	AWG 12 (µH; mΩ)	AWG 10 (µH; mΩ)	AWG 8 (µH; mΩ)
5	4.53; 1.71	4.29; 1.62	4.06; 1.53	3.83; 1.44
10	9.06; 3.41	8.58; 3.23	8.12; 3.06	7.66; 2.89
15	13.59; 5.12	12.87; 4.85	12.18; 4.59	11.49; 4.33
20	18.12; 6.83	17.16; 6.47	16.25; 6.12	15.32; 5.77
25	22.65; 8.53	21.45; 8.08	20.31; 7.65	19.15; 7.22

Table 3. Solar panel model parameters.

Parameter	Value
Maximum power P_{max}	250 W
Current at maximum power I_{mp}	8.31 A
Voltage at maximum power V_{mp}	30.1 V
Short circuit current I_{SC}	8.83 A
Open circuit voltage V_{OC}	37.4 V

3. Power Converters Modelling

As already mentioned, DCMG structure contains different PEC topologies that fulfill particular functions in the system. In this section, we presents modeling developments for converters that are part of the DCMG: boost converter, DC–DC full bridge converter, Cuk converter and DC-AC 1ϕ full bridge models are obtained. Each converter procedure of modeling appears in Sections 3.1–3.4.

3.1. Boost Converter Average Model

Each PV array has an interface boost converter before the connection with the corresponding bus. In this DCMG, boost converter preserves the traditional topology, but an input capacitor has been included to avoid PV array output voltage variations and a thermistor previous to the mentioned capacitor to limit the inrush current at the initial charge (see Figure 2).

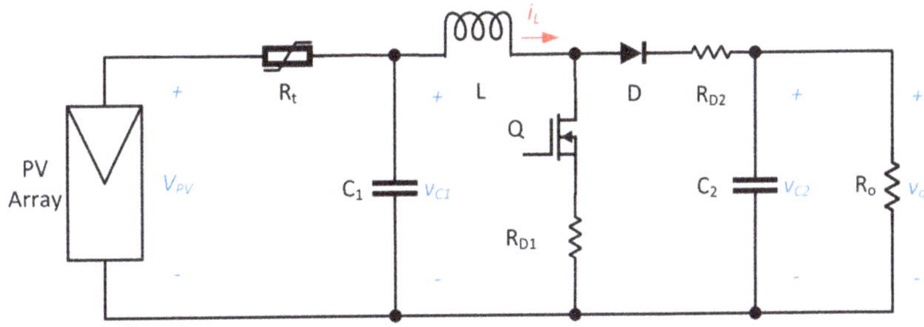

Figure 2. Boost converter.

In Figure 2, boost converter scheme has two power devices, a MOSFET (Q) and a diode (D); then, a switching function is defined for each in order to obtain the average model for the converter. In this way, S_1 and S_2 are switching functions for Q and D, respectively. If $S_1 = 1$, Q is in conduction state, else, if $S_1 = 0$, Q is not conducting. Same logic is used for S_2 and D, but S_1 and S_2 perform in complementary mode, so the two power devices operate in a complementary way. The following combinations are not allowed $S_1 = 1$ and $S_2 = 1$; $S_1 = 0$ and $S_2 = 0$. In order to have an improved resultant model of this converter, the conduction resistances for the switching devices have been included: R_{D1} for the MOSFET and R_{D2} for the diode. The Kirchhoff Voltage Law (KVL) is applied to inductor loop and Kirchhoff Current Law (KCL) is used in capacitors nodes in order to obtain the state equations for the switching model. The average of the switching function, $\overline{S}(t)$, corresponds to modulator signal duty cycle, $u(t)$, in this case, $\overline{S}(t) = u(t)$. Taking this deduction into account, it is possible to go from the switching model to the averaged model, which is given by:

$$\begin{aligned} \frac{dv_{C1}}{dt} &= -\frac{1}{R_t C_1} v_{C1} - \frac{1}{C_1} i_L + \frac{1}{R_t C_1} V_{PV} \\ \frac{di_L}{dt} &= \frac{1}{L} v_{C1} - \frac{[(R_{D1} - R_{D2})u + R_{D2}]}{L} i_L - \frac{(1-u)}{L} v_{C2} \\ \frac{dv_{C2}}{dt} &= \frac{(1-u)}{C_2} i_L - \frac{1}{R_O C_2} v_{C2} \end{aligned} \tag{1}$$

As mentioned before, the architecture of the DCMG has two main buses in different voltage level, so the model obtained can be used for both versions, the three 380 V boost converters linked to the primary bus and the 190 V boost converter in the secondary bus.

3.2. DC–DC Full Bridge Converter Average Model

DC–DC full bridge converters process power from one of the main buses to meet voltage and current requirements for final loads like LED luminaries, personal computers, etc. This topology is an important part of the system because there are several versions at different power ratings in the DCMG. This configuration allows an adequate output voltage regulation, provides isolation between the bus and the final load and gives reliability to the general system. This kind of isolated converter is able to manage a greater power rating than other isolated topologies. DC–DC full bridge converter is a two-stage topology, integrated by an inverter (Q_1–Q_4), a rectifier (D_1–D_4), linked by high-frequency transformer, and a LC output filter (Figure 3).

The DC–DC full bridge converter operates with phase shift pulse wide modulation. Two specific pair of power devices in the inverter are defined; Q_1 and Q_2 are the first pair and Q_3 and Q_4 are the second pair. The conduction period of time for each pair of MOSFETs must last less than a half of switching period. Diodes D_1, D_2, D_3 y D_4 conduct depending on its own polarization state in a specific time. A switching function S_1 is defined according to the on and off times in the devices Q_1, Q_2, D_1 and D_2 take the zero state and the one state simultaneously. In the same way, Q_3, Q_4, D_3 and D_4 synchronize their on and off state, and the corresponding switching function S_2 is assigned to this set; S_1 and S_2 operate with 180° phase. In order to obtain the switched model, v_x is defined as the rectifier voltage output. By applying Kirchhoff's laws, equations for switching model in time domain are obtained from the DC–DC full bridge switching equivalent circuit. Then, the average model is obtained by considering the modulating signal, $u(t)$, as the average of the selected commutation function; the resultant average model is described as:

$$\begin{aligned} \frac{dv_{C2}}{dt} &= -\frac{1}{R_o C_2} v_{C2} + \frac{1}{C_2} i_L \\ \frac{di_L}{dt} &= -\frac{1}{L} v_{C2} + \frac{1}{L} 2nuV_i \end{aligned} \tag{2}$$

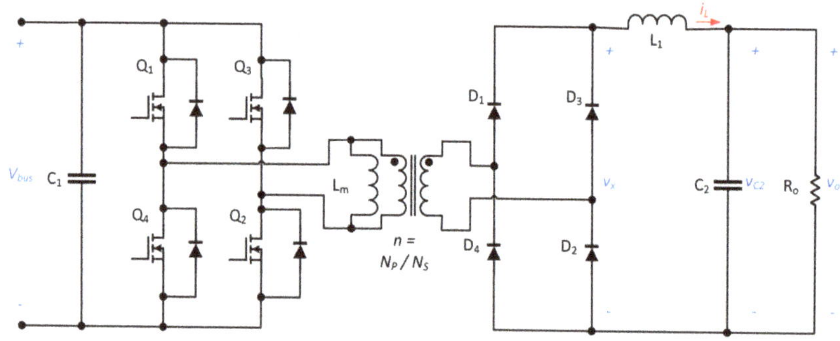

Figure 3. DC-DC Full Bridge Converter.

The average model in (2) can be used in any of the following versions: 380 V to 190 V at 1.2 kW to supply electronic loads in each cubicle, 190 V to 48 V at 400 W for supply drivers for LED luminaries. Two versions, 380 V to 190 V at 3.3 kW for power transfer between primary and secondary bus, in the other way, a 190 V to 380 V at 3.3 kW version. For charge–discharge bateries, a 380 V to 250 V at 2.5 kW and 250 V to 380 V at 2.5 kW for ESS at primary bus; finally, two more versions 190 V to 125 V at 1.25 kW and 125 V to 190 V at 1.25 kW for ESS in secondary bus.

3.3. *Cuk Converter Average Model*

In order to include LED luminaries as final load, a driver based on Cuk converter is designed. Cuk driver is the interface between 400 W DC-DC full bridge converter and LED luminaire; Cuk converter schematic is shown in Figure 4.

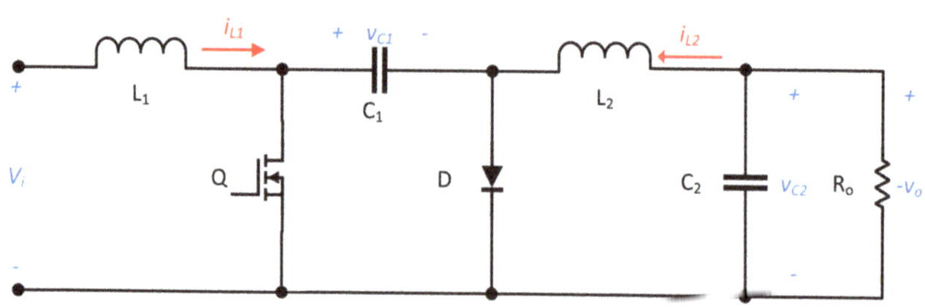

Figure 4. Cuk converter.

In Figure 4, two power devices, MOSFET (Q) and diode (D), can be distinguished. Switching functions S_1 and S_2 are defined for Q and D, respectively. Again, if $S_1 = 1$, Q is conducting, else, if $S_1 = 0$, the device is off state. The same logic is used for function S_2, and complementary operation is assumed. The following combinations are not allowed

$S_1 = 1$ with $S_2 = 1$ and $S_1 = 0$ with $S_2 = 0$. By applying Kirchhoff laws, state equations in time domain are obtained from the Cuk converter switching equivalent circuit; then, the averaged model is obtained by considering the modulating signal $u(t)$ as the average of the commutation function. The averaged model is described as:

$$\begin{aligned}
\frac{di_{L1}}{dt} &= -\frac{(1-u)}{L_1}v_{C1} + \frac{V_i}{L_1} \\
\frac{di_{L2}}{dt} &= \frac{u}{L_2}v_{C1} + \frac{1}{L_2}v_{C2} \\
\frac{dv_{C1}}{dt} &= \frac{1-u}{C_1}i_{L1} + \frac{u}{C_1}i_{L2} \\
\frac{dv_{C2}}{dt} &= -\frac{1}{C_2}i_{L2} + \frac{1}{R_oC_2}v_{C2}
\end{aligned} \tag{3}$$

Cuk output current flowing through LED array inside the luminary must be regulated to avoid unwanted lighting effects.

3.4. Single Phase DC-AC Full Bridge Bidirectional Converter Average Model

Interconnected systems includes one or more PECs that interacts with mains; the topology selected has bidirectional capability. In the first operation mode, the converter takes power from the mains and supplies loads in DC side, rectifier mode. In second operation mode, power harvested in PV arrays exceeds building power demand; then, it is possible to inject power to the grid, inverter mode. Then, the selected topology is a DC–AC full bridge converter (also called H4) [35,36] and has the capability to operate bidirectionally. This feature allows the converter to be integrated in the DCMG architecture to interacts with the mains. The converter schematic diagram appears in Figure 5.

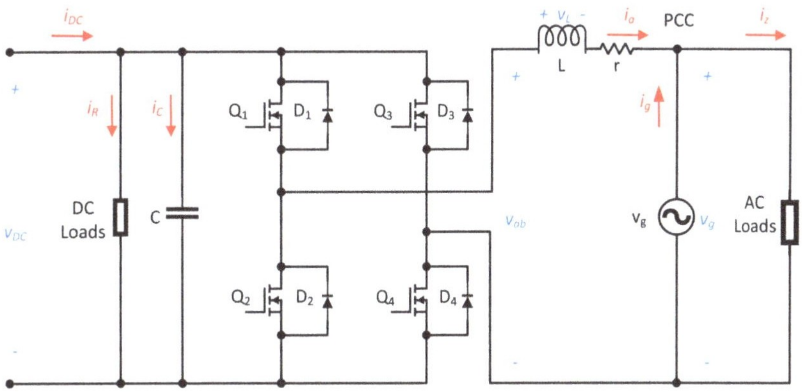

Figure 5. Single phase DC-AC full bridge converter.

A first pair of power devices defines is Q_1 and Q_4 and operates synchronized; the other pair Q_2 and Q_3 is synchronized as well but in a complementary way in reference with the first pair of power devices. Switching functions are defined as follows: if $S_1 = 1$, Q_1 and Q_4 are in the on state, else, $S_1 = 0$, Q_1 and Q_4 are in the off state. A second switching function, S_2, operates with Q_2 and Q_3 power devices. To obtain the corresponding switching simplified equivalent circuit, KVL is applied to the inductor side, and r is the series resistive value associated to the inductor and contributes with some damping to system. Then, KCL is applied to the PV side; the resultant set of equations represents the converter switching model. Finally, the average value of the switching function $\overline{S}(t)$ is defined as $u(t)$, the modulating signal.

$$\frac{di_a}{dt} = \frac{1}{L}(uv_{dc} - i_a r - v_g)$$
$$\frac{dv_{dc}}{dt} = \frac{1}{C}(ui_a - \frac{1}{R}v_{DC} + i_{DC}) \quad (4)$$

Two version of this converter are included in DCMG, 220 V_{RMS} to 190 V_{CD} at 3.3 kW and 127 V_{RMS} to 190 V at 1.5 kW. They are placed in primary and secondary bus, respectively. The 3.3 kW version in the primary bus is able to supply the ECVE.

4. Converters Design

First, some considerations to take into account for converters design are: for this development system, global efficiency is not a main issue; converters design and implementation includes an output bleeding resistor at 10% of nominal power in order to avoid no-load operation and keep the converter on performing in continuous conduction mode (CCM). DCMG is an interconnection of power converters do specific tasks; eventually, some of these converters can perform in a cascade connection, and high-frequency switching noise increases at every stage until reach a unsafe condition for the system. For this reason, a design criterion is established: current ripple must be less than 10% inductor nominal current value.

4.1. Boost Converter Design

Using the above criteria and following the design procedure in [37], the mathematical expressions to assign values to PEC passive elements are obtained. For 2.5 kW and 380 V voltage output boost converter, eventually, up to three of these converters set primary bus and supplies EVCE. The fourth boost converter at 2.5 kW and 190 V, voltage output sets secondary bus. Passive devices minimum obtained values for L_{min} and C_{min}, but about twenty times L_{min} value is selected in order to meet ripple current criteria and keep converter operating in CCM; parameter values are shown in Table 4.

Table 4. Parameter design values for both versions of boost converter.

Version (V_{out})	I_L (A)	R_o (Ω)	L (mH)	C_1 (μF)	C_2 (μF)	R_{D1} (mΩ)	R_{D2} (mΩ)	Δi_L (%)	Δv_{C2} (%)
380	14.25	57.76	1.3	82	150	65	45	9.03	0.12
190	31.78	14.44	0.65	82	150	65	45	7.75	0.52

4.2. DC–DC Full Bridge Converter Design

For all versions of this topology, the inductor current ripple criteria is applied to avoid high frequency components in the system. In the DCMC, the different DC–DC full bridge converter versions perform particular tasks. 400 W, 190 V input and 48 V output version supplies up to six tandems driver-LED luminary from the secondary bus. 1.2 kW , 380 V input and 190 V output version takes power from primary bus to supply electronic loads, like personal computer and others in teachers offices. To carry out the transfer between the primary bus and the secondary bus, there are two more versions of the DC–DC full bridge topology, both with a capacity of 3.3 kW. The first one performs a step-down function by transferring power from the primary bus at 380 V to the secondary bus at 190 V. The second of these versions has step-up capacity and performs the opposite operation, transfering energy from the secondary bus with a voltage of 190 V to the bus primary bus at 380 V.

On the other hand, the DCMR architecture is a multibus array, so each bus has at least one GD, one ESS and some loads. In the selected system, there are two main buses, so there is one ESSs in each bus. The ESS of the primary bus is made up of a serial arrangement of twenty batteries. To carry out the power management in the ESS batteries in the primary bus, two versions of the full bridge DC–DC converter are used, the first of these versions with a capacity of 2.5 kW is used to charge the SAE from the 380 V bus at 250 V output. The second version for this task, also in 2.5 kW to take energy from the ESS at 250 V and set the primary bus to 380 V. The ESS in the secondary bus is composed by a series arrangement of ten batteries. To carry out the power management in the ESS batteries, another couple of versions of the DC–DC full bridge converter are used; the first of these versions has a capacity of 1.25 kW and is used for the ESS load from the 190 V bus and output at 125 V. In addition, a 1.25 kW version is included in the system to take power from the ESS at 125 V and sets the secondary bus at 190 V. The design procedure is based in procedure given in [38]. All of the versions perform in 50 kHz switching frequency. The values calculated for all the versions mentioned appear in Table 5.

Table 5. Parameter design values for all the versions of DC–DC full bridge converter.

Version (kW)	V_i (V)	V_o (V)	I_L (A)	R_o (Ω)	L (mH)	C_1 (μF)	C_2 (C_2)	$\%\Delta i_L$ (%)	$\%\Delta v_{C2}$ (%)
0.40	190	48	8.33	5.76	380	1	82	9.03	0.12
1.2	380	190	31.78	14.44	650	1	150	7.75	0.52
3.3	380	190	17.37	10.94	550	1	100	6.09	0.0076
3.3	190	380	43.76	8.68	2.3	1	100	5.97	0.0016
2.5	380	250	10	25	1.3	1	100	6.09	0.0032
2.5	250	380	6.58	57.76	3.1	1	100	4.46	0.0011
1.25	190	125	10	12.5	700	1	100	5.43	0.0065
1.25	125	190	12.5	28.88	1.5	1	100	5.13	0.0022

For each version, passive elements minimum obtained values for this topology are L_{min} and C_{min}; then, three times L_{min} value to meet current inductor ripple criteria and to keep the converter performing in CCM operation.

4.3. Cuk Converter Design

Cuk driver is connected to one of the unidirectional tertiary buses at 48 V derived from a DC–DC full bridge converter connected to the secondary bus. The Cuk converter particular objective is to supply a LED luminary with specific voltage and regulated current level. The design is based in [38]; the resultant values appear in Table 6.

Table 6. Parameter values design Cuk converter.

P (W)	f_s (kHz)	L_1 (mH)	L_2 (mH)	V_{in} (V)	V_o (Ω)	C_1 (nF)	C_2 (μF)
60	100	1.2	1.1	−68	43.76	220	1

4.4. DC–AC Bidirectional Full Bridge Converter Design

This topology allows the interaction MG between secondary bus and mains, when solar irradiation increases until there is a power amount greater than building demand; then, the DC–AC converter is able to inject power into the grid through a single phase. Otherwise, if the PV array in not collecting power enough to fulfill loads demand inside the place, the converter sets the bus by taking power from the grid. The power stage design is based in [39]. There are two versions in the proposed DCMG. The first one is 1.2 kW to manage power transfer between secondary bus and the mains. The second version is a 3.3 kW converter. This converter transfers power between the primary bus and the mains and supplies the EVCE. The obtanied parameter values appear in Table 7.

Table 7. Parameter values design for single phase DC–AC full bridge converter.

Version (kW)	V_{ac} (V_{RMS})	V_{DC} (V)	R_o (Ω)	L (mH)	r (Ω)	C (μF)
3.3	220	380	43.76	7	0.27	7200
1.4	127	190	43.76	4.1	0.2	4576

Coupling inductor is calculated for maximum active and reactive power; DC capacitor performs two functions: to filter voltage ripple in DC bus and storage power to compensate lack of active power.

5. Hierarchical Control Development

As mentioned before, a hierarchical control structure is selected to manage the system. In a summarized way, two main parts are distinguished: local control and coordinated control. This kind of control strategy, keeps voltage buses regulated by feedback the complete PEC state. Local control stage is able to receive enable signals, operation mode selection and references. In this manuscript, the coordinated control stage is based in an EMS that prioritizes the consumption of the power collected in PV arrays.

5.1. Local Controllers Development

Local control is the lowest level control stage placed in the converter that governs the PEC with in situ variable measurements. Local control is able to receive signals like references, enable and operation mode selection from higher levels controllers in order to perform coordinated control. Obtained PEC models are based in state equations in the time domain; then, the linear control design strategies are suitable for boost converter and DC–DC full bridge converters. State feedback controller uses the complete state information given by average model to obtain gain values to use, and to generate a control signal definied as $u(t)$. Outstanding features of this control strategy are: state feedback controller brings zero error in state stable and robustness to transients and input voltage variations but the design and tuning procedure is not trivial.

Cuk converter and single phase DC–AC full bridge converter perform with controllers based in classical strategies; this kind of controller is robust and easy to tune. For Cuk converter, PI controller performs current regulation, and for DC–AC converter, a cascade structure based in PI controllers with a current inner loop and voltage external loop is used for regulation.

5.1.1. Boost Converters Controllers Development

Every boost converters in the system is able to perform in two modes: output voltage regulation and maximum power point tracking (MPPT). Voltage regulation strategy is used when a specific bus supplies a medium or heavy load condition. An MPPT strategy is used when there is a lot of power harvested in PV arrays and medium or light load condition is supplied by an specific bus.

A. Voltage Regulation Controller.

A control linear technique called state feedback is selected for local controllers design. Evaluating model boost converter obtained in (1) as a static model, i.e., with all the variables in steady state, the conclusion is the boost converter is a nonlinear system. It is necessary to obtain a linearized model in a fixed operation point. First step to linearization is to obtain matrix **A** and **B** through Jacobian for the average model evaluated in a fixed operation point. For matrix **A**:

$$\mathbf{A} = \left. \begin{vmatrix} \frac{\partial f_1}{\partial v_{C1}} & \frac{\partial f_1}{\partial i_L} & \frac{\partial f_1}{\partial v_{C2}} \\ \frac{\partial f_2}{\partial v_{C1}} & \frac{\partial f_2}{\partial i_L} & \frac{\partial f_2}{\partial v_{C2}} \\ \frac{\partial f_3}{\partial v_{C1}} & \frac{\partial f_3}{\partial i_L} & \frac{\partial f_3}{\partial v_{C2}} \end{vmatrix} \right|_{\substack{u = D \\ v_{C1} = V_{C1} \\ i_L = I_L \\ v_{C2} = V_{C2}}} \quad (5)$$

For matrix **B**:

$$\mathbf{B} = \left. \begin{vmatrix} \frac{\partial f_1}{\partial u} \\ \frac{\partial f_2}{\partial u} \\ \frac{\partial f_3}{\partial u} \end{vmatrix} \right|_{\substack{u = D \\ v_{C1} = V_{C1} \\ i_L = I_L \\ v_{C2} = V_{C2}}} \quad (6)$$

Particular structure of matrix **A** in linearized model obtained is:

$$\mathbf{A} = \begin{bmatrix} \frac{-1}{R_t C_1} & \frac{-1}{C_1} & 0 \\ \frac{1}{L} & \frac{[(R_{D1} - R_{D2})D + R_{D2}]}{L} & \frac{-(1-D)}{L} \\ 0 & \frac{(1-D)}{C_2} & \frac{-1}{C_2 R_O} \end{bmatrix} \quad (7)$$

and for matrix **B**

$$\mathbf{B} = \begin{bmatrix} 0 \\ -\frac{[(R_{D1} - R_{D2})D + R_{D2}]i_L}{L} + \frac{V_{C2}}{L} \\ \frac{-I_L}{C_2} \end{bmatrix} \quad (8)$$

Now, matrix **A** and **B** are incorporated to linealized state equation:

$$\hat{\dot{x}} = \mathbf{A}\hat{x} + \mathbf{B}\hat{u}$$

with:

$$\hat{\dot{x}} = \begin{bmatrix} \frac{\partial \hat{v}_{C1}}{\partial t} \\ \frac{\partial \hat{i}_L}{\partial t} \\ \frac{\partial \hat{v}_{C2}}{\partial t} \end{bmatrix}$$

and:

$$\hat{x} = \begin{bmatrix} \hat{v}_{C1} \\ \hat{i}_L \\ \hat{v}_{C2} \end{bmatrix}$$

Input vector $\hat{u}$, in this case, has a single element u, it corresponds to modulating signal. The output equation is given by:

$$\hat{y} = \mathbf{C}\hat{x} + \mathbf{D}\hat{u}$$

For this application, direct transfer matrix has not elements $\mathbf{D} = 0$, and matrix $\mathbf{C}$ is defined as:

$$\mathbf{C} = \begin{bmatrix} 0 & 0 & 1 \end{bmatrix}$$

State feedback controller structure for boost converter is shown in Figure 6.

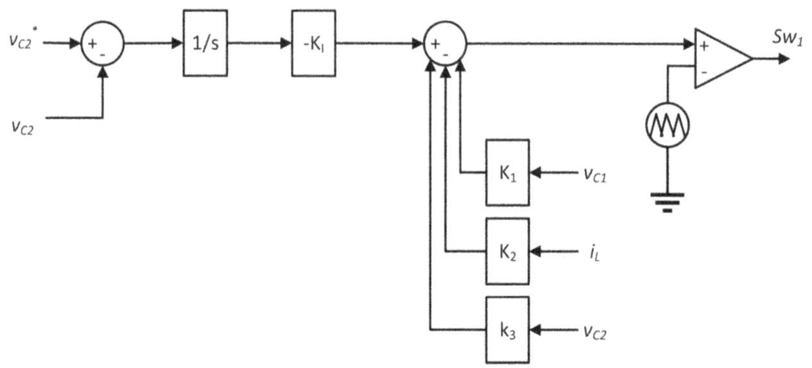

Figure 6. Feedback state controller structure.

Control scheme includes a gain-integrator trajectory to lead steady state error to zero and gain trajectories to feedback system state. The first step is to build up the augmented system, and the controllability matrix is required. For the second step, it is necessary to calculate gains values; the Ackermann method is applied to define gain values K_I, K_1, K_2 and K_3 as elements of the vector:

$$\mathbf{K_a} = \begin{bmatrix} K_1 & K_2 & K_3 & K_I \end{bmatrix}$$

Parameters such as overshoot percentage, $\%OS$, and settle time, T_s, are considered to calculate the desired polinomy. Two auxiliar poles are necessary to balance system order and the gain vector $\mathbf{K_a}$. The product of desired polynomial evaluated in terms of the augmented system and controllability matrix brings out spacific values for gains in matrix $\mathbf{K_a}$. Parameter values for desired polynomial of boost converter version 2.5 kW at 380 V output are: $\%OS = 0.1\%$ and $T_s = 12$ ms, two auxiliar poles are needed to balance the augmented system, the first one is placed one hundred times real part of the complex conjugate value of the dominant poles, the second pole is placed two hundred times real part of the complex conjugate value. The procedure is done again in order to obtain 190 V boost converter gain values. Selected parameters to calculate desired polynomial are: $\%OS = 0.1\%$ and $T_s = 8$ ms. Two auxiliar poles are needed to balance. The first one is placed seventy-five times real part of the complex conjugate value of the dominant poles. The second pole is placed one hundred and fifty times real part of the complex conjugate value of the dominant poles of the complex conjugate value.

B. MPPT Controller.

In specific periods of time during the day, particularly in sunny seasons of year, sun irradiance can increase; then, PV arrays harvest a surplus power. Under these conditions, it is possible to perform a better harness of this increased power collected by using a MPPT algorithm. There are many strategies to implement a MPPT algorithm, the most accepted of them are summarized in [40]. Nevertheless, the most popular algorithm and the one selected for this development is called perturb and observe ($P\&O$); this is a sampling technique that uses two variables measurement, current and voltage, to calculate power.

P&O can be implemented in analog version or in digital version and does not require a tuning procedure.

Voltage and current PV array output measurements are used to calculate a currente power data is calculated P_1; then, a voltage little variation (ΔV) or a duty cycle disturbance (ΔD) in boost converter leads to a new power value calculation, P_2. A comparision between P_1 y P_2 is performed. If P_2 is greater than P_1, disturbance moved in the right direction; else, disturbances must move in the opposite direction. In this way, it is possible to obtain the power peak value (P_{mpp}) and, consequently, voltage peak value (V_{mpp}). Algorithm flowchart of the MPPt controller is shown in Figure 7.

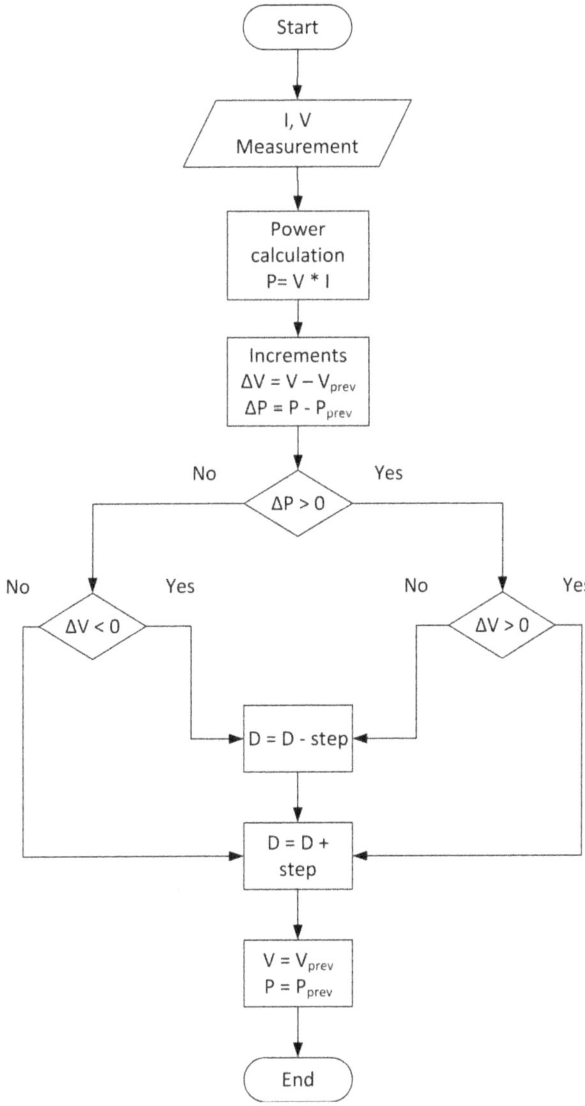

Figure 7. MPPT controller flowchart algorithm.

Behavior of boost converters, the two versions, performing in both modes: voltage regulation mode, in closed loop, facing load changes and MPPT controller perfomance are presented and commented in next section.

5.1.2. DC–DC Full Bridge Converters Controllers Development

The DC–DC full bridge converter model obtained in (2) is evaluated as a static model, i.e., with all the variables in steady state, and the obtained result is this kind of PEC is a linear system; then, a linearization procedure is not necessary. Once again, same control linear technique is used to design a feedback state controller, its structure is similar that shown in Figure 6, but, in this case, only two variables are feedbacked (v_{C2} and i_L). Controller structure includes a gain-integrator trajectory that leads steady state error to zero and other gain trajectories to feedback system state. Furthermore, it is necessary to calculate gain values K_I, K_1 and K_2; they are defined using the Ackermann method. Gains before mentioned are part of the gain vector given as:

$$\mathbf{K_a} = \begin{bmatrix} K_1 & K_2 & K_I \end{bmatrix}$$

The main parameters to design desired polynomial are overshoot percentage $\%OS$ and settle time T_s. Now, one auxiliar pole in needed to balance system order and gains vector $\mathbf{K_a}$. The product of desired polynomial evaluated in augmented system matrix and controllability matrix results values for gain matrix $\mathbf{K_a}$. The selected values for desired polynomial for DC–DC converter 400 W at 48 V are: $\%OS = 0.1\%$ and $T_s = 700$ μs. In order to obtain 1.2 kW DC-DC full bridge converter gain values, the selected parameters values are: $\%OS = 0.1\%$ and $T_s = 1.9$ ms. For bidirectional power transfer between buses, the step-down version 3.3 kW 380 V to 190 V DC-DC full bridge converter takes the following parameters for desired polynomial are: $\%OS = 0.1\%$ and $T_s = 1$ ms. The step-up version 3.3 kW 190 V to 380 V uses next parameters for this desired polynomial $\%OS = 1\%$ and $T_s = 2$ ms. In all cases, auxiliar pole is placed five times real part of the complex conjugate value.

To obtain the gains for the full bridge DC–DC converters for the charge–discharge of the ESS in the primary bus; the first one for battery charge, 2.5 kW and 380 V to 250 V, uses the following parameters to define the desired polynomial: $\%OS = 1\%$ and $T_s = 1.5$ ms. On the other hand, the converter that manages the battery discharge, 2.5 V and 250 V to 380 V, uses the following parameters to define the desired polynomial: $\%OS = 1\%$ and $T_s = 2.3$ ms. Finally, for the charge–discharge of ESS in the secondary bus. The first converter that manages charge cycles, 1.25 kW and 190 V to 125 V, uses the following parameters to define the desired polynomial: $\%OS = 1\%$ and $T_s = 1.1$ ms. The converter that perform discharge events for batteries uses the following parameters for desired polynomial: $\%OS = 1\%$ and $T_s = 1.6$ ms. Again, the auxiliary pole is placed five times the real part of the complex conjugate of the dominant pair. Results apperar in Tables 8 and 9.

Behavior of DC–DC full bridge converter, all versions, performing in close loop and facing load changes are presented and commented in next section.

Table 8. Gain values for both vesions of Boost converter.

Version	K_1	K_2	K_3	K_I
380 V	0.996579	0.205617	0.030030	12.038176
190 V	0.623042	0.119862	0.005269	17.391827

Table 9. Gain values for all the vesions of DC–DC full bridge bridge converter.

Version	V_{in}	V_{out}	K_1	K_2	K_I
400 W	190 V	48 V	0.065227	0.098410	364.003904
1.2 kW	380 V	190 V	0.020290	0.039492	35.027654
3.3 kW	380 V	190 V	0.032890	0.054304	117.792483
3.3 kW	190 V	380 V	0.008089	0.027744	5.407997
2.5 kW	380 V	250 V	0.013945	0.033055	31.541976
2.5 kW	250 V	380 V	0.009073	0.032140	12.959751
1.25 kW	190 V	125 V	0.029152	0.047667	83.196595
1.25 *KW*	125 V	190 V	0.0018966	0.047627	39.884327

5.1.3. Cuk Converter Controller Development

For Cuk converter controller design, classical control techniques are used. The 400 W DC-DC full bridge converter version is the previous stage to this converter and sets 48 V as input for the driver, which energizes the 60 W LED luminary at 68 V. The block diagram of the controller is shown in Figure 8.

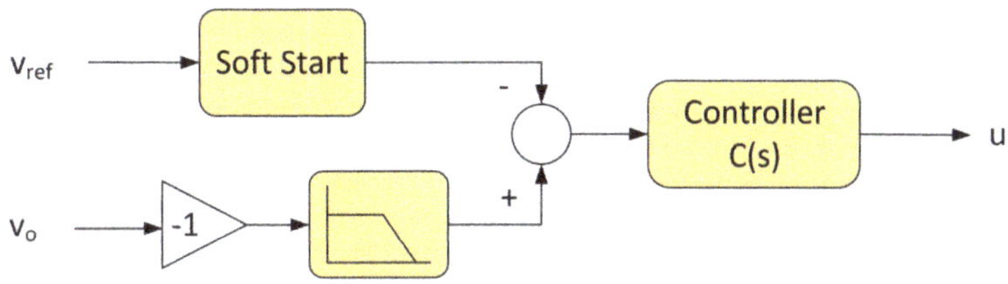

Figure 8. Cuk controller structure.

Due to the nature of the Cük converter, the output voltage is obtained with a negative polarity; for this reason, the measured voltage is multiplied by a gain value of −1. The measured voltage signal passes through a low-pass filter in order to make the control using the direct component of the converter output signal. On the other hand, the reference signal passes through a block that incorporates first order dynamics to achieve the effect of a soft start of the converter. Finally, the controller provides low-frequency gain, so it is proposed to place a zero at 400 Hz, so that the expression for the controller is given as:

$$C(s) = 1 + \frac{2\pi(400\ \text{Hz})}{s} \quad (9)$$

To validate the operation of the converter and the LED luminaire, the closed loop converter perform appears in results section.

5.1.4. DC–AC Bidirectional Full Bridge Converters Controllers Development

The controller for DC–AC bidirectional converter must be able to regulate DC bus and, at the same time, operates with unitary power factor, both tasks performing as rectifier. The other way, besides inject power to mains, converter is able to compensate reactive power and harmonics operating in inverter mode. Controller structure is composed by two stages, reference signal generation and voltage–current cascade controller, controller output is a set of sinusoidal signal u that is processed in a SPMW modulator to obtain a 3 ϕ switching pattern for converter power devices. The complete schematic is shown in Figure 9.

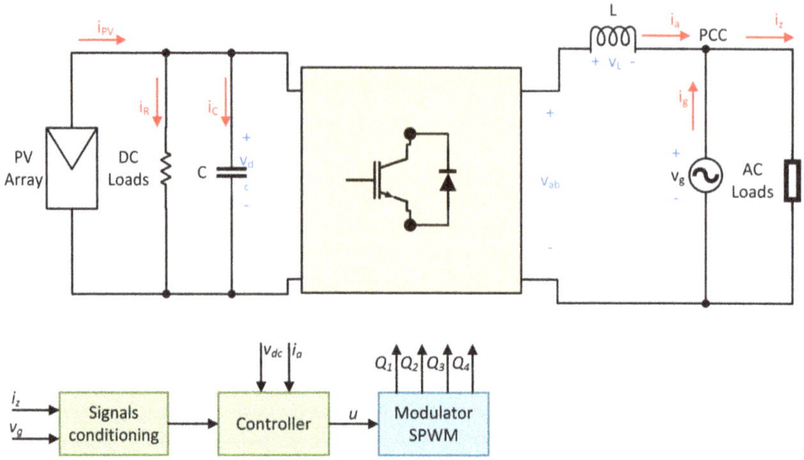

Figure 9. Converter and controller scheme.

Single phase dq transformation is applied to current circulating in AC side, i_z, to place the variable in a rotating reference frame (see Figure 10).

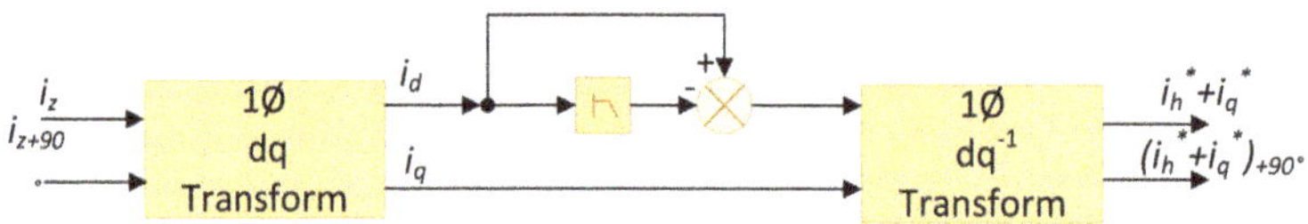

Figure 10. Reference signal generator.

The dq transform is applied to current circulating in AC side, i_z, to place it in a rotating reference frame, mathematically depicted as:

$$\begin{bmatrix} i_d \\ i_q \end{bmatrix} = \begin{bmatrix} sin\phi & -cos\phi \\ cos\phi & sin\phi \end{bmatrix} \begin{bmatrix} i_z \\ i_{z+90^\circ} \end{bmatrix}$$

In this rotating frame i_z current components can be separated in: I_d, term corresponds to active power and I_q referred to reactive power; the sum of the remaining terms meet current harmonics components, i_h. After transformation, i_d and i_q are:

$$\begin{aligned} i_d &= I_d + \tilde{i_d} \\ i_q &= I_q + \tilde{i_q} \end{aligned} \tag{10}$$

Resultating signal are composed by a constant levels of I_d and I_q and several frequencies components, $\tilde{i_d}$ y $\tilde{i_q}$; it can de expressed as (10). In order to do nonactive power compensation, I_d is filtered; this way, active power is separated. Then, the obtained signal is subtracted from i_d and takes back to the time reference frame through dq inverse transform; this mapping operation is defined by:

$$\begin{bmatrix} i_d^* + i_h^* \\ (i_d^* + i_h^*)_{+90^\circ} \end{bmatrix} = \begin{bmatrix} sen\phi & cos\phi \\ -cos\phi & sen\phi \end{bmatrix} \begin{bmatrix} i_d - I_d \\ i_q \end{bmatrix}$$

For this application, control objectives are: DC bus regulated and active power management. The control structure shown in Figure 11 includes two PI classical controllers in a cascade configuration. The external loop regulates DC bus to reference value v_{dc}^*, and its output is a part of current signal reference i_d^* referred to active power demanded by load.

The complete expression for inner loop reference is the sum of i_d^*, non-active terms $i_q^* + i_h^*$ and a compensation term $\frac{2}{V_g}$.

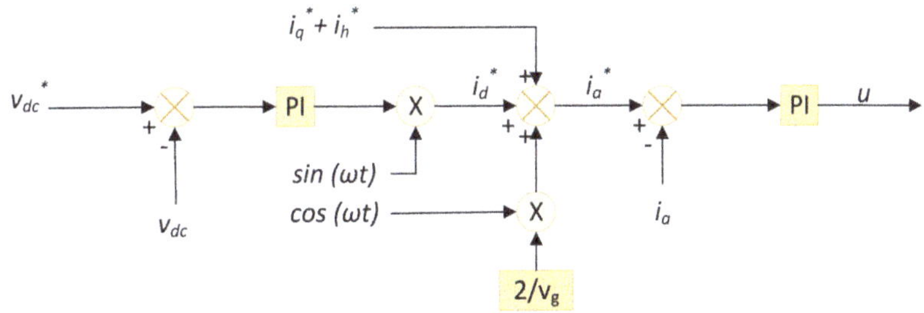

Figure 11. Cascade controller structure.

Considerations for controller design are: first, current inner loop is, at least, ten times faster than voltage external loop; second, for tuning, external loop PI controller gain must be selected carefully to avoid important current transients if an excessive value were selected. Parameter values for PI controllers for the 3.3 kW version at 220 V_{RMS}, and parameter values for PI controllers for the 1.5 kW version at 127 V_{RMS} appear in Table 10.

Table 10. Parameter values for DC–AC full bridge PI controllers.

Version	K_v	τ_v	K_i	τ_i
3.3 kW	10	0.12 s	0.08	0.03 s
1.5 kW	10	0.1 s	0.08	0.025 s

Behavior of DC–AC bidirectional single phase full bridge converter performing in close loop taking and injecting power to mains are presented and commented in next section.

5.2. Energy Management System Design

As supervisory control level, an EMS is included to control the multibus multivoltage DCMG. The main idea to design the EMS is to perform in a similar way to an industrial production system. The EMS is programmed in a central concentrator (CC) hardware; from here, enable signals are generated in order to harness all the energy harvested in PV arrays in various consumption points of the system. In order to reach harmonic performance of the DCMG, a list of loads priorities have to be met:

1. To supply loads into the building (lighting system and electronics loads)
2. To perform charge EV sequence or inject power to mains
3. To carry out the charge of the ESS batteries

Once the order of priorities is known, initially, EMS performs a logic startup to define an initial condition of the sistem called initial state of the system (μ_{ko}). A first sequential stage takes control of those PEC that can provide power to the system eventually, as a source condition. Another sequential stage controls the system load condition by enabling PECs that may consume power in certain periods of time. In order not to occupy too many resources of the CC, source condition and load condition changes are performed by one finite machine state (FMS) each.

The EMS design is based in Petri net theory. Petri nets applications are popular in industrial production items, logistics, resources management and others. In other hand, just a few developments are reported in power systems such as study of protection dynamic behavior [41] and hierarchical restoration after a fail event [42]. Nevertheless, due to discrete nature of the concept, it is possible to use a Petri net to read the new measurements, process them and update the system state of the corresponding transitions (enable signals) in a fixed period of time. The basic fundamentals of Petri net appear in a summarized way in [43].

EMS proposed consists of three stages: the startup test, source converters perform in closed loop at one tenth of the rated power to know how many and what are the PECs available to work. With the information collected, it is possible to set the DCMG initial state μ_{ko}. The second stage is a set of FMS to enable-disable source PECs; the most adequate FMS, in the current period, is selected to perform in concordance with the previous state of the system (μ_{k-1}). In the same way, the third stage is a set of FMS to enable–disable load PECs, the most adequate FMS; for the current period, it is selected to perform in concordance with the previous state of the system. An EMS block scheme is shown in Figure 12.

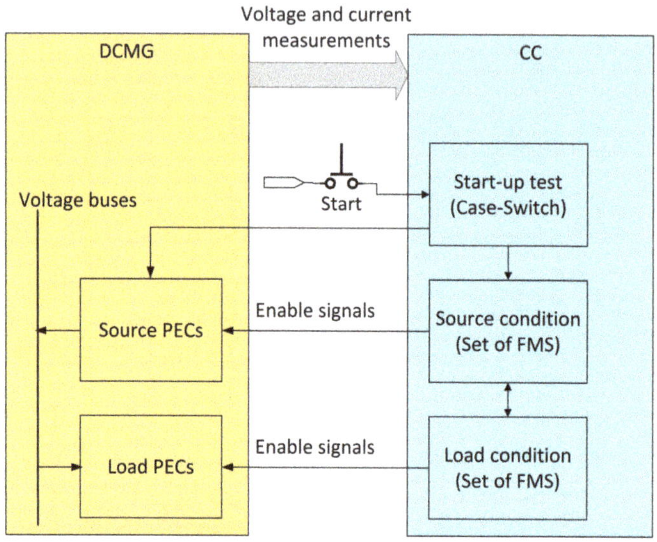

Figure 12. EMS stages block diagram.

After startup is performed and the initial state (μ_{ko}) is defined once, then, the two FMSs are selected to move to a new states (μ_k , μ_{k+1}, . . .). Voltage and current measurements are elements of the state vector, which is divided into two: the part associated with sources and the part associated with loads. In order to get the next period of time ($k + 1$), measurements are evaluated by the EMS and the FMSs activates the necessary enable signals (transitions) to move to a new state or remains in the same state for another period of time. The DCMG as a Petri net appers in Figure 13.

In Petri nets theory, four basic elements can be identify: place (circle), transition (rectangle), arc (arrow) and token (dot), they are associated with DCMG elements in a specific way. A place can represent as one of the following options: a tandem boost PV array and boost PEC ($APV1M$–$APV4M$), a DC-DC full bridge PEC and the electronic load (EC) or six LED luminaries (LED), ESSs battery charger PEC ($BAT1C$–$BAT2C$), also primary bus and secondary bus are considered ($BUSP$, $BUSS$) places, but they are not able to store power more than one period of time. A transition represents the enable signal of a PEC, since Petri net point of view, an active transition allows the power flow ($T0$–$T17$). An arc is the medium through power flows, it also indicates the direction of the flow. A token represents a power unit (W, kW, etc.). The places $APV1M$–$APVM4$ are connected to the respective bus through two trajectories arc and transition. One transition is activated if boost converter operates with feedback state controller in closed loop, voltage regulation mode. The other transition is turned on when this kind of PEC operates in MPPT mode.

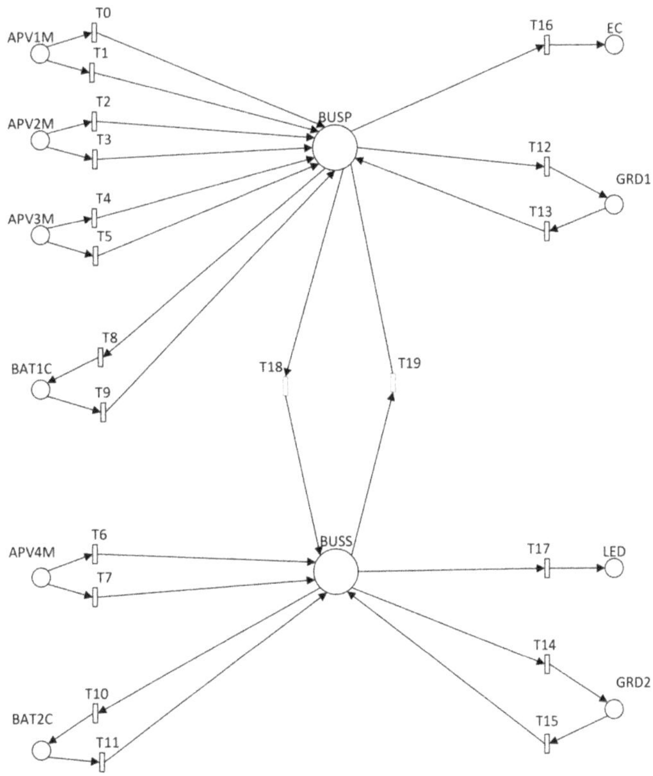

Figure 13. DCMG multibus multi-voltage Petri net scheme.

CC receives and quantifies the following measurements: PV arrays power output, the three linked to primary bus ($PV1$–$PV3$) and for the PV array related to secondary bus, $PV4$. Generally, if $i = 1,2,3,4$ then PVi is any of the PV arrays. If $PVi = 0$ the PV array is not working because is not connected or damaged. If $PV1 = 1$, low power is harvested form PV array, enough to set the bus, but not a lot of loads can be supplied. At $PVi = 2$, there is power enough to supply all the loads in the building and inject power to mains. If $PVi = 3$ there is surplus power harvested, taskes before mentioned can be done also a ESS batteries charge sequence is possible. This quantification is process in CC in binary form using two digits.

In a similar way, ESS batteries SoC is associated with parameters $BAT1$, for ESS in primary bus and, $BAT2$ is relaed to secondary bus. In general way, if $i = 1,2$ then $BATi$ is any of the ESS batteries. Now, if $BATi = 0$ ESS is not working for one of the following reasons: ESS is not connected or ESS is damaged. If $BAT = 1$, SoC is low and is not possible to keep the the bus regulated with some loads. When $BAT1 = 2$, SoC is at medium level is possible to supply loads and perform a charge sequence for an EV. If $BAT = 3$, SoC level value is high, then, all the loads into the building can be supplied, perform charge sequence for EV or inject power to mains. Again, $BATi$ value es represented with two binary digits.

Mains has two point of connection to DCMG, one through primary bus, $GRD1$, and the other via secondary bus, $GRD2$. Again, to assert in general way, $i = 1,2$, then $GRDi$ is any of mains voltage measurements. The quantification associated to these parameters is: if $GRDi = 0$ there is not voltage in mains, else, if $GRDi = 1$, there is voltage in point connection with mains. At this time, $GRDi$ is a binary number with two possible values.

All the paramemters before mentioned are concatenated to build the DCMG state vector. Depending on the state vector information, the needed transitions are activated to move to a more convenient state in the corresponding FMSs to reach a new source condition or load condition or maintain the current state.

5.2.1. Startup Stage

In the proposed CDMG there are eight PEC able to supply system loads, source converters. The startup test stage enables all the source PECs at one tenth of rating power in closed loop and in stand-alone operation. The aim of this test is to verify how many and what PECs are able to be connected to the corresponding bus through plug-and-play. Once the PEC is enabled, the test last a period of time enough to let pass the transient and reach steady state, in this condition output voltage measuremet is compared with output reference value in order to verify if measurement value is into $\pm 10\%$ reference value. If the test is pass by a specific PEC, the corresponding bit is set '1', else is set '0', and data is sent to CC. With the resulting eight bits information, CC takes a decision and set the system initial state of source condition. Let SC be the source active bit vector defined and ordained as shown in Table 11.

Table 11. SC vector.

$SGD2$	$SGD1$	$SBT2$	$SBT1$	$SPV4$	$SPV3$	$SPV2$	$SPV1$
$Bit7$ MSB	$Bit6$	$Bit5$	$Bit4$	$Bit3$	$Bit2$	$Bit1$	$Bit0$ LSB

In a detailed way, $SGD1$ and $SGD2$ are bits associated with the DC-AC full bridge converters condition, particularly when perform in rectifier mode as source PEC. If one of them passes the initial test or stays in on mode then $SGDi = 1$, else, $SGDi = 0$ if not passes the initial test or stays in off mode. $SBT1$ and $SBT2$ are bits associated with the step-up DC-DC full bridge converters at the ESSs, in batteries discharge mode. If one of them passes the inicial test or stays on mode $SBTi = 1$, else, $SBTi = 0$ if not passes the initial test or stays in off mode. Finally, $SPV1$, $SPV2$, $SPV3$ and $SPV4$ are bits associated with the boost converters condition, specially when performs in regulation mode. If $SPVi = 1$, PEC is able to operate or stays in on mode, else, $SPVi = 0$ if not able to operate or in off mode.

With eight PECs in the system, one bit each, there are 256 possible options to initialize the DCMG operation. In the start-up stage, the CC prioritizes the consumption of power collected in PV arrays, then ESS power is harness, finally, power from mains is taken. It means, only 22 combinations are available options to become initial DCMG state, most of the using PV arrays. If there is no PEC able to pass the start-up test, DCMG goes back to system off condition.

5.2.2. Source Condition Establishment Stage

After the startup test, the initial state is set in the DCMG already, a FMS is selected from a set. There are some options for FMS, depending on state vector v_k measurements. If V_k shows another sosource ready to provede power to system, vector SC change its value in the corresponding parameter, a new FMS is selected.

5.2.3. Load Condition Establishment Stage

In the proposed DCMG, there are six load PECs. The most critical loads are lighting system (LLED) and DC-DC full bridge converters to supply electronic loads (LEC). The rest of the load PECs are: two step-down DC-DC full bridge converters to charge batteries in ESSs and the before mentioned DC-AC full bridge converters, particularly in inverter mode. If the load PEC is demanding power, its bit value is set '1', else is set '0', and data is sent to CC. With the resulting six bits information, CC takes a decision and set the state of the FMS load condition, FMS source condition state is already known. Let be LC the load active bit vector defined and ordained as shown in Table 12.

Table 12. *LC* vector.

*LGD*2	*LGD*1	*LBT*2	*LBT*1	*LLED*	*EC*
*B*5 MSB	*B*4	*B*3	*B*2	*B*1	*B*0 LSB

In a detailed way, $LGD1$ and $LGD2$ are bits associated with the DC–AC full bridge converters condition when perform in inverter mode, as load PEC. If one of them is injecting power to mains $LGDi = 1$, else, $LGDi = 0$ if is not injecting to mains or stays in off mode. $SBT1$ and $SBT2$ are bits associated with the step-up DC–DC full bridge converters at the ESSs, in batteries charge mode. If one of them is in charging sequence $LBTi = 1$, else, $LBTi = 0$ if stays in off mode. Finally, $LLED$ and LEC are bits associated with the DC–DC full bridge converters, the first one supplies one to six cuk converter and LLED luminarie and the second powers electronic loads in a teachers office. Both, If $LLED = 1$ is suppling to lighting system, else, $LLED = 0$ if PEc is in off mode. Same logic is applied to LEC.

6. Results

In order to validate the novel architecture concept Psim simulation results are presented. The section is divided into two subsections: first, simulation results of converters performing in closed loop in single stage. Second, some trajectories of the multibus multi-voltage DCMG architecture supplied form PV arrays are tested to validate the correct operation of the system only with local control.

6.1. Performance of Converter in Stand-Alone

In order to validate models, designs and controllers of each converter version is presented performing in stand-alone, most of cases supplied from an ideal source to validate stand alone performance while faces load changes. Inductor current ripple criterion must be met to avoid high frequency components circulation during cascade connection events.

6.1.1. Boost Converter Closed Loop Performance

Boost converters in the DCMG are able to operate in two modes, output voltage regulation using and MPPT. Firstly, the both versions of boost converter in regulation mode behavior are verified with the same load sequence. The simulation test starts at 100% load condition, then, 22.5% load steps are removed until reach a final load at 10%. At 0.5 s, 22.5% load steps are connected back until meet 100% load condition again. First version, 380 V output voltage boost converter behavior appears in Figure 14.

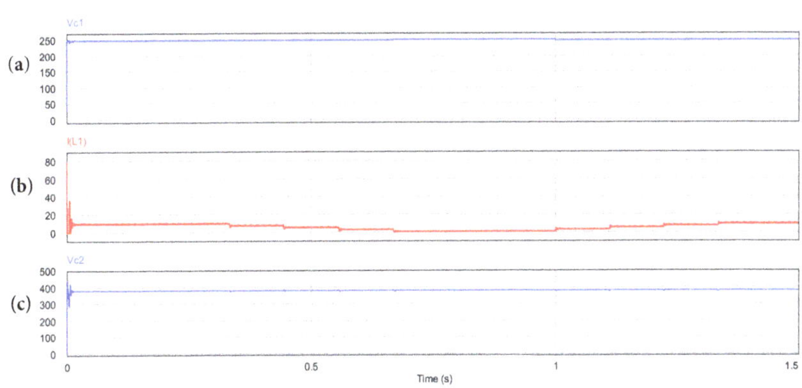

Figure 14. Boost converter 2.5 kW at 380 V with state feedback controller. (**a**) input voltage, (**b**) inductor current and (**c**) output voltage.

No significant dynamics is observed in capacitor input voltage at starting or during load steps, see Figure 14a. Inductor current ripple is not greater than 10% nominal value and converter performs in CCM all the simulation time, see Figure 14b. Overshot is noticed in DC bus voltage at the starting because not higher level of control is included, in steady state, transitions are smooth during load changes, see Figure 14c. Boost converter at 190 V output voltage simulation results appears in Figure 15.

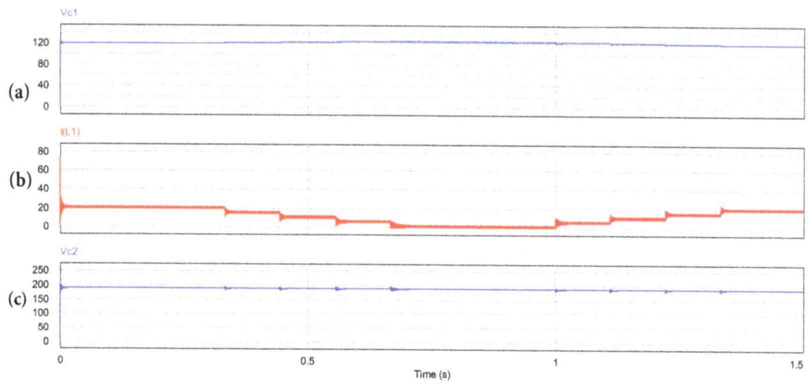

Figure 15. Boost converter 2.5 kW at 190 V with state feedback controller. (**a**) input voltage, (**b**) inductor current and (**c**) output voltage.

Again, no important transients are observed in capacitor input voltaje at starting neither during load transitions, see Figure 15a. Inductor current ripple keeps in criteria value and the converter performs in CCM, see Figure 15b. Starting oscillations take more time than those in 380 V version, they appear at starting and load transitions but overshot is not significant, see Figure 15c. A simulation to evaluateas a 380 V boost converter in MPPT mode is perform using *P&O* algorithm, Figure 16.

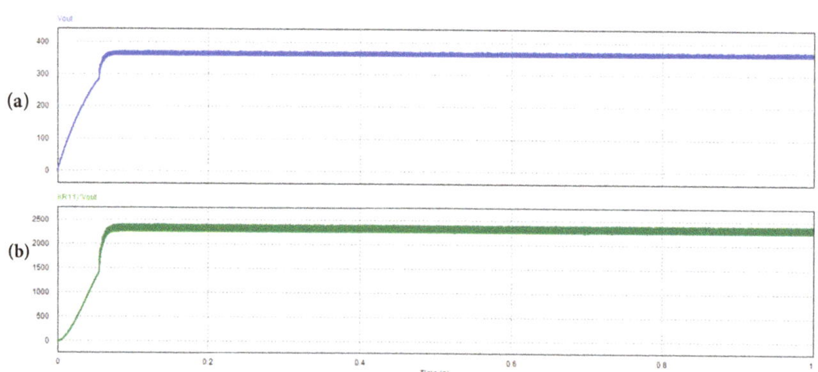

Figure 16. Sixth set of simulation results, (**a**) primary DC bus and (**b**) power harvested from PV array.

A sequence from 0% to 100% PV array rating power shows: in Figure 16a boost converter output voltage to set primary bus, mainly voltage value keeps around 380 V but, voltage ripple performing in this mode is larger than in regulation mode deu to. In Figure 16b waveform of power taaken form PV array in time domain.

6.1.2. DC-DC Full Bridge Closed Loop Performance

In the DCMG proposed there are several DC-DC full bridge converters versions, in this subsection some of this versions results are presented, PECs selected to test are 400 W at 48 V, 1.2 kW at 190 V and 3.3 kW at 190 V. First, 400 W at 48 V topology supplies six 60 W tandems Cuk driver-LED luminary and the simulation is relative to this dynamics. Simulation results are shown in Figure 17.

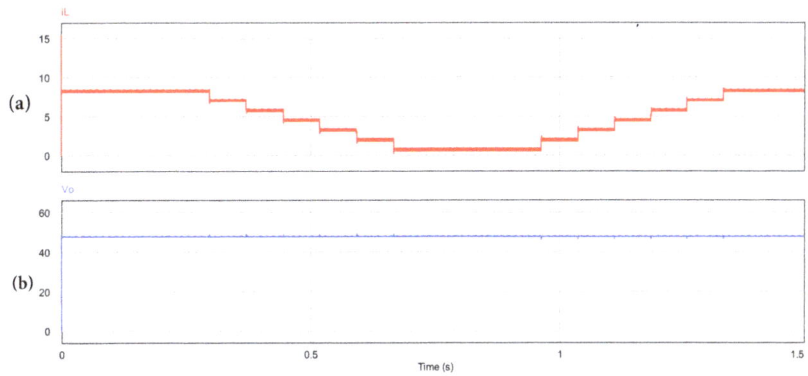

Figure 17. DC-DC full bridge converter 400 W at 48 V with state feedback controller. (**a**) inductor current, (**b**) output voltage.

Simulation test starts with 100% load condition, then tandems are disconnected one by one until converter remains at 10% load operation, only with bleeding resistor connected. Further, tandems are connected one by one to meet 100% load condition. In this case, inductor current and output current is the same. Inductor current ripple keeps its value according the critieria and converter keeps on CCM during the test, Figure 17a. There are not significant overshot in DC bus at luminaries connection/disconnection sequence, Figure 17b. Converter DC-DC full bridge 1.2 kW at 190 V output version behavior is tested, see Figure 18.

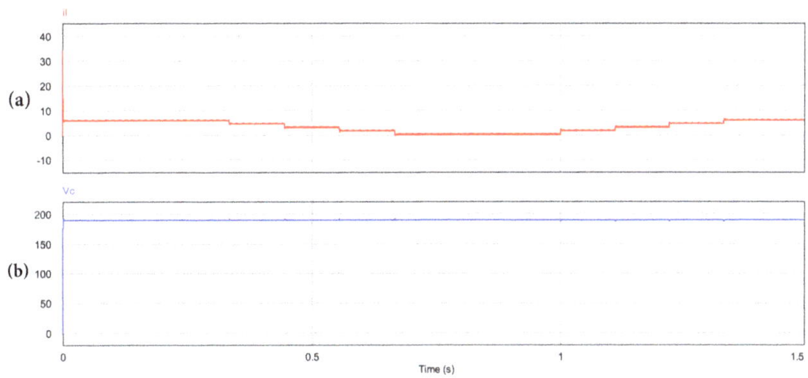

Figure 18. DC-DC full bridge converter 1.2 kW at 190 V with state feedback controller. (**a**) inductor current, (**b**) output voltage.

Again, simulation starts at 100% load condition, then 22.5% load steps are removed until 10% nominal load. At 1 s, 22.5%, load steps are connected back until meet 100% load condition again. Smooth transients presence is noticed in inductor current ripple and CCM condition remains during the test, see Figure 18a. No significant overshot appears in DC bus at load connection/disconnection sequence, Figure 18b. Finally , converter DC-DC full bridge 3.3 kW at 190 V version behavior is tested in the same way than before mentioned versions, see Figure 19.

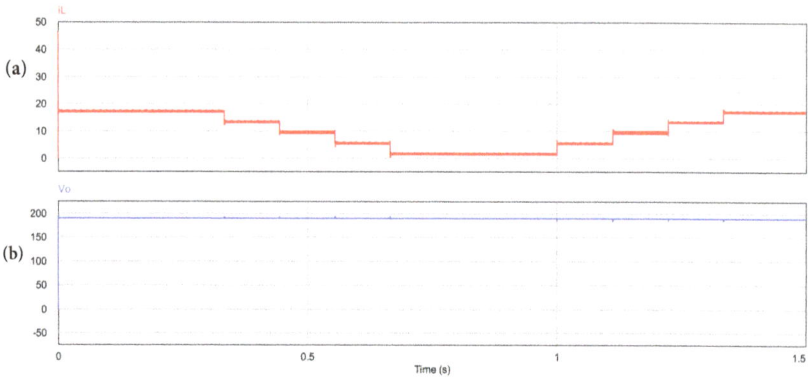

Figure 19. DC-DC full bridge converter 3.3 kW at 190 V with state feedback controller. (**a**) inductor current, (**b**) output voltage.

Test simulation starts at 100% load condition, then 22.5% load steps are disconnected until final load at 10% nominal power. At 1 s, 22.5% load steps are connected back until meet 100% load condition again. Current ripple increases at 55% load condition but ripple current criteria is met and CCM reamins during the test, see Figure 19a. Smooth transients apear in DC bus, with no overshot, at load connection/disconnection sequence, see Figure 19b. All the DC-DC full bridge converters versions remaining are tested in the same way, similar results were obtained in all cases.

6.1.3. Cük Converter

Cük converter closed loop behavior is validated with a simulations. Simulation time goes from soft start to steady state. No load changes are considered because a PV array with no variant features is always supplied by the driver. Four dynamics appear in model in (3), inductor one current and capacitor one voltage are shown in Figure 20.

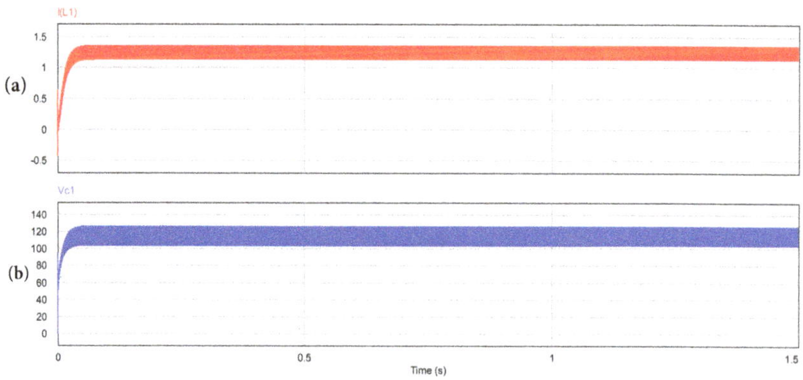

Figure 20. Cuk converter performing in closed loop. (**a**) inductor one current, (**b**) capacitor one voltage.

A 0.05 s soft start is perform at the beginning, no transient and well regulated variables are observed. The other pair of state variables, inductor two current and capacitor two voltage appear in Figure 21.

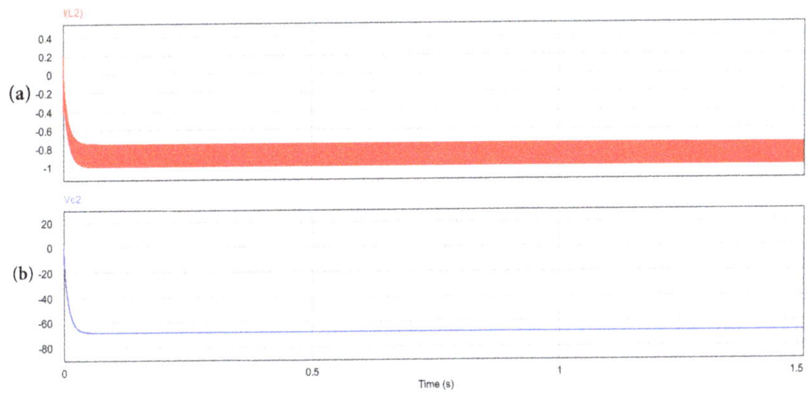

Figure 21. Cuk converter performing in closed loop. (**a**) inductor two current, (**b**) capacitor two voltage.

Inductor two curren ripple is 20% nominal value, this condition avoids undesirable effects in light spectrum. Voltage output, v_{C2}, performs a smooth start and reminds regulated during simulation time.

6.1.4. DC-AC Bidirectional Single Phase Full Bridge Converter

This converter features bidirectional capability, then, two simulations are performed in order to vaulate DC-AC single phase full bridge converter. The first one, in rectifier mode, 1.5 kW are transferred from AC to DC side, Figure 22 shows behavior in steeady state, negative sign in active power indicates power flows from the grid to DC side. Regulated DC bus with small voltage ripple can be seen in Figure 22a, mains current and voltage appear in Figure 22b, a 180° angle phase exists between grid voltage and grid current, this condition define the sign for active power; active and reactive power appears in Figure 22c reactive power is near to zero and converter operates with unitary power factor.

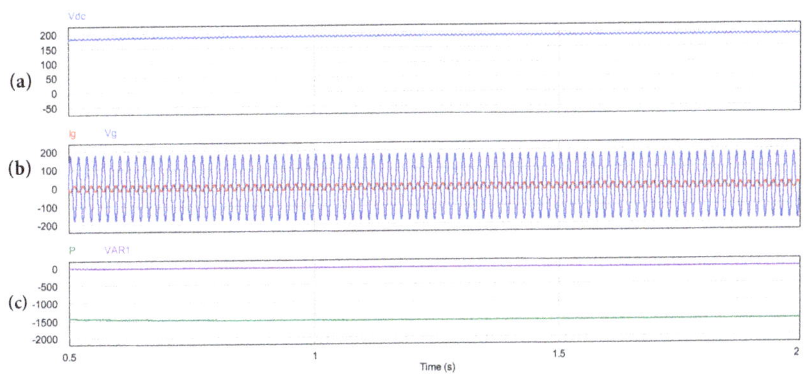

Figure 22. 1.5 kW DC-AC single phase full bridge converter, closed loop performing in rectifier mode. (**a**) DC bus voltage, (**b**) grid current *(red)* and grid voltage *(blue)*, (**c**) active *(green)* and reactive *(purple)* power.

Next simulaton corresponds to inverter operation, 1.5 kW are transfered from PV array to mains, Figure 23 shows behavior in stady state, positive sign in active power indicates power flows from DC side to the grid. Regulated DC bus with small voltage ripple can be seen in Figure 23a, mains current and voltage appear in Figure 23b, there is a zero phase angle between gird current and voltge, then, the positive sign is defined for active power, active and reactive power appears in Figure 23c; and there is no reactive power to compensate.

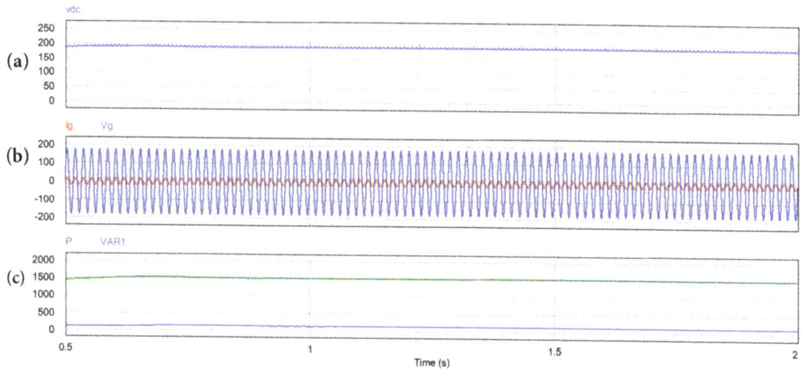

Figure 23. 1.5 kW DC-AC single phase full bridge converter, closed loop performing in inverter mode. (**a**) DC bus voltage, (**b**) grid current *(red)* and grid voltage *(blue)*, (**c**) active *(green)* and reactive *(purple)* power.

The second version in rectifier mode, 3.3 kW are transferred to DC side, Figure 24 shows behavior in steady state, negative sign in active power indicates power flows from the grid to DC side. Regulated DC bus with small voltage ripple can be seen in Figure 24a, mains current and voltage appear in Figure 24b, again a 180° angle phase exists between grid voltage and grid current, this condition define the sign for active power; active and reactive power appears in Figure 24c reactive power is near to cero and converter operates with unitary power factor.

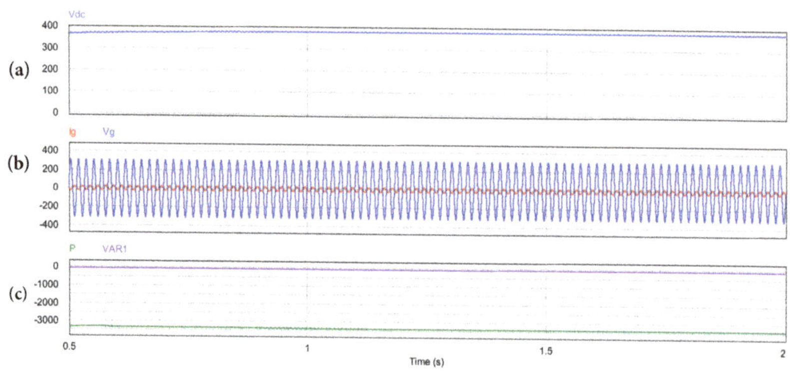

Figure 24. 3.3 kW DC-AC single phase full bridge converter, closed loop performing in rectifier mode. (**a**) DC bus voltage, (**b**) grid current *(red)* and grid voltage *(blue)*, (**c**) active *(green)* and reactive *(purple)* power.

This simulation corresponds to inverter operation, 3.3 KW are transfered from PV array to mains, Figure 25 shows behavior in stady state, positive sign in active power indicates power flows from DC side to the grid. Regulated DC bus with small voltage ripple can be seen in Figure 25a, mains current and voltage appear in Figure 25b, there is a zero phase angle between grid current and voltage, then, the positive sign is defined for active power, active and reactive power appears in Figure 25c; and there is no reactive power compensation.

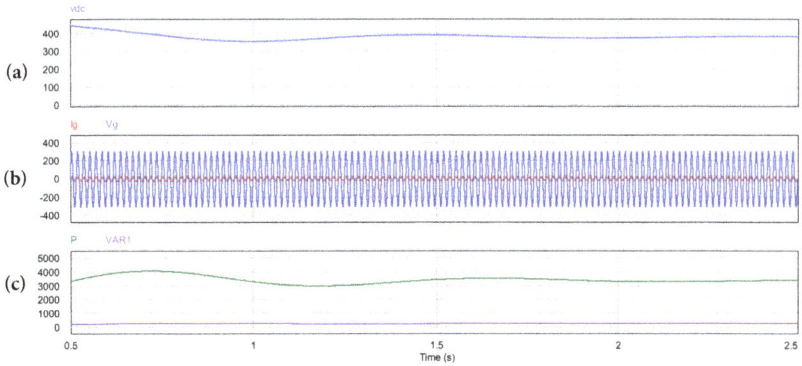

Figure 25. 3.3 kW DC-AC single phase full bridge converter, closed loop performing in inverter mode. (**a**) DC bus voltage, (**b**) grid current *(red)* and grid voltage *(blue)*, (**c**) active *(green)* and reactive *(purple)* power.

DC-AC bidirectional converter results confirm that is possible to have an DCMG able to interact with mains, it feature gives several possibilities for energy administration in the system.

6.2. Performance of DCMG Multibus Muti-Voltage

In order to validate multibus multi-voltage DCMG architecture concept two trajectories in the architecture are tested by simulation. Only state feedback controller as local control stage is included, no higher level controllers are considered. PECs are enabled by an digital signal and the output voltage reference signal. First trajectory to evaluate appear in Figure 26.

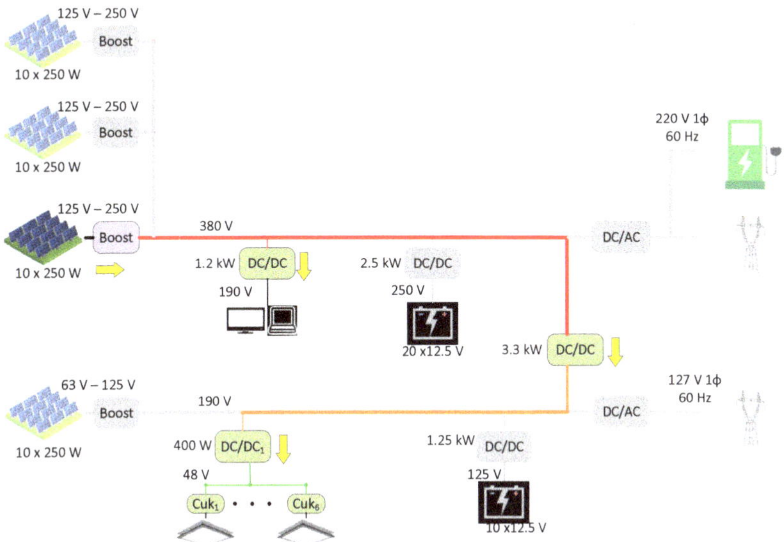

Figure 26. First test trajectory for DCMG multibus multi-voltage operating with only one boost converter.

Primary DC bus is generated by a single boost converter and supplies the following elements: a single 1.2 kW converter that supplies a 100 W electronic load; a 3.3 kW converter to generate secondary DC bus that supplies a 400 W converter to supply six tandems Cuk driver-LED luminaire as final load. Simulation results are shown in Figure 27.

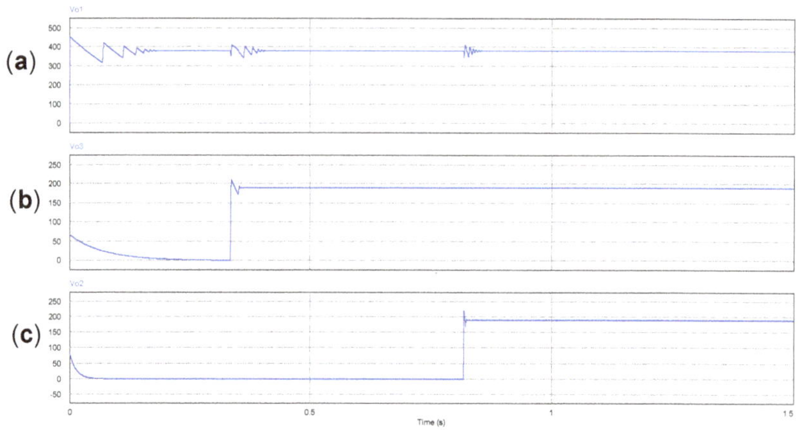

Figure 27. First set of simulation results, (**a**) primary DC bus, (**b**) 1.2 KW converter output voltage and (**c**) secondary DC bus.

Significant oscillations appear at the beginning of the simulation in primary bus, see Figure 27a; this kind of transient can be overcome using a secondary control loop in a centralized communication configuration for DCMG. Then, reduced amplitud oscillations are produced in primary bus by input capacitors in load converters being connected to primary or secondary buses. At 0.3 s, 1.2 kW converter is enabled and a smooth transient is produced in the primary bus Figure 27b. At 0.8 s secondary bus is established by 3.3 kW converter and corresponding transitory can be noticed in primary bus, again a negative exponential voltage dynamics can be noticed at the starting of simulation, see Figure 27c. Some results of the same test are presented in Figure 28.

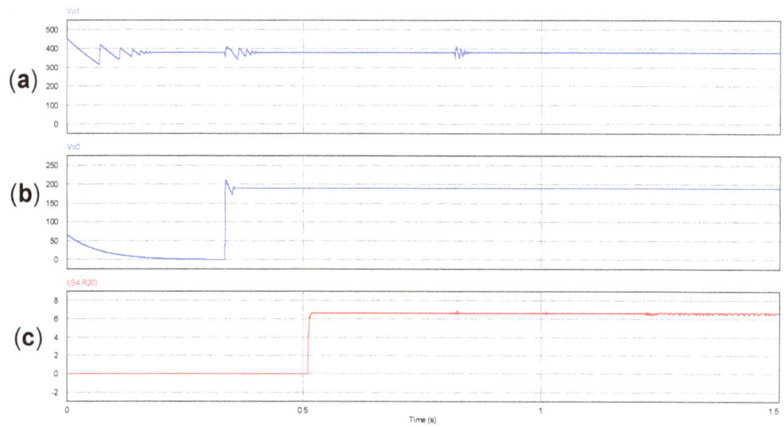

Figure 28. Second set of simulation results, (**a**) primary DC bus, (**b**) 1.2 kW converter output voltage and (**c**) electronic load input current.

Once 1.2 kW is enabled at 0.3 s, further at 0.5 s a personal computer is turned on, Figure 28. Waveforms are: in Figure 28a, principal bus voltage and Figure 28b, 1.2 kW converter output voltage are used like time references; Figure 28c shows current input electronic load enabled at 0.5 s, some ripple can be seen caused by loads dynamic in secondary bus. More results are presented in Figure 29.

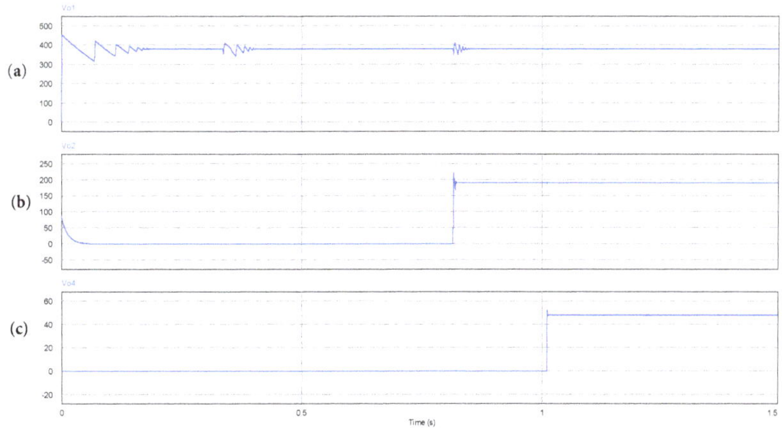

Figure 29. Third set of simulation results, (**a**) primary DC bus, (**b**) 1.2 kW converter output voltage and (**c**) terciary lighting bus.

Waveforms are: Figure 29a, primary bus voltage; and Figure 29b, 1.2 kW converter output voltage are used like time refrences; Figure 29c represents a tertiary lighting bus enabled at 1 s by 400 W converter, well regulated 48 V bus can be seen despite connection/disconnection of tandem driver Cuk-LED luminaries. The furthest functional trajectory for DCMG is from boost converter to lighting tandem driver-luminary, Figure 30 shows this power path.

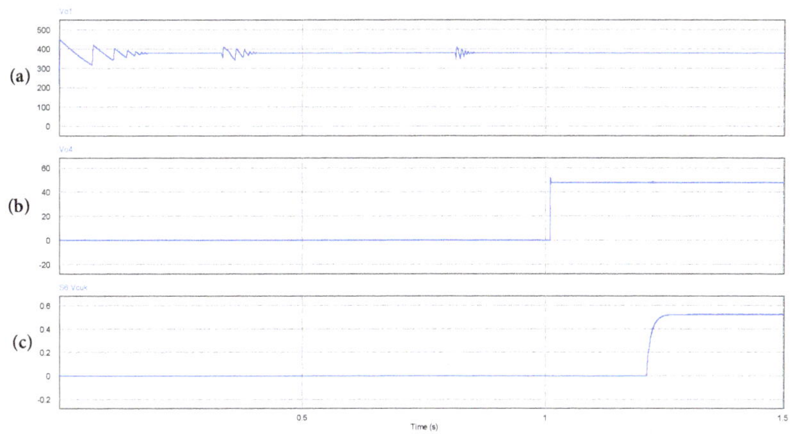

Figure 30. Fourth set of simulation results, (**a**) primary DC bus, (**b**) 1.2 kW converter output voltage and (**c**) Cuk driver reference signal.

Waveforms are: in Figure 30a, primary bus voltage as time reference; 400 W voltage output converter, Figure 30b, finally, Cuk convertert voltage output to supply one of the six luminaries on the set, see Figure 30c. The same longest trajectory is reviewed again, but luminaries connection appear in Figure 31.

In Figure 31a, primary bus voltage as time reference is shown; secondary bus is shown in Figure 31b, in Figure 31c, 400 W converter current output suppling six tandem Cuk driver-LED luminary connection sequence, current ripple increases when another luminary is connected, at full load condition current ripple is about 10% of nominal inductor current.

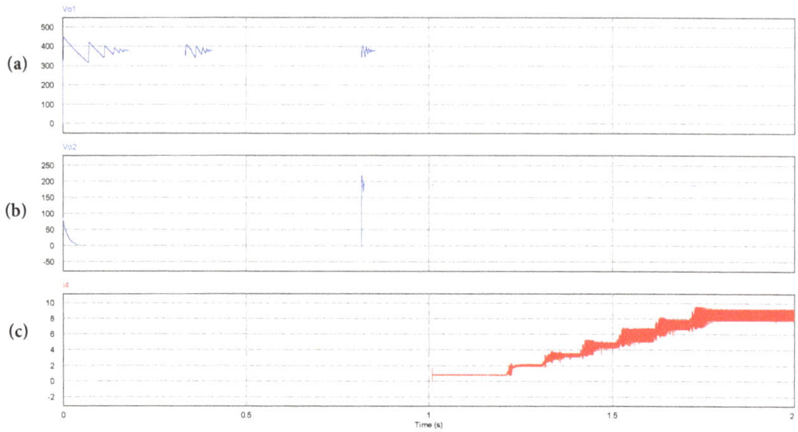

Figure 31. Fifth set of simulation results, (a) primary DC bus, (b) 1.2 kW converter output voltage and (c) 400 W converter current output.

A second test is done, now all three boost converters generate the primary bus, but the rest of the converters in the DCMG remain the same, again the aim is to supply an electronic load in primary bus and group of six tandem driver-luminary from secondary bus. An scheme of the new trajectory is presented in Figure 32.

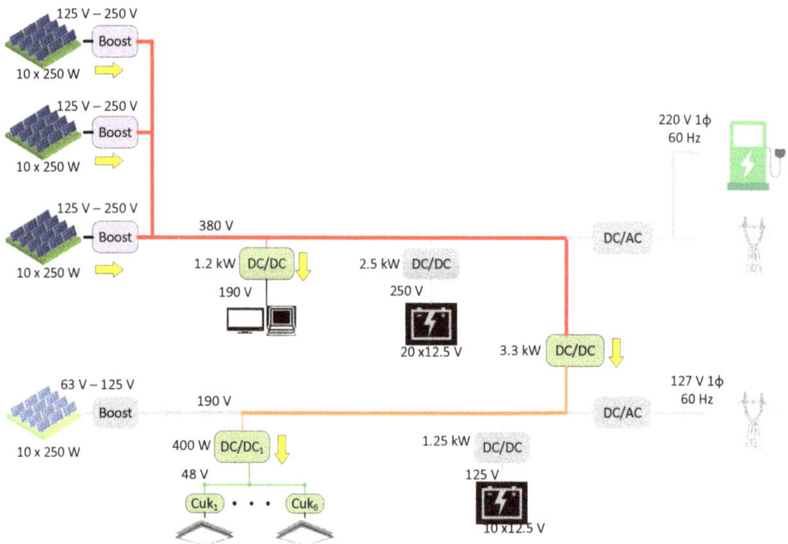

Figure 32. Second test trajectory for DCMG multibus multi-voltage performing with three boost converters.

In this case, it is necessary to verify current sharing using feedback state as local control strategy. The most important differences against test one appear in Figure 33, primary bus voltage presents starting transient with longer settle time than test one, see Figure 33a. Figure 33b shows total current from three boost converter and it is important to realize current in Figure 33c, one boost converter current output is same waveform that total current but divided by three.

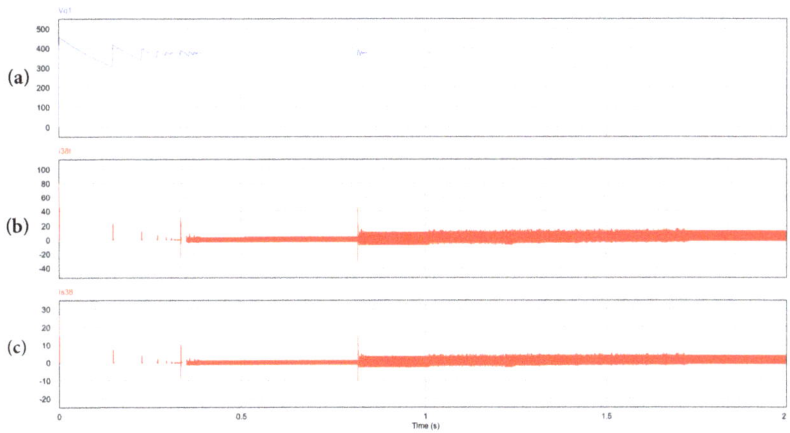

Figure 33. Sixth set of simulation results, (**a**) primary DC bus, (**b**) total current from three boost converters array and (**c**) boost converter number one output current.

6.3. Performance of DCMG with EMS

The EMS start-up stage let a pre-charge in the boost converter input capacitor then closed loop operation at one tenth of PEC rating power is perform, a simulation of this sequence using one of the 380 V output boost is shown in Figure 34.

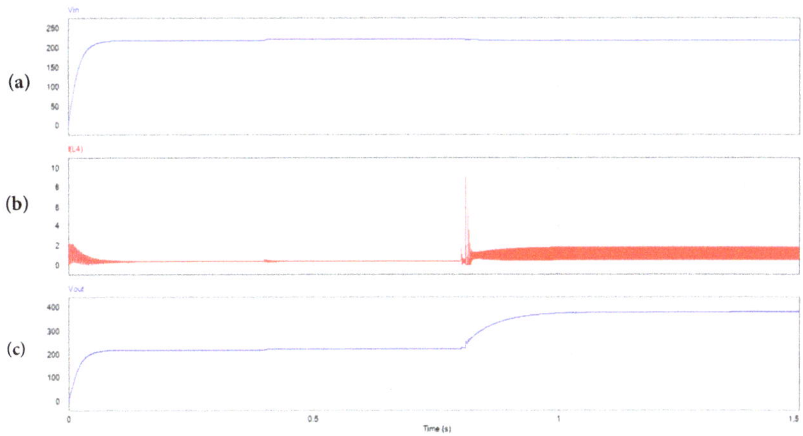

Figure 34. Start-up sequence with one 380 V output boost convertert, (**a**) input capacitor voltage, (**b**) inductor current (**c**) boost converter voltage output.

In Figure 34a input capacitor is charge through the thermistor at low temperature resistance value (10 Ω), at 0.4 s thermistor reduce its resistance value by temperature increase (0.22 Ω). Figure 34b, at the begining a transient current flows through inductor to charge output capacitor, at 0.8 s a second transient take place because the enable signal is set. In Figure 34c since the simulation starts, voltage output has the same value than input voltage by converter particular behavior. When converter is enabled reference value moves smoothly from pre-charge value to rating value, 0.8 s–1 s.

The star-up test stage considers eight source PECs, then 256 possible combination exist. Nevertheless, EMS prioritizes the usage of harvested power on PV array, then, as many times as possible, DCMG starts by using power collected in PV arrays through boost converters. Next simulation uses an incremental counter as input to EMS start-up to get the adequate signal output after the test, see Figure 35.

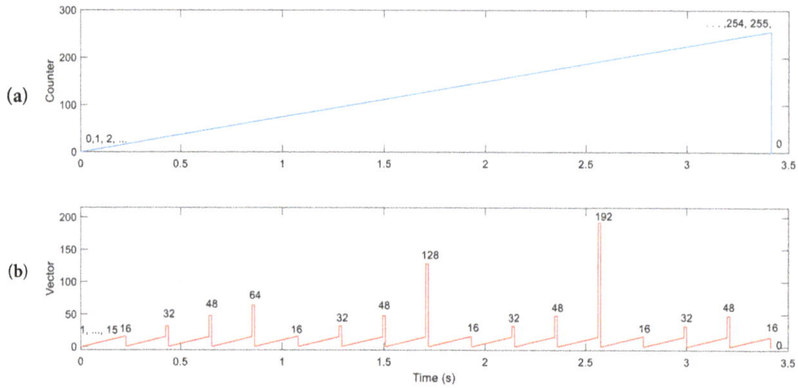

Figure 35. Sixth set of simulation results, (**a**) Incremental counter input and (**b**) start-up test ouput value.

According to SC vector, in Table 12, given values from 1 to 15 represents system start with PV arrays. If PV arrays do not pass the test, algorithm assumes as result one of the following: 16 (primary bus ESS batteries), 32 (secondary bus ESS batteries), 48 (both buses ESSs), 64 (power incoming from mains through primary bus), 128 (power incoming from mains through secondary bus), 192 (power incomng from mains through both buses). If there is no source able to supply power to the system output value is 0, it means, DCMG remains in off state and the building should be powered in a typical way, as an ultimate emergency action. A complete start-up test sequence is shown in Figure 36.

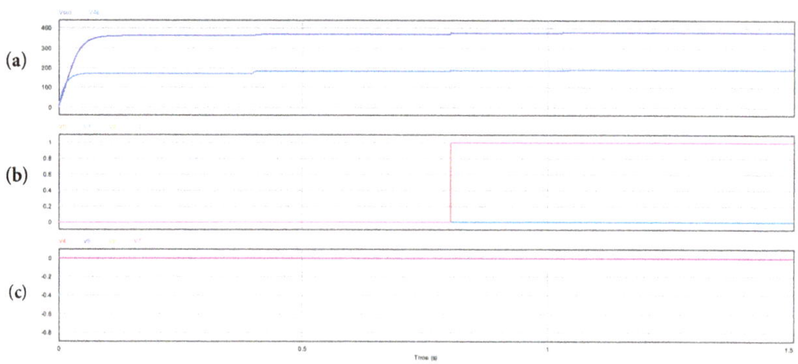

Figure 36. Sixth set of simulation results, (**a**) 380 V boost converter ($SPV1$, blue) and 190 V boost converter ($SPV2$, cyan), (**b**) SC vector, the least significant four bits (**c**) SC vector, the most significant four bits.

In Figure 36a, primary and secondary buses are established by one boost converter each, it is assumed that only this two PECs passed the initial test. In Figure 36b, $v_0 = SPV1$ and $v_3 = SPV4$ takes '1' value and the other two least significant bit renains in '0', finally, in Figure 36c the most significant four bits, related to ESSs batteries and DC-AC converters keeps on '0'.

7. Conclusions

In this paper a novel concept for DCMG architecture classified as multibus multi-voltage is presented, this development is suitable for building already in operation where different kinds of loads with a variety of current and voltage requirements must be supplied. This concept of architecture main advantage is reliability for final users, according to typical multibus schemes.

A hierarchical control structure is proposed to govern the system. Relative to local control strategy: average models of all PECs topologies included in the proposed DCMG are obtained by using switching functions. Passive elements values for each converter are obtained using well known procedures. Finally, feedback state controllers are design, then they are tuned from the desired polynomial based on overshoot percentage and settle time, further Ackemann method is used to obtain gains values for boost and DC-DC full bridge converters, all versions. PI classic controllers are used in Cük an DC-AC converters.

For higher supervisory control level, a EMS based in Petri net theory is proposed as coordinated control stage. EMA consist of three parts, a start-up test to define the initial state of the system. After initial state is set, connection-disconnection of source PECs is performed by set of FMSs which enable the needed transitions for the required state. In same manner, *SC* vector sets an specific FMSs to go through its states to add or shed loads PECs.

Simulation results for every converter in stand-alone operation are presented in order to validate each version performance in closed loop with feedback state controller as local control stage, here inductor ripple criterion is verified. Some trajectories of the DCMG architecture are under test to verify system operation and validate the concept. Some EMS operations are presented to validate supervisory control.

With results presented is possible to visualize that all the trajectories in DCMG are operative. In further stages of this development: Petri net EMS can be modified to reach another environmental aims or economic goals. Experimental PECs are under implementation to validate the DCMG in a phisycal way. A digital communications network can be considered in order optimize hardware CC resources. The proposed DCMG architecture is flexible and can be implemented in several kind of buildings considering peak power consumption and site loads demand.

Author Contributions: Methodology, J.M.-N.; Validation, R.O.-G.; Investigation, E.R.-S.; Writing—original draft, H.R.-E.; Writing—review & editing, C.G.-T. All authors have read and agreed to the published version of the manuscript.

Funding: This research received no external funding.

Institutional Review Board Statement: Not applicable.

Informed Consent Statement: Not applicable.

Data Availability Statement: Not applicable.

Conflicts of Interest: The authors declare no conflict of interest.

References

1. Ipakchi, A.; Albuyeh, F. Grid of the future. *IEEE Power Energy Mag.* **2009**, *7*, 52–62. [CrossRef]
2. Lasseter, R.H. MicroGrids. In Proceedings of the 2002 IEEE Power Engineering Society Winter Meeting (Conference Proceedings (Cat. No.02CH37309)), New York, NY, USA, 27–31 January 2002, Volume 1; pp. 305–308.
3. Peterson J.L. *Petri Net Theory and the Modeling of Systems*; Independently Published: Chicago, IL, USA, 2019; ISBN: 1080591176, 9781080591176.
4. Justo, J.J.; Mwasilu, F.; Lee, J.; Jung, J.W. AC-microgrids versus DC-microgrids with Distributed Energy Resources: A Review. *Renew. Sustain. Energy Rev.* **2013**, *24*, 3878–405. [CrossRef]
5. Guerrero, J.M.; Vasquez, J.C.; Matas, J.; De Vicuna, L.G.; Castilla, M. Hierarchical Control of Droop-Controlled AC and DC Microgrids - A General Approach Toward Standardization. *IEEE Trans. Ind. Electron.* **2011**, *58*, 158–172. [CrossRef]
6. Xu, L; Chen, D. Control and Operation of a DC Microgrid With Variable Generation and Energy Storage. *IEEE Trans. Power Deliv.* **2011**, *26*, 2513–2522. [CrossRef]

7. Kakigano, H.; Nomura, M.; Ise, T. Loss Evaluation of DC Distribution for Residential Huoses Compared with AC System. In Proceedings of the 2010 International Power Electronics Conference—ECCE ASIA, Sapporo, Japan, 21–24 June 2010; Volume 1; pp. 480–486.
8. Che, L.; Shahidehpour, M. DC Microgrids: Economic Operation and Enhancement of Resilience by Hierarchical Control. *IEEE Trans. Smart Grid* **2014**, *5*, *5*, 2517–2526.
9. Salomonsson, D.; Soder, L.; Sannino, A. An Adaptive Control System for a DC Microgrid for Data Centers. *IEEE Trans. Ind. Appl.* **2008**, *44*, 1910–1917. [CrossRef]
10. Kakigano, H.; Miura, Y.; Ise, T. Distribution Voltage Control for DC Microgrids Using Fuzzy Control and Gain-Scheduling Technique. *IEEE Trans. Power Electron.* **2013**, *28*, 2246–2258. [CrossRef]
11. Sun, K.; Zhang, L.; Xing, Y.; Guerrero, J.M. A Distributed Control Strategy Based on DC Bus Signaling for Modular Photovoltaic Generation Systems With Battery Energy Storage. *IEEE Trans. Power Electron.* **2011**, *26*, 3032–3045. [CrossRef]
12. Schonbergerschonberger, J.; Duke, R.; Round, S.D. DC-Bus Signaling: A Distributed Control Strategy for a Hybrid Renewable Nanogrid. *IEEE Trans. Ind. Electron.* **2006**, *53*, 1453–1460. [CrossRef]
13. Karlsson, P.; Svensson, J. DC bus voltage control for a distributed power system. *IEEE Trans. Power Electron.* **2003**, *18*, 1405–1412. [CrossRef]
14. Dragicevic, T.; Lu, X.; Vasquez J.C.; Guerrero, J.M. DC Microgrids—Part I: A Review of Control Strategies and Stabilization Technique. *IEEE Trans. Power Electron.* **2016**, *31*, 4876–4891. [CrossRef]
15. Dong, D.; Cvetkovic, I.; Boroyevich D.; Zhang, W.; Wang, R.; Mattavelli, P. Grid-Interface Bidirectional Converter for Residential DC Distribution Systems—Part One: High-Density Two-Stage Topology *IEEE Trans. Power Electron.* **2013**, *28*, 1655–1666. [CrossRef]
16. Guerrero, J.M.; Vasquez, J.C.; Matas, J.; Castilla, M.; De Vicuna, L.G. Control Strategy for Flexible Microgrid Based on Parallel Line-Interactive UPS Systems. *IEEE Trans. Ind. Electron.* **2009**, *56*, 726–736. [CrossRef]
17. Kwasinski, A.; Onwuchekwa, C.N. Dynamic Behavior and Stabilization of DC Microgrids with Instantaneous Constant-Power Loads. *IEEE Trans. Power Electron.* **2011**, *26*, 822–834 [CrossRef]
18. Guo, L.; Hung, J.Y.; Nelms, R.M. Evaluation of DSP-Based PID and Fuzzy Controllers for DC–DC Converters. *IEEE Trans. Ind. Electron.* **2009**, *56*, 2237–2248.
19. Onwuchekwa, C.N.; Kwasinski, A. Analysis of Boundary Control for Buck Converters with Instantaneous Constant-Power Loads. *IEEE Trans. Power Electron.* **2010**, *25*, 2018–2032. [CrossRef]
20. Lu, X.; Sun, K.; Guerrero, J.M.; Vasquez, J.C.; Huang, L. Double-Quadrant State-of-Charge-Based Droop Control Method for Distributed Energy Storage Systems in Autonomous DC Microgrids. *IEEE Trans. Smart Grid* **2015**, *6*, 147–157. [CrossRef]
21. Anand, S.; Fernandes, B.G.; Guerrero J. Distributed Control to Ensure Proportional Load Sharing and Improve Voltage Regulation in Low-Voltage DC Microgrids. *IEEE Trans. Power Electron.* **2013**, *18*, 1900–1913. [CrossRef]
22. Lu, X.; Guerrero, J.M.; Sun, K.; Vasquez J.C. An Improved Droop Control Method for DC Microgrids Based on Low Bandwidth Communication With DC Bus Voltage Restoration and Enhanced Current Sharing Accuracy. *IEEE Trans. Power Electron.* **2014**, *29*, 1800–1812. [CrossRef]
23. Mudaliyar, S.R.; Mishra, S.; Sharma, R.K. Coordinated controller design and stability analysis of DC microgrid with constant power load. In Proceedings of the 2017 IEEE Second International Conference on DC Microgrids (ICDCM), Nuremburg, Germany, 27–29 June 2017; pp. 322–329.
24. Bhavsar, Y.S.; Joshi, P.V.; Akolkar, S.M. Energy management in DC microgrid. In Proceedings of the 2015 International Conference on Energy Systems and Applications, Pune, India, 30 October–1 November 2015; pp. 209–213.
25. Varghese, A.; Chandran, L.R.; Rajendran, A. Power flow control of solar PV based islanded low voltage DC microgrid with battery management system. In Proceedings of the 2016 IEEE 1st International Conference on Power Electronics, Intelligent Control and Energy Systems (ICPEICES), Delhi, India, 4–6 July 2016; pp. 1–6.
26. Zhang, Y.; Li, Y. Energy Management Strategy for Supercapacitor in Droop-Controlled DC Microgrid Using Virtual Impedance. *IEEE Trans. Power Electron.* **2017**, *32*, *4*, 2704–2716. [CrossRef]
27. Chen, Y.K.; Wu, Y.C.; Song, C.C.; Chen, Y.S. Design and Implementation of Energy Management System With Fuzzy Control for DC Microgrid Systems. *IEEE Trans. Power Electron.* **2013**, *28*, *4*, 1563–1570. [CrossRef]
28. Schweizer, M.; Kolar, J.W. Design and Implementation of a Highly Efficient Three-Level T-Type Converter for Low-Voltage Applications. *IEEE Trans. Power Electron.* **2013**, *28*, 899–907. [CrossRef]
29. De Doncker, R.W.A.A.; Kolar, J.W. A Three-Phase Soft-Switched High- Power-Density DC/DC Converter High- Power Applications. *IEEE Trans. Ind. Appl.* **1991**, *27*, 63–73. [CrossRef]
30. Jain, A.K.; Ayyanar, R. PWM Control of Dual Active Bridge: Comprehensive Analysis and Experimental Verification. *IEEE Trans. Power Electron.* **2011**, *26*, 1215–1227. [CrossRef]
31. Balog, R.S.; Krein, P.T. Bus Selection in Multibus DC Microgrids. *IEEE Trans. Power Electron.* **2011**, *26*, *3*, 860–867. [CrossRef]
32. Zhao, Z.; Hu, J.; Liu, J. Hierarchical coordinated control of multi bus-based DC microgrid considering fault occurred in buses. In Proceedings of the 2016 China International Conference on Electricity Distribution (CICED), Xi'an, China, 10–13 August 2016; Volume 1; pp. 1–6.
33. Rashidirad, N.; Hamzeh, M.; Sheshyekani, K.; Afjei, E. An Effective Method for Low-Frequency Oscillations Damping in MultiBus DC Microgrids. *IEEE J. Emerg. Sel. Top. Circuits Syst.* **2017**, *7*, 403–412. [CrossRef]

34. Pereira, A.; Freitas, A.A.; Pereira E. 3-Transmission Line Parameter Calculation. In *Transmission of Electrical Energy*; CRC Press Taylor and Francis Group: Boca Raton, FL, USA, 2020; pp. 99–170.
35. Patrao, I.; Figueres, E.; González-Espín, F.; Garcerá, G. Transformerless topologies for grid-connected single-phase photovoltaic inverters. *Renew. Sustain. Energy Rev.* **2011**, *15*, 3423—3431. [CrossRef]
36. Wang, J.; Ji, B.; Zhao, J.; Yu J. From H4, H5 to H6—Standardization of full-bridge single phase photovoltaic inverter topologies without ground leakage current issue. *IEEE Energy Convers. Congr. Expo.* **2012**, *1*, 2419–2425.
37. Hart, D. *Power Electronics*; WCB/McGraw-Hill: New York, NY, USA, 2010; ISBN:9781283450485.
38. Rashid, M.H. *Power Electronics: Devices, Circuits, and Application*; Pearson: London, UK, 2014; ISBN:9780133125900; LCCN: 2013011062 .
39. Gonzalez, O.; Perez-Ramirez, J. Beristain, J. Photovoltaic Power Injection to the Grid with Reactive Power and Harmonic Compensation Using a Simple H Bridge Converter. In Proceedings of the IEEE Workshop on Power Electronics and Power Quality Applications (PEPQA), Bogotá, Colombia, 2–4 June 2015; Volume 6; pp. 1–6.
40. Subudhi, B.; Pradhan, R. A Comparative Study on Maximum Power Point Tracking Techniques for Photovoltaic Power Systems. *IEEE Trans. Sustain. Energy* **2013**, *4*, 89–98. [CrossRef]
41. Beniuga, O.; Beniuga, R.; Leca, M.; Sarbu, G.C. Using Petri Net Tool to Study the Dynamic Behaviour of Power Systems Protections. In Proceedings of the International Conference and Exposition on Electrical And Power Engineering (EPE), Iasi, Romania, 18–19 October 2018; pp. 436–419.
42. Fountas, N.A.; Hatziargyriou, N.D.; Valavanis, K.P. Hierarchical time-extended Petri nets as a generic tool for power system restoration. *IEEE Trans. Power Syst.* **1997**, *12*, 837–843. [CrossRef]
43. Gonzalez, RO.; Gonzalez, G.G.; Escobar, J.; Barazarte, R.Y. Applications of Petri Nets in Electric Power Systems. In Proceedings of the IEEE Central America and Panama Convention (CONAPAN XXXIV), Panama City, Panama, 12–14 November 2014; Volume 1; pp. 1–6.

applied sciences

MDPI

Review

Recent Advancements in Anaerobic Digestion and Gasification Technology

KeChrist Obileke *, Golden Makaka and Nwabunwanne Nwokolo

Department of Physics, Faculty of Science and Agriculture, University of Fort Hare, Private Bag X1314, Alice 5700, South Africa; gmakaka@ufh.ac.za (G.M.)
* Correspondence: kobileke@ufh.ac.za

Abstract: In recent times, there has been a growing demand for the use of biomass as an alternative energy due to its sustainable nature. At present, anaerobic digestion and gasification has been proven as a promising technology for exploiting this energy from biomass. This review aims to provide a comprehensive and detailed analysis of the combination of anaerobic and gasification technology as a hybrid system for sustainable waste-to-energy generation. This review reveals that both anaerobic digestion and biomass gasification have been successfully demonstrated as technologies for energy recovery. However, to improve the conversion efficiency in both technologies, the utilization of an intensifier, additive, and enhancer will be required. Moreover, temperature has been identified as a major factor affecting the technologies and should be considered. The bibliometric study conducted revealed that China is the leading country and has set the pace for other countries to follow suit. Subsequently, waste-to-energy research could be easily implemented on a global scale. This study recommends an experimental study of anaerobic digestion and gasification as a hybrid system.

Keywords: anaerobic digestion; biogas; gasification; syngas; temperature

Citation: Obileke, K.; Makaka, G.; Nwokolo, N. Recent Advancements in Anaerobic Digestion and Gasification Technology. *Appl. Sci.* **2023**, *13*, 5597. https://doi.org/10.3390/app13095597

Academic Editor: Maria Vicidomini

Received: 10 April 2023
Revised: 24 April 2023
Accepted: 28 April 2023
Published: 1 May 2023

1. Introduction

According to the statistical review of world energy (2021), the COVID-19 pandemic was reported to have had a dramatic effect on the energy market. This contributed to the fastest rate of decline in primary energy and carbon emissions since the Second World War. Despite this report, the good news is that renewable energy continued to grow. For recording purposes, the current rate of primary energy consumption and carbon emission fell by 4.5% and 6.3%, respectively. This decline was the largest for primary oil since 1945 and the lowest level since 2011 for carbon emissions [1]. Renewable energy sources, such as wind, solar, and hydroelectricity, all grew despite the decline in overall energy demand. Statistically, renewable energy sources with the inclusion of biofuel and the exclusion of hydro rose by 9.7%, lower than the 10-year average of 13.4% per annum but with an increase to 2.9 Exajoules (EJ) in terms of energy. Solar and wind energy rose by 1.3 EJ and 1.5 EJ, respectively, thereby contributing to the largest renewable energy growth. Similarly, hydroelectricity in China improved by 0.4 EJ [1]. From the authors' point of view, it is worrisome that biomass energy was not mentioned specifically, as it accounts for more than one-third of primary energy.

To this effect, biomass energy is regarded as renewable and sustainable energy. It is interesting that approximately 32% of the country's overall energy supply comes from biomass, and 70% of the national population is said to depend on it. In addition to natural biomass, the use of conventional waste biomass as a source of energy is another promising option [2]. One major way and option for achieving waste-to-energy focusing on biogas and syngas is through anaerobic digestion and gasification technology [3]. These technologies are regarded as mature biochemical and thermochemical processes and are currently receiving research interest. Its efficiency in the substrate and product separation as well as the microbes' ecology of anaerobes test contribute to its recent attention. Without a doubt, this review will promote and contribute to the development of the practical application of both technologies as a hybrid system.

2. Methodology

Amongst the existing types of literature reviews, which include systematic review, narrative review, traditional review, critical review, and state-of-the-art review, a conceptual review was employed [4]. The choice was based on the objective of the study, which aims to synthesize the area of conceptual knowledge, thereby contributing to a better understanding of a particular topic. Regarding the study, a conceptual review is most applicable, as it provides sensitive areas of the review based on recent advances in anaerobic digestion and biomass gasification. Furthermore, as a detailed and broad review containing a vast amount of information, the conceptual type of review was deemed to be the most suitable type to use. Findings were obtained from reputable databases, such as Springer, ScienceDirect, SCOPUS, Web of Science, and Google Scholar. In addition, substantial information from conference proceedings and industry reports was employed. The authors employed recent studies and literature in the review.

3. Anaerobic Digestion Technology

The term "anaerobic digestion" (AD) refers to the degradation of biomass under anoxic conditions in both natural and engineered systems [5]. Akhiar et al. [6] defined anaerobic digestion as a process wherein microorganisms degrade organic matter in the absence of oxygen to produce biogas. This technology is regarded as the earliest approach toward achieving a circular bioeconomy. Moreover, due to increasing industrial interest, studies on anaerobic microbial processes have greatly improved in identifying the important process parameter and factors that can promote methane production from all types of organic matter, such as manure and slurry [5]. Methane is said to substitute natural gas, especially as it tends to build up opportunities with regard to traditional petrochemistry [7]. Furthermore, anaerobic digestion has proven to be an ideal end-of-pipe technology for the treatment of waste. It serves as a process in the recovery of various biotechnological applications because of the wide substrate spectrum [5]. Such a biotechnological application, which includes the microbial electrochemical process and production of medium-chain fatty acid by anaerobic fermentation, can be coupled with anaerobic digestion. Anaerobic digestion has received immense interest in recent times because it addresses the major challenges faced by developing countries regarding environmental pollution resulting from the use of fossil fuels [8].

4. Recent Studies on Anaerobic Digestion Focusing on Biogas Production

Globally, approximately 50 million small-scale domestic digesters, 132,000 commercial digesters, and 700 upgrading biogas digesters have been installed. These digesters are fully functional and operate using anaerobic digestion technology. For instance, in Europe, to be specific, 15,000 commercial biogas digesters are in full operation [9]. In terms of biogas production, up to 31.24 billion cubic meters (bcm) were reported in 2015 from Europe. This number is predicted to increase by 70 bcm by 2030 because of the readily available resources [10]. On another note, the total global energy production resulting from biogas was approximately 2.8×10^5 TJ in 2000; fourteen years later, it increased to 1.3×10^6 TJ [NBD 2016]. According to the world biogas association report released in 2019, 10.5 GW of electricity generation came from the 17,783 biogas digesters in Europe. Although the specific time was not disclosed, it was indicated that Germany is leading with 10, 971 biogas digesters, followed by Italy (1655), France (742), Switzerland (632), and the United Kingdom 613 [11].

Pramanik et al. [12] provided a comprehensive review of anaerobic digestion as an effective technology for biogas. The review finds that anaerobic digestion achieves a higher energy production, especially the co-digestion process, whereas the two-stage anaerobic digestion process helps in increasing methane production. Hence, the authors recommended the use of food waste as a suitable substrate for biogas digester because of its excellent biodegradability and high-water content. In another study, Connor et al. [13] emphasized that the European agricultural sector is known for small-scale anaerobic digesters. This is because of the small-scale individual farm sizes and land productivity, which are insufficient to meet the feedstock requirement of large and medium-scale plants. The study reveals that there is a need for national support which will aid in adopting small-scale anaerobic digesters within the European countries. Nevertheless, there is still an overriding obstacle hindering the full commercialization of small-scale digesters, and this has to do with upfront investment costs. In addition to this, substrate type as well as a mode of utilization is another factor of consideration for efficient biogas production. Mono-digestion is characterized by imbalanced nutrients, lack of microbial communities, and fluctuation in operational parameters. In a bid to address these challenges, Ngan et al. [14] conducted a study focusing on the anaerobic digestion of rice straw with livestock manure. Rice straw is high in cellulose and, hence, requires pre-treatment before being fed into the biogas digester. Evidently from the study, it was revealed that rice straw when co-digested with biological waste (livestock manure) for anaerobic digestion showed a better degradability potential. Therefore, the condition for optimum pre-treatment, operational parameters, and mixture proportion with the substrate improves methane production, according to a study by Ngan et al. [14].

Various factors, such as anticipated toxicity, lower pH (6.6), and low carbon C:N ratio (12:2), are said to be responsible for the challenges faced in using only textile waste (sludge) for energy and nutrient recovery. To this effect, Kumar et al. [15] conducted a study on the anaerobic co-digestion of textile aerobic sludge with food waste and cow dung. This was assessed through a biochemical methane test under a mesophilic temperature of 35 °C. From the result, cow dung mixed with textile sludge in a ratio of 1:1 generated 524.4 mL/g of biogas, whereas using textile sludge only did not produce any biogas. In addition, the mixture of food waste and textile sludge resulted in biogas production of 288.3 mL/g, which is lower compared with that of cow dung with textile sludge. The findings from the study provided an efficient method for textile sludge utilization in biogas production.

The potential of using chicken manure as the sole feedstock for biogas production under anaerobic digestion was studied by Jurgutis et al. [16]. The choice of chicken manure in the study was because of its high biogas potential yield using a biochemical methane test. A methane yield of 508 mL CH_4 g^{-1} VS was recorded throughout the 217 days of monitoring. The novelty of the study is associated with the adaptation of the same operational conditions (organic loading rate and retention time) during the fermentation process. To this end, approximately 93% of chicken manure was digested, with an organic loading rate of 3.14 kg $VSL^{-1}d^{-1}$ and total ammonium nitrogen of 5.5 g L^{-1}. This proves that it is possible to digest chicken manure as a stand-alone feedstock as long as the ammonium concentration is carefully controlled, which is achieved via the freshwater dilution process.

In terms of the design parameters, Obileke et al. [17] mentioned that the biogas digester dimension and material of construction are important factors of consideration. The authors developed a design equation that considered the volume of the digester, inlet, and outlet chamber dimensions, as well as the shape of the digestion chamber. In terms of construction, the understudied biogas digester was fabricated using high-density polyethylene (HDPE) plastic, whereas the inlet and outlet chamber was constructed using bricks/cement. The developed plastic biogas digester and the appropriate dimension were found to be suitable for anaerobic digestion for biogas production (methane and carbon dioxide reported to be 60% and 30%, respectively) using cow dung as the substrate.

To stabilize organic waste while generating biogas, the anaerobic digestion of poultry litter was carried out by Chaump et al. [18]. The study originated as a result of technical challenges faced, which included seasonality of litter production, low carbon/nitrogen ratios, and limited digestibility of bedding. From the study, biomethane potential of 0.24–0.30 L/gVS and 0.15–0.16 L/gVS was obtained for litter leachate and whole litter, respectively, whereas the insoluble bedding material left after leaching was reported to be 0.08–0.13 L/gVS. The information from the results indicates that poultry litter leachate can be amended for digestion despite having a low carbon/nitrogen ratio. Hence, litter can be stored before digestion, although this leads to biomethane losses. The authors opined that the loss is because of the decomposition of organics during storage.

5. Intensification of the Anaerobic Digestion Process

As stated by Sathyan et al. [8], the need to improve anaerobic digestion has been of great interest in recent years. The use of different types of additives or accelerants has been explored in several synergies, thereby enhancing the methane yield and quality of digestate, which are products of anaerobic digestion. It is important to know that the use of the additive in enhancing the performance of anaerobic digestion reduces the challenges associated with slow start-up, substrate-induced inhibition, and low energy output [8]. Some of these additives include microbes, enzymes, microelements, biological metabolites, adsorbents, deforming agents, and chelating agents. Therefore, Table 1 presents these additives, their substances, and the effects from various studies.

Table 1. Additives used in anaerobic digestion enhancement.

Types of Additives	Functions	References
Enzymes (Cellulose, Amylase, Lipase, etc.)	Facilitate the activities of hydrolysis of the particulate compounds	[19]
Nutrients (Nitrogen, Phosphorus, Potassium, etc.)	Essential for the growth of biogas fermentation microorganisms	[20]
Microelements (Iron, Cobalt, Nickel, etc.)	Essential for methanogens activities	[21]
Inhibitors (Heavy-metal ions)	Promote the biological activity of the system at a low concentration	[22]
Forming agent (Polyester deformer)	Help to eliminate formation during anaerobic digestion	[23]
Absorbent (Aluminium powder)	Help to increase biogas content and methane content to a different degree; decrease BOD and COD content	[24]
Chelating agent	Increase the growth rate of methanogens and gas production of anaerobic digestion	[25]
Accelerants	Increase biogas yield or production, COD, and buffer capacity as well as degradation rate	[20,21,26,27]
Carbon-based materials (Carbon nanotube, Graphene, and Carbon cloth)	Promote microbial growth and microbial immobilization; provide nutrients to microbes, resulting in faster methane production; stimulate the mechanism of methanogens as well as improve the efficiency and stability of the anaerobic process, thereby increasing methane yield	[28–32]

Some of these additives are cost-effective and have been widely studied to test their efficacy. In this case, their findings and results were successful in enhancing gas production, favourable digestion condition, and superior organic degradation [21,33].

Carbon-based materials intensify the process of anaerobic digestion by promoting the growth of microorganisms, providing required nutrients to microorganisms and improving the activities of enzymes. These results in quick methane production and content as well as improved quality digestate [31]. Previous studies conducted by Capson et al. [29], Lin et al. [30], Mostafa et al. [31], and Yang et al. [32] revealed that carbon material is necessary for anaerobic digestion because it enhances its efficiency and stability, thereby increasing the methane yield.

The properties of graphene in relation to anaerobic digestion include electric conductivity, high mechanical strength, and large surface area. To improve the stability and efficiency of biogas digester through anaerobic digestion, graphene is required and recommended. The introduction of graphene during anaerobic digestion improves the methanogens, which is the microorganism responsible for methane production in the low-temperature range of 10–20 °C [27,34]. Furthermore, graphene results in the change in syntrophic microbes between methane production archaea and acidogenic microbes, which increases the rate of methane production by 28% to be specified through 71.1% *Methanobacterium* and *Methanosarcina* (11.3%). Graphene indeed intensifies and influences microbial community dynamics during anaerobic digestion.

6. Feedstock for Anaerobic Digestion

The nutritional composition is regarded as a major factor that determines the suitability of any feedstock or substrate for biogas production. This nutritional composition affects the biogas yield, methane content, biodegradability, and degradation of the biomass kinetic involved [35]. These nutrients include carbohydrates, protein, and fats, as shown in Table 2. Interestingly, any feedstock for the anaerobic digestion process should be readily biodegradable and should not contain any toxic components that would impair the activities of microbes [36]. Table 2 shows the nutrient compositions of feedstocks with their methane yields and methane compositions.

Table 2. Reported studies on nutritional composition for AD with methane yield and composition [37,38].

Composition of Nutrient	Methane Yield (m^3/kgVS)	Methane Composition (%)
Carbohydrate	0.42	50
Protein	0.50	50
Lipids	1.01	70

From Table 2, it is seen that lipids produced the highest methane yield of 1.01 m^3/kgVS as well as a methane percentage composition of 70% compared with protein and carbohydrates. This implies that substrates that are rich in lipids hold great potential for methane production. Nevertheless, the use of such substrate should be closely monitored, as a drop in pH can be toxic to the microbial community because of degradation that releases long-chain fatty acids. Considering the feedstocks that are protein-rich, Nwokolo et al. [39] mentioned that these feedstocks have a high potential for methane yield. However, they increase the alkalinity of anaerobic digestion because of the degradation process that releases ammonium. This process enhances the digestate and inhibits the activities of the methanogens. Further, Nwokolo et al. [39] stated that this inhibition occurs because of the shift in equilibrium from ammonium to ammonia, usually in a concentration value of 53–1450 mg/L. On a broader scale, substrates or biogas production are classified into agricultural waste, industrial organic waste, and municipal waste.

6.1. Agricultural Waste (AW)

Agricultural wastes are widely used due to their availability, and they include animal waste and lignocellulosic biomass in the form of residue from crops and forest and energy crops. However, the authors chose to focus on animal waste as a major feedstock for biogas production. This is driven by its availability, biodegradability, and global usage as compared with others [39]. These animal wastes are characterized by high nutrient content and organic matter concentration, neutral pH, different microbes, etc. Hence, they produce a lower rate of biogas production because of their pre-digestibility by the animal intestine. The conversion of animal waste using anaerobic digestion provides benefits such as effective waste management techniques and sustainable biogas energy generation. Mostly, animal wastes, such those from pigs, goats, sheep, cattle, and poultry, are widely studied as viable substrates to produce biogas. Regarding characteristics of the animal wastes, parameters such as total solids (TS), volatile solids (VS), pH, and carbon-to-nitrogen ratio (C/N) play a vital role in the enhancement of biogas yield using animal waste. To this effect, the recommended ranges of these parameters for the different types of animal waste are summarised in Table 3.

Table 3. Optimum ranges of substrate parameters for enhanced biogas production.

Animal Waste	Total Solids (TS)	Volatile Solids (VS)	pH	Carbon/Nitrogen Ratio	References
Pig manure	6.4–7.5	6.2–82.8	6.4–7.5	5.7–13.5	[40–43]
Goat manure	33.7–55.5	27.7–89.4	7.9	18.0	[44,45]
Sheep manure	22.3–40.0	18.7–72.7	7.16–8.1	11.3–14.7	[26,44]
Cattle manure	14.5–22.7	11.9–72.0	7.1–8.6	14.59–18.9	[26,40,44]
Poultry manure	20.0–92.6	18.3–84.1	6.9–7.4	7.5–9.75	[22,40,46,47]

From Table 3, it can be observed that different animal wastes have different parameter conditions (TS, VS, pH, and C/N ratio). The discrepancies in the physiochemical characteristics of these animal wastes can be attributed to their origin, the difference in their diet, variation in digestion, and intestinal micro-organism [48]. Depending on the pH, the buffering capacity protects the process of digestion against acidification linked with high total solids. Furthermore, animal waste with the required pH provides a suitable environment for the optimum functioning of the microorganism during anaerobic digestion. Total solids are responsible for the contribution of water towards the growth of microorganisms, thereby facilitating the dissolution and transport of nutrients. As the amount of water decreases or increases, it reduces the activities of microorganisms because of the total solids [49].

6.2. *Industrial Waste (IW)*

Industrial organic waste for anaerobic digestion consists of by-products, residues, and waste from numerous activities in the industries. Some of these industries include the food and beverages industry, pulp and paper industry, petrochemical industry, and textile industry. Selected wastes from industries include food and beverage waste, sugar, starch, and fruit processing wastes produced in large quantities. According to Lup et al. [50], industrial organic wastes are characterized by a high content of hydrocarbons, fats, and protein. Industrial wastes are usually not used as feedstock for anaerobic digestion, especially for biogas production, because of their low degradability (30–50%) and recalcitrant chemical properties [39]. This particular property does not affect food waste, which is an example of industrial waste. Hence, they are the most widely used substrate among industrial waste for anaerobic digestion. The pulp and paper industry generates waste of high biological oxygen demand (BOD), high dye content of 300–2800 m/g, and 1200–6500 colour units [20]. To improve the biodegradability of industrial waste for anaerobic digestion, it is recommended that they are co-digested with other substrates because of their low total solid content of <1%. However, industrial waste for anaerobic digestion employed various technologies which are specific for biogas production via anaerobic digestion. Some of these technologies are shown in Table 4.

Table 4. Various technologies employed in industrial waste for anaerobic digestion [36].

Technologies for Industrial Waste	Description
Up-flow anaerobic sludge blanket (UASB)	This involves the pumping of waste into the sludge bed, consisting of granular bacteria and a sludge blanket. The UASB essentially separates gas and prevents bacteria sludge from flowing into the waste.
Anaerobic fixed-film (AFF) system	The AFF employs the use of a fermentation tank to provide a biological firm as an intermediary on which to fix bacteria, thereby reducing bacteria losses from the system.
Continuous stirred tank reactor (CSTR) system	This technology uses a stirring machine characterized by high efficiency of digestion to reduce the hold-up time within the anaerobic digester.
Modified covered lagoon (MCL) system	The MCL system is covered by high-density polyethylene (HDPE) plastic sheet as a container for biogas production. This system increases the contact area of bacteria sludge with the waste.

6.3. Municipal Solid Waste (MSW)

Municipal solid waste is waste generated from household waste, commercial waste, and waste from cleaning. Precisely, it consists of a fraction of household and garden waste. The composition of MSW includes paper waste, plastic waste, glass waste, metal waste, organic waste, and textile waste. For efficient digestion of these solid wastes, impurities, such as plastics, glass, and metals, need to be removed, as they affect the operation of the process, especially for biogas production [51–54]. To lower greenhouse gas emissions and lessen the dependency on the use of external energy, the utilization of source-separated municipal organic waste is needed [55]. In considering the anaerobic digestion of MSW, particle size is one factor that must be considered. The particle size impacts the surface area, which increases and influences the hydrolysis rate during anaerobic digestion and lag phases. These are known as the limiting stage of adaptation of microorganisms to the substrate. [56]. Towards this, Parra-Orobio et al. [56] mentioned that a particle size of <2 mm is recommended and effective for methane production, having an electrical energy of 2960.4 kWh per week. The recommended particle size of <2 mm is 19% higher than in reactors with larger particles. Thus, the <2 mm particle size indicates lower design cost and maintenance. This is because the smallest particle enables a shorter retention time. Despite the recommended value of the particle size, an optimal range for the particle size still has not been defined. Notwithstanding, the size variability (from the crushing process) is another parameter capable of affecting the kinetics of the anaerobic digestion of MSW. Approximately1.3 tonnes of MSW are generated globally every year. This MSW consists of 46% organic waste, 17% paper waste, 10% plastic waste, 5% glass waste, 4% metal waste, 3% textile waste, 13% inert waste, and 2% miscellaneous waste [57].

At the continental level, Africa generates the highest organic composition of municipal waste at 66%, followed by Europe (54%), South America (53%), and then Asia (47%). North America and Australia are regarded as the countries with the least composition of municipal waste generation, producing 26% and 25%, respectively. The variation in the organic composition of this waste by continent might be attributed to the rates of the organic portion of waste from low-income countries (50–70%) versus high-income countries (20–40%) [58]. Interestingly, anaerobic digestion of municipal waste provides many benefits, such as the separation of organic biodegradable fractions from total municipal waste, generation of renewable energy, reduction of landfilling, and mitigation of pollution.

A grouping of the waste category of the feedstock required for anaerobic digestion was discussed earlier, and this is presented in Table 5.

Table 5. Feedstock of anaerobic digestion by waste classification.

Feedstock for Anaerobic Digestion	Source of Waste
Animal waste	Energy crop, livestock manure, remains from harvest and farm mortality, etc.
Industrial waste	Dairy product waste, processing of food/beverage, pharmaceutical industries, waste from a slaughterhouse, residues from agro-processing, etc.
Municipal waste	Sewage sludge, yard trimmings, food waste from restaurants, and organic fraction of MSW.

7. Gasification Technology

The thermochemical conversion of biomass materials to produce syngas, coal, tar, and some low-molecular-weight hydrocarbons in the presence of the oxidizing agent is known as gasification [59]. Similar to anaerobic digestion (biogas), the major end-product of gasification is syngas, which consists of carbon monoxide, hydrogen, methane, carbon dioxide, and traces of other gases (flammable gases). Although the end application of gasification products is mainly for electricity and heat generation, it can be used as transportation fuel [60]. The biomass gasification system consists of the following individual units: a feeding system, gasifier, cyclone, and gas scrubber. These units are said to contribute to the overall efficiency of the gasification process [61]. During gasification, the oxidizing agent, consisting of oxygen, air, steam, or mixtures thereof, is usually provided in stoichiometric amounts to maintain partial oxidation of the biomass material. The choice of any of the oxidizing agents is guided by the expected quality of the syngas in terms of the heating value and cost [62]. To lower the cost, the air is usually used as an oxidizing agent; however, it results in syngas with low heating values (4–6 MJ/Nm^3) [63]. This is said to be attributed to the higher concentration of nitrogen present in syngas. Using oxygen and steam as oxidizing agents increases the heating value of syngas. It was reported that the heating value of 9–10 MJ/Nm^3 and 17–18 MJ/Nm^3 was obtained for oxygen and steam, respectively. To maintain the partial oxidation of biomass material during gasification, there is usually the provision of sub-stoichiometric quantities [59].

8. Recent Studies on Gasification Focusing on Syngas Production

This section presents the selected recent studies conducted by various scholars and researchers on biomass gasification. This is necessary to have an overview and track the trend of research in this regard. On the positive side, the development of the biomass gasification process has contributed to high potential energy efficiency that is environmentally friendly and inexpensive, mostly as a waste-to-energy technology. Therefore, its contributions cannot be neglected, especially for the production of syngas. Currently, on the gradual progress made so far on biomass gasification, it is interesting to mention that some high-profile gasifier units have been installed and are operating in a different part of the world. Table 6 shows a few of the gasification plants installed in selected parts of the world.

Table 6. The recent operating gasifier units with their capacity and locations [64].

Gasifier Unit	Location	Capacity
TCG global steam reforming	USA	136,000 t/year
Two-stage thermal process	Norway	78,000 t/year
Fixed-bed dry bottom	USA	16,000 t/day
General Electric	China	1500 t/day

Reviewing the recent studies on biomass gasification from the literature, the gasification of eucalyptus wood chips in a downdraft gasifier to produce syngas in South Africa was conducted by Nwokolo et al. [59]. The authors sourced the eucalyptus wood chip from the sawmill industry. The eucalyptus wood was gasified using air as the oxidizing agent, and the produced syngas were measured. With the aid of a non-dispersive infra-red sensor and palladium/nickel (Pd/Ni) gas sensor., the volumetric composition of the syngas produced was as follows; hydrogen (22.3–22.5%), carbon monoxide (22.3–24.3%), methane (1.9–2.1%), carbon dioxide (9.8–10.7%), and nitrogen (41.5–42.9%). These syngas produced a higher heating value (HHV) of 6.08 MJ/Nm^3, obtained from hydrogen, carbon monoxide, and methane composition. The study provided evidence that gasification is an alternative for the waste-to-energy conversion.

To predict and optimise the rate of syngas production, Dang et al. [65] developed a robust biomass gasification process model. This was possible by incorporating updated biomass reaction kinetics and dense bed hydrodynamics into the Aspen Plus platform. The model was developed using 864 datasets of four input parameters and demonstrated good agreement and best line of fit under a range of operating conditions. This was achieved when comparing the model prediction with the experimental result. An optimal syngas yield from 61.4% to 78.5% by volume was obtained on a dry N_2-free basis. According to the optimization model result, a maximum syngas production of 78.6 vol% occurred at a temperature of 900 °C, equivalent ratio (ER) of 0.23, steam-to-biomass (S/B) ratio of 0.21, and moisture content (MC) at 30 wt%.

Li et al. [26] carried out an experimental study on the gasification of cotton stalks to enhance the Boudouard reaction and carbon monoxide production. The parameters studied included carbon dioxide flow rate (50–400 mL/min), carbon dioxide concentration (0–100%), temperature (750–950 °C), and feedstock particle size of 0.1–8 mm. From the study, it was revealed that high temperature had a positive effect on gas yield and gasification efficiency (GE) due to the Boudouard reaction. As the particle sizes increased, GE first increased and then decreased due to the different heating conditions of the particle, with the highest GE of 88% at 0.8–1 mm obtained. The yield of CO and GE improved with an increase in the flow rate of CO_2 and CO_2 concentration. However, the conversion rate of CO_2 decreased at a large CO_2 flow rate, and the growth rate of GE decreased for CO_2 concentrations greater than 50%. Therefore, the maximum gasification efficiency (GE) reached up to 125% at the recommended optimum temperature, particle size, and CO_2 concentration of 850–900 °C, 50–100%, and 1mm, respectively. Gasification efficiency (GE) is defined as the ratio of the low heating value (LHV) of gases to that of biomass feedstock.

Synthesis gas composition was obtained from the gasification of biomass involving a mixture of carbon black, resulting from waste tires and sewage sludge. Varank et al. [66] modelled the process via the response surface method, while optimization involved the use of Box–Behnken design. For the syngas composition, the determination coefficient obtained for H_2, CO, CH_4, and heating values was 92.86%, 95.40%, 96.15%, and 96.80%, respectively. These were the indicators used for the modelling using a second-degree polynomial equation. The high determination coefficient reported was an indication that the polynomial equation conformed to a sufficient level with the experimental data. Based on the validation result, the percentages of the H_2, CH_4, and CO were 12.75%, 8.07%, and 7.87%, respectively, whereas the heating value was 1420.3 $kcal/m^3$. The findings from the study confirmed that the gasification process was suitable for the production of high-quality syngas from waste material and recommended the use of the Box–Behnkan design for the optimization of the process.

To produce syngas, Peng et al. [67] employed the use of a catalyst during the gasification of wood residue in a fluidized bed gasifier while using a combination of air and steam as an oxidizing agent. During the study, the following parameters were considered; catalyst loadings (20%, 30%, and 40%) at residence time (20, 40, and 60 min) and temperature (750, 825, and 900 °C). The authors revealed that a high temperature of 900 °C, catalyst loading of 40%, and residence time of 60 min was found most suitable to produce syngas of high quality.

To facilitate the optimization of energy efficiency and economic viability of biomass gasification, Yao et al. [68] developed a gasification model. The model aimed to predict the rate of production and quality of syngas. It was mentioned that the three regions of the velocity profiles (natural convection region, mixed convection region, and forced convection region) were used for the development and building of the model. From the study, it was observed that the developed model provided accurate predictions against experimental data; nonetheless, there was a deviation of approximately 10%. Moreover, an equivalence ratio of 0.25 was found suitable for maximizing the economic benefits of biomass gasification.

Air steam gasification of pine sawdust with calcined cement as a catalyst was investigated for the production of syngas. The study was motivated by the operational problems relating to the formation of tar during gasification. As a result, tar was minimized in the producer gas due to catalyst cracking. According to Sui et al. [69], the employment of cement as a catalyst enhanced the hydrogen concentration and lowered the tar yield.

9. Intensification of Gasification Process

As mentioned earlier, in anaerobic digestion, intensification focuses on ways to enhance and improve biomass gasification. To optimize the process of gasification, Paulauskas et al. [70] investigated the efficacy of residual char gasification in a gasifier. The study used char made from wood and sewage sludge under different operating conditions in the gasification set-up. The findings showed that the intensifier and additive caused an increase in air flux from 0.11 kg/m^2s to 0.32 kg/m^2s as well as in the gasification rate from 0.8 to 2.61 g/min. Consequently, a 72% decrease in the duration of wood char gasification was recorded. This implies that wood char and sewage sludge char are suitable for the intensification of biomass gasification.

A continuous fly ash circulation was used as an intensifier or enhancer in industrial dual-fluidized gasification systems. This was evidenced in the study conducted on the effect of ash circulation on the performance of a dual fluidized bed gasification system. In the study, Pissot et al. [71] mentioned that ash-forming elements are released from fuel during the gasification of biomass. Hence, these elements are known to have a positive impact on the quality of the gas produced. To assess the impact on the gas quality, the coarse ash was intensified by an injection of untreated olivine in fine particle size and then recirculated. This resulted in a decrease in the tar concentration. From the study, it was revealed that the effect of ash circulation increased with time at a higher rate than in the control experiment.

In another study, Chen et al. [72] explored the use of calcium oxide in the steam gasification of sewage sludge to enhance the quality of syngas and hydrogen yield. The calcium oxide (CaO) was in situ to capture CO_2 released directly from the biomass matrix, thereby causing a steam reformation reaction. This process drives the conversion of biomass process in the production of more hydrogen. Thus, the quality of hydrogen in the gas stream was enhanced as high as 70–80% due to CO_2 sorption.

Similarly, Yang et al. [73] developed a two-stage sorption that relies on the integration of a CaO-based CO_2-carrying cycle in the gasification of sewage sludge, thereby intensifying the sludge carbon. During this process, the CO_2 was first captured and stored in the form of $CaCO_3$ at a higher temperature to gasify the sludge char for carbon monoxide (CO) production. The proposed intensification process resulted in 284.7 NmL of syngas per gram of dry sludge, gross CO/H_2 molar ratio of 2.3, and hydrogen and carbon-rich gas steam at a temperature of 550 °C and 750 °C, respectively. The study recommends CaO-based CO_2 as an efficient option for improving carbon during the gasification of sewage sludge. On a similar note, the use of calcium oxide improved the mole ratio of hydrogen from 0.0 to 0.70 and the yield of syngas from 10.70 vol% to 49.40 vol% [74].

As observed, the catalyst enhanced syngas yield because of its effect on the activation energy [21]. To improve the formation of hydrogen, the study used dolomite ($MgCO_3{\cdot}CaCO_3$) as a catalyst, which was calcinated for 4 h at a temperature of 900 °C. Dolomite, being a natural mineral-based catalyst, enhanced the formation of hydrogen. One advantage of dolomite is that it is inexpensive, abundantly available, and can aid in the removal of tar from producer gas [75]. Findings from a study by Guan et al. [76] revealed that the application of dolomite in MSW gasification reduced the content of tar from 9.71 weight% to 0.0 weight% and increased the syngas production from 0.620 to 1.470 m^3/kg feedstock. Regarding the content of hydrogen in the syngas, this also increased from 30.6 to 50.2%. In a similar development, transition-metal-based catalysts (copper, iron, and cobalt) are said to be efficient and effective in the steam reformation of tar. For high tar content in syngas, these transition-based catalysts are easily deactivated during carbon deposition [74]. Nickel is another suitable catalyst that aids in deactivation and sintering during gasification reactions and reducing fouling. For instance, Wang et al. [77], mentioned that the addition of nickel oxide (NiO) in MSW gasification caused an increase in the rate of syngas production from 0.78 to 1 m^3/kg MSW and in hydrogen production from 21.9 to 80.7 g H_2/kg MSW.

10. Feedstock for Gasification Technology

Biomass gasification entails the use of various types of feedstocks, which include but are not limited to, woody biomass (chips, wood pellet, sawdust, and tree bunches); agricultural and food biomass (rapeseed, jatropha, miscanthus, sugar cane bagasse, nutshell, rice husk, wheat straw, and oil production waste); and industrial and animal biomass [64]. Before gasification, feedstocks are characterized to determine their suitability; Table 7 presents the characterization of some selected feedstocks for gasification.

Table 7. Characteristics of feedstock for gasification.

Types of Feedstocks	Moisture Content (wt%)	Volatile Matter Content (wt%)	Ash Content (wt%)	Heat of Combustion (MJ/kg)	References
Woody Biomass					
Wood pellet	6.7	84.3	0.8	20.8	[78]
Saw dust	7.4	-	0.6	-	[79]
Chips (pine wood)	4.5	78.4	2.6	16.1	[80]
Root of mango tree (tree bunches)	5.73	67.87	3.91	18.52	[81]
Eucalyptus prophyll	11.37	75.34	0.27	17.16	[82]
Agricultural and Food Biomass					
Miscanthus	10.67	65.65	5.34	17.00	[81]
Oil production waste	5.9	80.1	7.6	20.1	[78]
Sugar cane bagasse	-	87.4	2.9	-	[83]
Wheat straw	2.50–7.00	60.0–75.0	5.64–15.0	16.12 ± 0.19	[78,84,85]
Rice husk	8.02	61.43	18.02	-	[84]
Pea nutshell	-	84.1	1.4	-	[83]
Cocoa nutshell	7.16	68.58	2.26	-	[84]
Jatropha seed cake	-	73.5	7.3	-	[83]
Animal and Industrial Biomass					
Microalgae	3.0–10.0	70–81	5.0–10.0	20.0–25.0	[86,87]
Brewer's spent grain	3.97	83.3	3.22	21.6	[88]
Spent coffee grounds	2.66	80.44	1.25	-	[85]
Natural rubber	1.71	89.98	3.60	45	[89]

For a successful study of the gasification process, the feedstock and its properties are necessary for syngas production [90]. These properties (moisture content, heating value, content of ash, alkali metals, and lignin ratio) are briefly discussed, as they affect the composition and yield of the syngas [91,92]. A study conducted by Tian et al. [34]

revealed that feedstock with a high composition of cellulose and hemicellulose resulted in the production of syngas with a higher concentration of carbon monoxide and methane, whereas feedstock with a high composition of lignin generated syngas with a higher concentration of hydrogen.

Having provided the summary of the properties of some selected feedstock for gasification in Table 7, it is important to state the optimum ranges for these feedstocks according to their classifications. Hence, Table 8 presents the summary of the recommended operating parameter of biomass waste from the literature.

Table 8. Summary of recommended operating parameters of biomass waste for gasification technology [64].

Classification of Biomass	Moisture Content (wt%)	Volatile Matter (wt%)	Ash Content (wt%)	Heat of Combustion (MJ/kg)
Woody biomass	4.0–12.0	60.0–85.0	0–5	16.0–20.0
Agricultural and food biomass	2.0–12.0	64.0–85.0	0–18	15–17
Animal and industrial biomass	3.0–10.0	70–81	5–10	24–25

11. Bibliometric Analysis of the Study

In this section, the data employed were obtained from the Web of Science (WoS) core collection. In total, 207 and 801 results from the WoS the core collection were obtained for anaerobic digestion and biomass gasification, respectively. The choice of WoS was to maintain consistency in the study and avoid any overlap in the data collection. The period considered for the search was set to be 10 years from 2011 to 2021, focusing on journal types, publishers, and countries. Original articles, review papers, and proceeding papers were the document types considered; these are presented in Table 9a,b. This search was carried out to investigate the trend in research on anaerobic digestion and biomass gasification. For the successful analysis of the bibliometric, the keywords criteria used were *"anaerobic digestion"*, *"biogas digester"*, and *"biogas"* for anaerobic digestion, while for biomass gasification, the keywords used were *"biomass gasification"*, *"gasifier"*, *and "syngas"*.

Although the entire period set for bibliometric analysis was 2011 to 2021, the authors also considered the recent years from 2017 to 2021. It was observed that there is a fluctuation in the total number of publications in the two research fields, implying that the research in the two fields has not been consistent. Hence, future study is recommended. Regarding the journals, Energy and Fuel are the most active journals in terms of the number of publications in anaerobic digestion and biomass gasification, with 83 and 587, respectively. The Energy and Fuel journals are published by Elsevier. For the anaerobic digestion and biomass gasification research, Elsevier published a total number of 80 and 498 publications (original article, review paper, and proceeding papers). Moreover, it can be deduced that Energy, Fuel and Green Sustainable Science Technology are present and common in research or study. In addition, in terms of publishers, Elsevier, Springer Nature, and Taylor and Francis are known as the popular publishers present in both journals. For countries researching anaerobic digestion and biomass gasification, China is regarded as the leading country (See Table 9a,b). This might be because of the country's interest in funding renewable energy and alternative energy research, especially as it relates to anaerobic digestion for biogas production and biomass gasification for syngas production.

Table 9. (**a**) Bibliometric analysis of anaerobic digestion research. (**b**) Bibliometric analysis of biomass gasification research.

(a)

	Publication year considered	Document type considered	Journals	Publishers	Countries
Anaerobic digestion	2017 (18)	Original articles (169)	Energy and Fuel (83)	Elsevier (80)	India (30)
	2018 (21)		Environmental Science (56)	Springer Nature (34)	China (28)
	2019 (17)	Review papers (21)	Biotechnology Applied Microbiology (48)	Wiley (11)	Germany (24)
	2020 (15)	Proceedings papers (18)	Green Sustainable Science Technology (30)	Taylor and Francis (10)	Denmark (20)
	2021 (23)		Agriculture Engineering (28)	MDPI (8)	South Africa (16)

(b)

	Number of publications per year	Document Type	Journals	Publishers	Countries
Biomass gasification	2017 (18)	Original articles (764)	Energy and Fuel (587)	Elsevier (498)	China (147)
	2018 (21)		Engineering Chemical (293)	American Chemical Society (58)	USA (100)
	2019 (17)	Proceedings papers (47)	Thermodynamic (150)	Springer Nature (50)	Italy (93)
	2020 (15)		Green Sustainable Science Technology (83)	MDPI (47)	Malaysia (43)
	2021 (23)	Review papers (37)	Environmental Science (77)	Taylor and Francis (29)	Sweden (43)

(**a**) Source: https://www.webofscience.com/wos/woscc/summary/211f5dbe-3eb0-4982-b0eb-72df5c6c1513-569f8899/relevance/1 (accessed on 30 June 2022); (**b**) Source: https://www.webofscience.com/wos/woscc/summary/df29a3ed-2d94-4f12-9c07-ae972f7233b5-3d84b141/relevance/1 (accessed on 30 June 2022).

12. Challenges Facing Anaerobic Digestion and Biomass Gasification

The recent challenges and future concerning anaerobic digestion and biomass gasification are discussed. For anaerobic digestion, though additives play an essential role during the intensification and enhancement of the anaerobic digestion process as earlier seen, not all of them (microbes) can stimulate complex organics, such as municipal sludges. Therefore, the need for favourable conditions capable of enhancing healthy growth over competing microorganisms should be developed. To achieve this, a biogas digester with two stages of anaerobic digestion has the tendency and capability to enrich the bacteria. To meet the energy demands with fewer carbon emissions than conventional energy systems, new technologies for generating this must be flexible. In this regard, carbon material as an enhancer or additive plays a critical role in this area. However, continued research in this direction is necessary in enhancing methane production in the future. Therefore, on the basis of the experimental studies, it can be seen that more attention and interest should be given to additives and their effect on continuous biogas digesters. [93–95]. Furthermore, regarding feedstock variability, Uddin and Wright [36] mentioned that the wide variability in the properties (physical and chemical) is a problem in choosing the proper technology. That is, the physical properties (feedstock size and moisture content) need to be compatible with the gasification technology. In the Akinbomi et al. [96] study, the challenges relating to anaerobic digestion focus on environmental and operational factors. The study revealed anaerobic digestion was inhibited by fluctuations in temperature, pH, and ammonia concentration. Nonetheless, the composition of the feedstock, position, handling, pumping, and mixing of the high-solids waste also play a role. According to Akinbomi et al. [96], anaerobic digestion is affected by temperature, pH, and feedstock composition (with regard to C/N). These are referred to as environmental factors. The authors recommended that the condition of the parameter of these environmental factors must be met for a stable, optimal, and efficient anaerobic digestion system. Accumulating intermediate products, such as ammonia, volatile fatty acid, and long-chain fatty acid, during anaerobic digestion is usually detrimental for the microbes, thereby causing process inhibition. For instance, NH_3 consumes VFA, which is responsible for the lowering of methane production via methanogens. As a result of this, the accumulation of VFA reduces the pH of the biomass digester. The higher organic loading rate is caused as a result of an accumulation of volatile fatty acids [37].

For biomass gasification, the thermodynamics of its operation is not well understood. The limitation affecting biomass gasification can simply be understood on the basis of a technical issue that the gasification technology is facing with poor mixing and heat transfer within the gasifier [97]. According to the authors, this makes it a bit challenging to achieve an even distribution of fuel as well as temperature across the geometry of the gasifier. Therefore, it is reported that the scaling-up of the gasifier is difficult. This is attributed to the characteristic of the design (gasifier geometry and air inlet velocity) [97]. Furthermore, the operation of the gasification process is said to be complex. This is a result of the need to simultaneously control air supply, bed material, and feedstock during the operation of the gasifier [98]. This factor prompts the high particulate observed, which circulates and causes equipment erosion during the gasification of syngas. Therefore, the gasifier requires high-pressure conditions for successful operation. Hence, this leads to low volumetric gas flow rate, condensation during compression, and other complications (defluidization). Although the process of biomass gasification has been in existence all this time, the fundamental principles underpinning its operation are still vague. This is especially true concerning the type of material suitable as feedstock. Due to the high temperature and pressure of the gasification process, there is deposited liquid ash in the gasifier wall because of the turning of fuel oxygen mixtures into the turbulent flame of the dust. This development brings about a technical issue when analysing the ash melting behaviours of the material used as feedstock in the gasification technology. As a result of its operating conditions, a specific type of material is used as a feedstock because of its complexity [97].

13. Conclusions and Recommendation

Among the various sources of renewable energy, biomass energy is regarded as the most widespread and readily available renewable energy source globally. Anaerobic digestion and biomass gasification have been identified as suiting technologies for their conversion to valuable products. Therefore, this research study reviewed recent advances in these technologies (2017–2021): intensification, feedstock, bibliometric analysis of anaerobic digestion, and biomass gasification. The review concluded by providing the challenges affecting both technologies. Having successfully established these headings under the present review, the following key findings in the field of anaerobic digestion and biomass gasification have been highlighted.

The system design for the biogas digester and gasifier depends on the dimension and material used for construction and fabrication. In terms of the performance of both technology, temperature affects both anaerobic digestion and biomass gasification. Therefore, the employment of intensifiers, additives, and enhancers is a promising approach to use and is required for improving the process of anaerobic digestion and biomass gasification, although co-digestion is recommended for anaerobic digestion due to its contribution to enhancing methane production. From the review, despite the various feedstock or substrates used in both technologies (agricultural and food waste and industrial waste), it was observed that animal waste and woody biomass are used mostly as feedstock or substrate for anaerobic digestion and biomass gasification, respectively. Considering the academic trend of the study, China is seen as the leader in anaerobic digestion and biomass gasification research. Based on this, the country will see a growth of biogas and syngas production in the coming years. Energy, Fuel, and Green Sustainable Science Technology are the most active journals on anaerobic digestion and biomass gasification. The review recommends the necessity to conduct a comprehensive and detailed bibliometric analysis of the status and trend of research in anaerobic digestion and biomass gasification technologies. To widen the scope of the technologies, more experimental studies are needed, focusing on intensification to enhance and improve both anaerobic digestion and biomass gasification. Future works are required, focusing on the development of a hybrid system in combining anaerobic digestion and gasification for the recovery of energy from animal waste and woody biomass.

Author Contributions: K.O. contributed to the conception of the study; K.O. and N.N. organized the database and performed the analysis of the review; G.M. supervised and provided resources for the study; K.O. wrote the first draft of the manuscript; K.O., N.N. and G.M. wrote various sections of the manuscript. All authors contributed to the manuscript revision, read, and approved the submitted version. All authors have read and agreed to the published version of the manuscript.

Funding: This research received no external funding.

Institutional Review Board Statement: Not applicable.

Informed Consent Statement: Not applicable.

Data Availability Statement: The data that support the findings of this study are available from the corresponding author upon reasonable request.

Acknowledgments: The authors would like to acknowledge the financial support received from the Govan Mbeki Research Development Centre (GMRDC), University of Fort Hare.

Conflicts of Interest: The authors declare no conflict of interest.

References

1. Dudley, B. *Statistical Review of World Energy*, 70th ed.; Whitehouse Associates Press: London, UK, 2021.
2. Benti, N.E.; Gurmesa, G.S.; Argaw, T.; Aneseyee, A.; Gunta, S.; Kassahun, G.S.; Aga, G.S.; Asfaw, A. The current status challenges and prospects of using biomass energy in Ethiopia. *Biotechnol. Biofuels* **2021**, *14*, 209. [CrossRef]
3. Victorin, M.; Davidsson, Å.; Wallberg, O. Characterization of mechanically pre-treated wheat straw for biogas production. *BioEnergy Res.* **2020**, *133*, 833–844. [CrossRef]
4. Votteler, R.G.; Brent, A.C. A literature review on the potential of renewable electricity sources for mining operations in South Africa. *J. Energy S. Afr.* **2016**, *27*, 1–21. [CrossRef]
5. Nikolausz, M.; Kretzschmar, J. Anaerobic digestion in the 21st century. *Bioengineering* **2020**, *7*, 157. [CrossRef] [PubMed]
6. Akhiar, A.; Zamri, M.F.; Torrijos, M.; Shamsuddin, A.; Battimelli, A.; Roslan, E.; Mohamad, H.M.; Carrere, H. Anaerobic digestion industries progress throughout the world. In *IOP Conference Series: Earth and Environmental Science*; IOP Publishing: Bristol, UK, 2020; Volume 476, p. 012074. [CrossRef]
7. Adnan, A.I.; Ong, M.Y.; Nomanbhay, S.; Chew, K.W.; Show, P.L. Technologies for Biogas Upgrading to Biomethane: A Review. *Bioengineering* **2019**, *6*, 92. [CrossRef] [PubMed]
8. Sathyan, A.; Haq, I.; Kalamdhad, A.; Khwaraikpam, M. Recent advancements in anaerobic digestion: A novel approach for waste to energy. *Adv. Organic. Waste Manag.* **2022**, 233–246. [CrossRef]
9. Zheng, L.; Chen, J.; Zhao, M.; Cheng, S.; Wang, L.P.; Mang, H.P.; Li, Z. What could China give to and take from other countries in terms of the development of the biogas industry? *Sustainability* **2020**, *12*, 1490. [CrossRef]
10. Xue, S.; Song, J.; Wang, X.; Shang, Z.; Sheng, C.; Li, C.; Zhu, Y.; Liu, J. A systematic comparison of biogas development and related policies between China and Europe and corresponding insights. *Renew. Sustain. Energy Rev.* **2020**, *117*, 109474. [CrossRef]
11. World Biogas Association. 2019. Available online: https://www.worldbiogasassociation.org/wp-content/uploads/2019/09/WBA-globalreport-56ppa4_digital-Sept-2019.pdf (accessed on 11 November 2022).
12. Pramanik, S.; Suja, F.; Md Zain, S.; Pramanik, B. The anaerobic digestion process of biogas production from food waste: Prospects and constraints. *Bioresour. Technol. Rep.* **2019**, *8*, 100310. [CrossRef]
13. Connor, S.O.; Ehimen, E.; Pillai, S.C.; Black, A.; Tormey, D.; Bartlett, J. Biogas production from small-scale anaerobic digestion plants on European farms. *Renew. Sustain. Energy Rev.* **2021**, *139*, 110580. [CrossRef]
14. Ngan, V.; Mervin, F.S.; Nam, T.; Van Theo, H.; Detras, M.; Hung, D.V.; Cuong, D.; Hung, N.V. Anaerobic digestion of rice straw for biogas production. *Sustain. Rice Straw Manag.* **2020**, 65–92. [CrossRef]
15. Kumar, P.; Samuchiwal, S.; Malik, A. Anaerobic digestion of textile industries wastes for biogas production. *Biomass Convers. Biorefinery* **2020**, *10*, 715–724. [CrossRef]
16. Jurgutis, L.; Slepetiene, A.; Volungevicius, J.; Amaleviciute-Volunge, K. Biogas production from chicken manure at different organic loading rates in a mesophilic full scale anaerobic digestion plant. *Biomass Bioenergy* **2020**, *141*, 105693. [CrossRef]
17. Obileke, K.; Mamphweli, S.; Meyer, E.; Makaka, G.; Nwokolo, N. Design, and fabrication of a plastic biogas digester for the production of biogas from cow dung. *J. Eng.* **2020**, *2020*, 1848714. [CrossRef]

18. Chaump, K.; Preisser, M.; Saravanan, R.; Prasad, R.; Adhikari, S.; Higgins, B. Leaching, and anaerobic digestion of poultry litter for biogas production and nutrient transformation. *Waste Manag.* **2019**, *84*, 413–422. [CrossRef]
19. Parawira, W. Enzyme research and applications in biotechnological intensification of biogas production. *Crit. Rev. Biotechnol.* **2012**, *32*, 172–186. [CrossRef]
20. Zhang, J.; Zhao, W.; Zhang, H.; Wang, Z.; Fan, C.; Zang, L. Recent achievement in enhancing anaerobic digestion with carbon-based functional materials. *Bioresour. Technol.* **2018**, *266*, 555–567. [CrossRef]
21. Yun, S.; Zhang, C.; Wang, Y.; Zhu, J.; Wei, Y. Synergistic effects of Fe salts and composite additives on anaerobic digestion of dairy manure. *Int. Biodeterior. Biodegrad.* **2019**, *136*, 82–90. [CrossRef]
22. Li, X.; Yun, S.; Zhang, C.; Fang, W.; Huang, X.; Du, T. Application of nano-scale transition metal carbides as accelerants in anaerobic digestion. *Int. J. Hydro. Energy* **2018**, *43*, 1926–1936. [CrossRef]
23. Liu, W.; Zheng, H.; Wu, Z.; Wang, Y. Effects of pH profiles on nisin fermentation coupling with foam separation. *Appl. Microbiol. Biotechnol.* **2010**, *85*, 1401–1407. [CrossRef]
24. Patel, V.; Patel, A.; Datta, M. Effects of adsorbents on anaerobic digestion of water hyacinth-cattle dung. *Bioresour. Technol.* **1992**, *40*, 179–181. [CrossRef]
25. Gaddy, J.L. Performance of Anaerobic Digesters. U.S. Patent US5783081A, 21 July 1998.
26. Li, J.; Xie, Y.; Zeng, K.; Flamant, G.; Yang, H.; Yang, X.; Zhong, D.; Du, Z.; Chen, H. Biomass gasification in molten salt for syngas production. *Energy* **2020**, *210*, 118563. [CrossRef]
27. Wu, B.; Lin, R.; Kang, X.; Deng, C.; Xia, A.; Dobson, A.D.; Murphy, J.D. Graphene addition to digestion of thin stillage can alleviate acidic shock and improve biomethane production. *ACS Sustain. Chem. Eng.* **2020**, *8*, 13248–13260. [CrossRef]
28. Abbas, Y.; Yun, S.; Wang, Z.; Zhang, Y.; Zhang, X.; Wang, K. Recent advances in bio-based carbon material for anaerobic digestion. *Renew. Sustain. Energy Rev.* **2021**, *135*, 110378. [CrossRef]
29. Capson-Tojo, G.; Moscoviz, R.; Ruiz, D.; Santa-Catalina, G.; Trably, E.; Rouez, M.; Crest, M.; Steyer, J.P.; Bernet, N.; Delgenès, J.P.; et al. Addition of granular activated carbon and trace elements to favour volatile fatty acid consumption during anaerobic digestion of food waste. *Bioresour. Technol.* **2018**, *260*, 157–168. [CrossRef]
30. Lin, R.; Cheng, J.; Zhang, J.; Zhou, J.; Cen, K.; Murphy, J.D. Boosting biomethane yield and production rate with graphene: The potential of direct interspecies electron transfer in anaerobic digestion. *Bioresour. Technol.* **2017**, *239*, 345–352. [CrossRef]
31. Mostafa, A.; El-Dissouky, A.; Fawzy, A.; Farghaly, A.; Peu, P.; Dabert, P.; Le Roux, S.; Tawfik, A. Magnetite/ graphene oxide nanocomposite for enhancement of hydrogen production from gelatinaceous wastewater. *Bioresour. Technol.* **2016**, *216*, 520–528. [CrossRef]
32. Yang, Y.; Zhang, Y.; Li, Z.; Zhao, Z.; Quan, X.; Zhao, Z. Adding granular activated carbon into anaerobic sludge digestion to promote methane production and sludge decomposition. *J. Clean. Prod.* **2017**, *149*, 1101–1108. [CrossRef]
33. De Vrieze, J.; Regueiro, L.; Props, R.; Vilchez-Vargas, R.; Jáuregui, R.; Pieper, D.H.; Lema, J.M.; Carballa, M. Presence does not imply activity: DNA and RNA patterns differ in response to salt perturbation in anaerobic digestion. *Biotechnol. Biofuels* **2016**, *9*, 244. [CrossRef]
34. Tian, T.; Qiao, S.; Li, X.; Zhang, M.; Zhou, J. Nano-graphene induced positive effects on methanogenesis in anaerobic digestion. *Bioresour. Technol.* **2017**, *224*, 41–47. [CrossRef]
35. Atelge, M.; Krisa, D.; Kumar, G.; Eskicioglu, C.; Nguyen, D.D.; Chang, S.W.; Atabani, A.; Al-Muhtaseb, A.H.; Unalan, S. Biogas production from organic waste: Recent progress and perspectives. *Waste Biomass Valor.* **2020**, *11*, 1019–1040. [CrossRef]
36. Uddin, M.; Mba, W.M. Anaerobic digestion fundamentals, challenges, and technological advances. *Phys. Sci. Rev.* **2022**, *19*, 1–19. [CrossRef]
37. Schnürer, A. Biogas production: Microbiology and technology. In *Anaerobes in Biotechnology*; Springer: Berlin/Heidelberg, Germany, 2016; pp. 195–234.
38. McGenity, T.J.; Timmis, K.N.; Fernández, B.N. *Hydrocarbon and Lipid Microbiology Protocols*; Springer: Berlin/Heidelberg, Germany, 2016.
39. Nwokolo, N.; Mukumba, P.; Obileke, K.; Enebe, M. Waste to energy: A focus on the impact of substrate type in biogas production. *Processes* **2020**, *8*, 1224. [CrossRef]
40. Shen, J.; Zhao, C.; Liu, Y.; Zhang, R.; Liu, G.; Chen, C. Biogas production from anaerobic co-digestion of durian shell with chicken, dairy, and pig manures. *Energy Convers. Manag.* **2019**, *198*, 110535. [CrossRef]
41. Wang, Z.; Jiang, Y.; Wang, S.; Zhang, Y.; Hu, Y.; Hu, Z.-H.; Wu, G.; Zhan, X. Impact of total solids content on anaerobic co-digestion of pig manure and food waste: Insights into shifting of the methanogenic pathway. *Waste Manag.* **2020**, *114*, 96–106. [CrossRef]
42. Ning, J.; Zhou, M.; Pan, X.; Li, C.; Lv, N.; Wang, T.; Cai, G.; Wang, R.; Li, J.; Zhu, G. Simultaneous biogas, and biogas slurry production from co-digestion of pig manure and corn straw: Performance optimization and microbial community shift. *Bioresourc. Technol.* **2019**, *282*, 37–47. [CrossRef]
43. Duan, N.; Zhang, D.; Lin, C.; Zhang, Y.; Zhao, L.; Liu, H.; Liu, Z. Effect of organic loading rate on anaerobic digestion of pig manure: Methane production, mass flow, reactor scale and heating scenarios. *J. Environ. Manag.* **2019**, *231*, 646–652. [CrossRef]
44. Achinas, S.; Li, Y.; Achinas, V.; Euverink, G.J.W. Influence of sheep manure addition on biogas potential and methanogenic communities during cow dung digestion under mesophilic conditions. *Sustain. Environ. Res.* **2018**, *28*, 240–246. [CrossRef]

45. Imeni, S.M.; Pelaz, L.; Corchado-Lopo, C.; Busquets, A.M.; Ponsá, S.; Colón, J. Techno-economic assessment of anaerobic co-digestion of livestock manure and cheese whey (Cow, Goat & Sheep) at small to medium dairy farms. *Bioresourc. Technol.* **2019**, *291*, 121872.
46. Scarlat, N.; Fahl, F.; Dallemand, J.F.; Monforti, F.; Motola, V. A spatial analysis of biogas potential from manure in Europe. *Renew. Sustain. Energy Rev.* **2018**, *94*, 915–930. [CrossRef]
47. Cheong, D.Y.; Harvey, J.T.; Kim, J.; Lee, C. Improving Biomethanation of Chicken Manure by Co-Digestion with Ethanol Plant Effluent. *Int. J. Environ. Res. Public Health* **2019**, *16*, 5023. [CrossRef] [PubMed]
48. Caruso, M.C.; Braghieri, A.; Capece, A.; Napolitano, F.; Romano, P.; Galgano, F.; Altieri, G.; Genovese, F. Recent updates on the use of agro-food waste for biogas production. *Appl. Sci.* **2019**, *9*, 1217. [CrossRef]
49. Orhorhoro, E.K.; Ebunilo, P.O.; Sadjere, E. Experimental Determination of Effect of Total Solid (TS) and Volatile Solid (VS) on Biogas Yield. *Am. J. Mod. Energy* **2017**, *3*, 131–135. [CrossRef]
50. Lup, D.T.; Stroe, A.M.; Chezan, P.M.; Pica, E.M. The Importance of Waste Incineration. *Stud. Univ. Babes-Bolyai Ambient.* **2018**, *63*, 43–48. [CrossRef]
51. Deng, L.; Liu, Y.; Wang, W. *Biogas Technology*, 1st ed.; Springer: Singapore, 2020.
52. Singh, A. Managing the uncertainty problems of municipal solid waste disposal. *J. Environ. Manag.* **2019**, *240*, 259–265. [CrossRef]
53. Rao, M.N.; Sultana, R.; Kota, S.H. *Chapter 2—Municipal Solid Waste BT—Solid and Hazardous Waste Management*, 1st ed.; Elsevier: Amsterdam, The Netherlands, 2017.
54. Salomone, R.; Saija, G.; Mondello, G.; Giannetto, A.; Fasulo, S.; Savastano, D. Environmental impact of food waste bioconversion by insects: Application of Life Cycle Assessment to process using Hermetia Illucens. *J. Clean. Prod.* **2017**, *140*, 890–905. [CrossRef]
55. Goossensen, M. Anaerobic Digestion of Municipal Organic Waste in Amsterdam; An Explorative Technical Feasibility Study and Scale Comparison. Master's Thesis, Wageningen University and Research, Wageningen, The Netherlands, 2017.
56. Parra-Orobio, B.; Torres-Lozade, P.; Marmolejo-Rebellon, F. Anaerobic digestion of municipal biowaste for the production of renewable energy: Effect of particle size. *Braz. J. Chem Eng.* **2017**, *34*, 481–491. [CrossRef]
57. Asamoah, B.; Nikiema, J.; Gebrezgabher, S.; Odonkor, E.; Njenga, M. *A Review on Production, Marketing and Use of Fuel Briquettes;* CGIAR Research Program on International Water Management Institute (IWMI): Anand, India, 2016.
58. Dlamini, S.; Simatele, M.D.; Serge Kubanza, N. Municipal solid waste management South Africa: From waste to energy recovery through waste-to-energy technologies in Johannesburg. *Local Environ.* **2019**, *24*, 249–257. [CrossRef]
59. Nwokolo, N.; Mukumba, P.; Obileke, K. Gasification of eucalyptus wood chips in a downdraft for syngas production in South Africa. *Int. J. Renew. Energy Res.* **2020**, *10*, 663–668.
60. Luo, X.; Wu, T.; Shi, K.; Song, M.; Rao, Y. Biomass Gasification: An Overview of Technological Barriers and Socio-Environmental Impact. In *Gasification for Low-Grade Feedstock;* Yun, Y., Ed.; InTechOpen: London, UK, 2018.
61. Kirsanovs, V.; Blumberga, D.; Karklina, K.; Veidenbergs, I.; Rochas, C.; Vigants, E.; Vigants, G. Biomass Gasification for district heating. *Energy Procedia* **2017**, *113*, 217–223. [CrossRef]
62. Solarte-Toro, J.; Chacón-Pérez, Y.; CardonaAlzate, C. Evaluation of biogas and syngas as energy vectors for heat and power generation using lignocellulosic biomass as raw material. *Electron. J. Biotech.* **2018**, *33*, 52–62. [CrossRef]
63. Gallucci, F.; Liberatore, R.; Sapengo, L.; Volponi, E.; Venturini, P.; Rispoli, F.; Pris, E.; Carnevale, M.; Colanttoni, A. Influence of Oxidant Agent on Syngas Composition: Gasification of Hazelnut Shells through an Updraft Reactor. *Energies* **2020**, *13*, 102. [CrossRef]
64. Glushkov, D.; Nyashina, G.; Shvets, A.; Pereira, A.; Ramanathan, A. Current Status of the Pyrolysis and Gasification Mechanism of Biomass. *Energies* **2021**, *14*, 7541. [CrossRef]
65. Dang, Q.; Zhang, X.; Zhou, Y.; Jia, X. Prediction, and optimization of syngas production from a kinetic based biomass gasification process model. *Fuel Process. Technol.* **2020**, *212*, 106604. [CrossRef]
66. Varank, G.; Ongen, A.; Guvenc, S.; Ozcan, H.; Can-Guven, E. Modelling, and optimization of syngas production from biomass gasification. *Int. J. Environ. Sci. Technol.* **2022**, *19*, 3345–3358. [CrossRef]
67. Peng, W.; Wang, L.; Mirzaee, M.; Ahmadi, H.; Esfahani, M.; Fremaux, S. Hydrogen and syngas production by catalytic biomass gasification. *Energy Convers. Manag.* **2017**, *135*, 270–273. [CrossRef]
68. Yao, Z.; You, S.; Ge, T.; Wang, C.W. Biomass gasification for syngas and biochar co-production: Energy application and economic evaluation. *Appl. Energy* **2018**, *209*, 43–55. [CrossRef]
69. Sui, M.; Li, G.; Guan, Y. Hydrogen, and syngas production from steam gasification of biomass using cement as catalyst. *Biomass Convers. Biorefinery* **2020**, *10*, 119–124. [CrossRef]
70. Paulauskas, R.; Zakarauskas, K.; Striūgas, N. An Intensification of Biomass and Waste Char Gasification in a Gasifier. *Energies* **2021**, *14*, 1983. [CrossRef]
71. Pissot, S.; Vilches, T.; Thunman, H.; Seemann, M. Effect of ash circulation on the performance of a dual fluidized bed gasification system. *Biomass Bioenergy* **2018**, *115*, 45–55. [CrossRef]
72. Chen, S.; Sun, Z.; Zhang, Q.; Hu, J.; Xiang, W. Steam gasification of sewage sludge with CaO as CO2 sorbent for hydrogen-rich syngas production. *Biomass Bioenergy* **2017**, *107*, 52–62. [CrossRef]
73. Yang, X.; Tian, S.; Kan, T.; Zhu, Y.; Xu, H.; Strezov, V.; Nelson, P.; Jiang, Y. Sorption-enhanced thermochemical conversion of sewage sludge to syngas with intensified carbon utilization. *Appl. Energy* **2019**, *254*, 113663. [CrossRef]

74. Maitlo, G.; Ali, I.; Mangi, K.H.; Ali, S.; Maitlo, H.A.; Unar, I.N.; Pirzada, A.M. Thermochemical Conversion of Biomass for Syngas Production: Current Status and Future Trends. *Sustainability* **2022**, *14*, 2596. [CrossRef]
75. Islam, M.W. A review of dolomite catalyst for biomass gasification tar removal. *Fuel* **2020**, *267*, 117095. [CrossRef]
76. Guan, Y.; Luo, S.; Liu, S.; Xiao, B.; Cai, L. Steam catalytic gasification of municipal solid waste for producing tar-free fuel gas. *Int. J. Hydrogen Energy* **2009**, *34*, 9341–9346. [CrossRef]
77. Wang, J.; Cheng, G.; You, Y.; Xiao, B.; Liu, S.; He, P.; Guo, D.; Guo, X.; Zhang, G. Hydrogen-rich gas production by steam gasification of municipal solid waste (MSW) using NiO supported on modified dolomite. *Int. J. Hydrogen Energy* **2012**, *37*, 6503–6510. [CrossRef]
78. Riaza, J.; Mason, P.; Jones, J.M.; Gibbins, J.; Chalmers, H. High temperature volatile yield and nitrogen partitioning during pyrolysis of coal and biomass fuels. *Fuel* **2019**, *248*, 215–220. [CrossRef]
79. Schneider, J.K.; da Cunha, M.E.; dos Santos, A.L.; Maciel, G.P.; Brasil, M.C.; Pinho, A.R.; Caramão, E.B. Comprehensive two-dimensional gas chromatography with fast-quadrupole mass spectrometry detector analysis of polar compounds extracted from the bio-oil from the pyrolysis of sawdust. *J. Chromatogr. A* **2014**, *1356*, 236–240. [CrossRef]
80. Nguyen, T.S.; He, S.; Raman, G.; Seshan, K. Catalytic hydro-pyrolysis of lignocellulosic biomass over dual Na2CO3/Al2O3 and Pt/Al_2O_3 catalysts using n-butane at ambient pressure. *Chem. Eng. J.* **2016**, *299*, 415–419. [CrossRef]
81. He, J.; Strezov, V.; Kumar, R.; Weldekidan, H.; Jahan, S.; Dastjerdi, B.H.; Zhou, X.; Kan, T. Pyrolysis of heavy metal contaminated Avicennia marina biomass from phytoremediation: Characterisation of biomass and pyrolysis products. *J. Clean. Prod.* **2019**, *234*, 1235–1245. [CrossRef]
82. Pighinelli, A.L.M.T.; Boateng, A.A.; Mullen, C.A.; Elkasabi, Y. Evaluation of Brazilian biomasses as feedstocks for fuel production via fast pyrolysis. *Energy Sustain. Dev.* **2014**, *21*, 42–50. [CrossRef]
83. Yi, L.; Liu, H.; Li, S.; Li, M.; Wang, G.; Man, G.; Yao, H. Catalytic pyrolysis of biomass wastes over Org-CaO/Nano-ZSM-5 to produce aromatics: Influence of catalyst properties. *Bioresour. Technol.* **2019**, *294*, 122186. [CrossRef] [PubMed]
84. Akubo, K.; Nahil, M.A.; Williams, P.T. Pyrolysis-catalytic steam reforming of agricultural biomass wastes and biomass components for production of hydrogen/syngas. *J. Energy Inst.* **2019**, *92*, 1987–1996. [CrossRef]
85. Liu, X.; Luo, Z.; Yu, C.; Xie, G. Conversion mechanism of fuel-N during pyrolysis of biomass wastes. *Fuel* **2019**, *246*, 42–50. [CrossRef]
86. Chang, Z.; Duan, P.; Xu, Y. Catalytic hydro-pyrolysis of microalgae: Influence of operating variables on the formation and composition of bio-oil. *Bioresourc. Technol.* **2015**, *184*, 349–354. [CrossRef]
87. Chernova, N.I.; Kiseleva, S.V.; Larina, O.M.; Sytchev, G.A. Manufacturing gaseous products by pyrolysis of microalgal biomass. *Int. J. Hydrogen Energy* **2020**, *45*, 1569–1577. [CrossRef]
88. Borel, L.D.M.S.; Lira, T.S.; Ribeiro, J.A.; Ataíde, C.H.; Barrozo, M.A.S. Pyrolysis of brewer's spent grain: Kinetic study and products identification. *Ind. Crop. Prod.* **2018**, *121*, 388–395. [CrossRef]
89. Ahmad, N.; Abnisa, F.; Wan Daud, W.M.A. Liquefaction of natural rubber to liquid fuels via hydrous pyrolysis. *Fuel* **2018**, *218*, 227–235. [CrossRef]
90. Alptekin, F.; Celiktas, M. Review on catalytic biomass gasification for hydrogen production as a sustainable energy form and social, technological, economic, environmental, and political analysis of catalysts. *ACS Omega* **2022**, *7*, 24918–24941. [CrossRef]
91. Chuayboon, S.; Abanades, S.; Rodat, S. Insights into the Influence of Biomass Feedstock Type, Particle Size and Feeding Rate on Thermochemical Performances of a Continuous Solar Gasification Reactor. *Renew. Energy* **2019**, *130*, 360–370. [CrossRef]
92. Yu, H.; Wu, Z.; Chen, G. Catalytic Gasification Characteristics of Cellulose, Hemicellulose and Lignin. *Renew. Energy* **2018**, *121*, 559–567. [CrossRef]
93. Liu, Y.; Yun, S.; Zhou, X.; Hou, Y.; Zhang, T.; Li, J.; Hagfeldt, A. Intrinsic origin of superior catalytic properties of tungsten-based catalysts in dye-sensitized solar cells. *Electrochim. Acta* **2017**, *242*, 390–399. [CrossRef]
94. Fan, Q.; Zhang, W.; Yun, S.; Xu, J.; Song, Y. III-Nitride polymorphs: XN (X=Al, Ga, In) in the Pnma phase. *Chem. Eur. J.* **2018**, *24*, 17280–17287. [CrossRef] [PubMed]
95. Liu, Y.; Yun, S.; Zhou, X.; Hou, Y.; Zhang, T.; Li, J.; Hagfeldt, A. The role of conductive materials in the start-up period of thermophilic anaerobic system. *Bioresour. Technol.* **2017**, *239*, 336–344.
96. Akinbomi, J.G.; Patinvoh, R.J.; Taherzadeh, M.J. Current challenges of high-solid anaerobic digestion and possible measures for its effective applications: A review. *Biotechnol. Biofuels* **2022**, *15*, 52. [CrossRef]
97. Mohammadi, A.; Anukam, A. The Technical Challenges of the Gasification Technologies Currently in Use and Ways of Optimizing Them: A. Review. In *Energy Recovery*; IntechOpen: London, UK, 2022. [CrossRef]
98. Basu, P. *Biomass Gasification, Pyrolysis and Torrefaction: Practical Design and Theory*; Academic Press: London, UK, 2018.

applied sciences

MDPI

Article

Dynamic Modelling and Energy, Economic, and Environmental Analysis of a Greenhouse Supplied by Renewable Sources

Francesco Calise, Francesco Liberato Cappiello, Luca Cimmino and Maria Vicidomini *

Department of Industrial Engineering, University of Naples Federico II, P.le Tecchio 80, 80125 Naples, Italy; frcalise@unina.it (F.C.); francescoliberato.cappiello@unina.it (F.L.C.); luca.cimmino@unina.it (L.C.)
* Correspondence: maria.vicidomini@unina.it

Abstract: This paper regards the design and dynamic modelling of a greenhouse coupled with renewable energy technologies to obtain a hybrid renewable energy plant as an optimal solution in the green farm framework. The considered technologies are PV panels, solar thermal collectors, and a biomass auxiliary heater. The system is also coupled with a pyrogasifier, supplied by wood and agricultural waste in the framework of a biocircular economic approach. To supply the investigated user, with a "green farm" located in Castelvolturno (Naples, South of Italy) reducing the energy consumption and operating costs, all of the main components of the plant were suitably designed. The operation of the designed components was simulated by a dynamic simulation model developed by TRNSYS software and validated by means of the literature results. A comprehensive energy, economic, and environmental analysis of the greenhouse is presented. The main results suggest that the investigated renewable plant reduces the total equivalent CO_2 emissions by 148.66 t/y. Considering the current high increases in energy prices as a result of the energy crisis due to the war, the system shows very significant profitability with a simple payback of only 1.7 years.

Keywords: dynamic simulation; green farm; PV panels; solar thermal collectors

Citation: Calise, F.; Cappiello, F.L.; Cimmino, L.; Vicidomini, M. Dynamic Modelling and Energy, Economic, and Environmental Analysis of a Greenhouse Supplied by Renewable Sources. *Appl. Sci.* **2023**, *13*, 6584. https://doi.org/10.3390/app13116584

Academic Editor: Davide Astiaso Garcia

Received: 4 May 2023
Revised: 17 May 2023
Accepted: 23 May 2023
Published: 29 May 2023

1. Introduction

Renewable energy sources (RES) [1] can be coupled with several energy systems to produce the heat or power needed for different plants in order to reduce the primary energy demand of the included components. RES systems may be integrated into agricultural waste conversion processes [2,3] to provide the energy required for the process and significantly reduce the energy conversion of the process. The exploitation of biomass as a renewable source is widely investigated in the literature [4]. In fact, biomass can be used to produce biogas by means of the anaerobic digestion process [5], and the biogas may be upgraded into biomethane and injected into the gas grid [6]. Furthermore, digestate is a by-product of the anaerobic digestion process which can be used as a fertilizer in agriculture [7]. Agriculture biomass conversion is a pivotal solution for farmers considering that their agricultural activities can be modified in order to obtain energy-efficient plants [8].

In this framework, the waste-to-energy (WTE) approach is considered as a viable strategy for the management of waste considering that in comparison to other methods, this approach provides renewable energy outputs and end products [9]. According to this strategy, by using low/negative-value organic waste, the production of renewable heat or power as well as fuel is possible. This allows for a minimal environmental impact to be made and also a significant decrease in organic waste volume [10]. Implementing this paradigm, agricultural waste (or other sources of biomass) can be globally recognized as resources, more than simply waste, producing renewable outputs in an efficient manner. This environmental-friendly approach also implies a reduction in fossil fuel dependence [11]. Thanks to agricultural waste conversion, it is possible to produce useful energy for specific users such as farms, single buildings, and groups of both.

However, for the efficient and economic use of agricultural waste for the production of energy vectors, clear guidelines and possible obstacles to be overcome in order to use agriculture biomass in the coming years [8] have to be identified. In particular, solar systems—namely, photovoltaic (PV) panels [12], solar thermal collectors (STC) [13], or PV/thermal (PVT) collectors [14]—can be absolutely coupled with greenhouse systems [15]. This option seems very smart for the adoption of natural gas boilers and electricity from the national grid. For example, the installation of PV panels for power production [16] can easily be a sustainable energy source to supply water pumps for irrigation farming purposes; solar thermal collectors can be employed to provide heat for the heating system of greenhouses with the aim of maintaining the operating temperature of the greenhouse within the designed temperature range [17]. For prediction and the increase in solar power plants, the tracking of solar power data can assist in determining the rate of performance loss in their systems. In addition, artificial intelligence is one strategy that has recently grown in popularity [18].

Several authors have investigated this issue. For example, a non-linear integrated controlled environment agriculture model was developed in order to define a correlation between several factors and the conditions of the crop being grown [19]: humidity and temperature control, the impact of weather disturbances, irrigation, and fertilization. The simulation results are related to a renewable-energy-powered semi-closed greenhouse situated in Ithaca, New York. Tomato growing was considered. The authors suggest that a model able to control the environment where the crops grow can be useful to increase the efficiency of renewable energy utilization from 4.7% to 127.5%. In the work of ref. [20], a mathematical model to simulate a greenhouse was developed and validated vs. experimental data. The equations regard four elements of the greenhouse, i.e., the inside air, cover, bare soil surface, and canopy surface. The model was used to simulate the operation of the research farm of Punjab Agricultural University, Ludhiana. The model solved using the Gauss–Seidel iteration method confirms good agreement with the measured data related to the winter operation for a tomato crop. A dynamic greenhouse environment simulator was developed in ref. [21] to predict, with different configurations, greenhouse dynamic behavior. The model was implemented in a web-based interactive application that allowed for the selection of the greenhouse design, operational strategies, and weather conditions (four seasons at four geographical locations). To forecast the hourly demand for heating the conventional greenhouse a quasi-steady state time-dependent thermal model was developed [22]. The model considers the lumped estimation of the heat transfer parameters of the greenhouse. The model takes into account the indoor environmental control parameters of the greenhouse, the thermal and physical properties of crops, construction materials, and several hourly weather parameters, such as cloud cover, relative humidity, temperature, and wind speed. The heat gain from environmental control systems and the heat loss for plant evapotranspiration are further phenomena considered in this model. The authors highlight the importance of the environmental control systems to reduce the total heating requirements over the year by about 13–56%. A Chinese-style solar greenhouse for Canadian prairies modeled by a detailed TRNSYS model is presented in ref. [23]. The model, that is used to predict the transient heating requirement of the greenhouse, was also validated by a new heating simulation model. The same model developed in ref. [23] was improved in ref. [24] and validated adopting the data collected from a solar greenhouse in Manitoba, Canada. The yearly simulation reveals that in the coldest month (January), the daily average heating could be twice as high (6.3 MJ/m^2·day) as the value obtained in a mild month such as March (3.4 MJ/m^2·day). Comparing this solar greenhouse with a traditional, local one, the heating cost is approximately 55% lower.

The open literature regarding dynamic modelling and multi-source integration is very wide [25,26]. In ref. [27], solar, geothermal, and biomass systems were integrated into a novel renewable trigeneration plant for the production of power, heat, and cool, respectively. In particular, a 193 m^2 PV, a 159 kWh lithium-ion battery, a 30 kWe organic Rankine cycle, a 350 kWth biomass auxiliary heater, a geothermal well at 96 °C, and a 80 kW single-stage H_2O/LiBr absorption chiller were the included technologies in this plant. The model of the plant was applied to the case study, a residential building in Campi Flegrei (Naples, South Italy). A primary energy saving of 139% was achieved due to the excess energy delivered to the grid. The achieved payback period was about 19 years, mainly due to the high capital cost of the used technologies. A novel ultra-highly efficient solar power system combining concentrating photovoltaic/thermal (CPVT) solar collectors coupled with an organic Rankine cycle (ORC) was investigated in ref. [28]. The CPVT collectors produced the power and heat used to supply the ORC unit. The solar field consists of a dish concentrator and a triple-junction PV layer, and it is equipped with a double axis tracking system. Both CPVT and ORC models are integrated in a more complex dynamic simulation model, developed in a TRNSYS environment. However, the results showed that the novel (ORC+CPVT) system produces only 6% more electrical energy than the system with only CPVT collectors. In ref. [29], two solar polygeneration plants designed to produce space heating and cooling energy, power, and fresh water are presented. These plants are designed for two small Mediterranean islands, the Favignana and Salina islands (South Italy). In the first layout, the CPVT collectors produced electric and thermal energy. The thermal energy supplies both a single-stage lithium bromide/water absorption chiller for cooling energy production and a multi-effect distillation unit for freshwater production. An auxiliary biomass-fired heater is also considered. In the second layout, a PV field was coupled with heat pumps for space heating, cooling, and domestic hot water production and a reverse osmosis unit for freshwater production. Both plants were modelled in a TRNSYS environment. In addition, in these models, the detailed estimation of the load of the buildings located on the islands was also determined using type 56 of TRNSYS. The best economic indexes were obtained for the layout using electricity-driven technologies, with a payback period of about 6 years. Evacuated flat-plate solar collectors and PV panels were integrated into two renewable polygeneration plants for the production of power, heat, and cool in ref. [30]. In the first renewable polygeneration system, a 6 kWe organic Rankine cycle (ORC), a 17-kW single-stage H_2O/LiBr absorption chiller, a geothermal well at 96 °C, a 200 kWt biomass auxiliary heater, a 45.55 kWh lithium-ion battery, and a 25 m^2 solar field were included. In the second layout, the solar thermal collectors were replaced by PV panels. A ground-cooled condenser was used to reduce the ORC condensation temperature. All of the included units were simulated in detail in a TRNSYS environment. The economic results showed that the layout with the PV panels and evacuated solar collectors obtained payback periods of 13 years and 15 years, respectively. CPVT collectors were used to match the heating and power demand of an anaerobic digestion plant supplied by the organic fraction of municipal solid waste by a dynamic simulation model in ref. [14]. This plant also includes a biomass auxiliary heating system and an upgraded section for biomethane production. For such hybrid solar–biomass systems, a detailed model, which is able to calculate the time-dependent production of biomethane, was developed in MATLAB® 2022a (Matrix Laboratory, MathWorks, Natick, MA, USA)and then integrated into the TRNSYS simulation model of the whole plant. The achieved primary energy saving was 24%, and a promising payback time of about 3 years was estimated.

Aims of the Work and Its Novelty

This work aims to increase renewable energy technology usage in the agricultural sector. With respect to the literature review, the work aims to show how hybrid renewable energy plants can be an optimal solution in the framework of the green farm and biocircular economy approach. In particular, even if in the current literature the dynamic simulation of a greenhouse has already been investigated [23], the previous paper did not include in its analysis the integration of the model of the greenhouse into a whole renewable energy plant, as occurs in the presented paper. In addition, in this work, a detailed economic analysis aimed at evaluating the profitability of the presented green farm is also presented. This analysis considers in detail the economic capital costs and maintenance costs of all of the components included in this renewable plant. The economic analysis also shows the effect of the current energy crisis on the spread of renewable energy technologies due to the sharp increase of energy prices. This aspect has never been investigated in the literature in the case of comprehensive dynamic simulations of renewable systems coupled with a greenhouse. The presented approach based on the dynamic simulation of a complex plant, including more different technologies, is very reliable and affordable for performing preliminary analysis and components design. In fact, the dynamic simulation allows us to mimic the real operating conditions of the plant, for example, the operating temperature of the greenhouse. This suggests a first and useful preliminary indication of the design phase. In fact, by knowing the operating temperature of the greenhouse during the year, the designer can indicate suitable vegetable species for such operating conditions. In addition, the simulation model can also be useful to optimize the operation of the plant according to a specific optimization function before the installation of the plant itself.

In order to address this aim, the following points were implemented:

1. The development of a greenhouse dynamic simulation model in a TRNSYS environment
2. The validation of the greenhouse dynamic simulation model by the literature values;
3. The evaluation of the hourly energy demand of the greenhouse and the whole company to simulate the electrical and thermal loads of the company itself;
4. The greenhouse model is integrated into a comprehensive dynamic simulation model, including several renewable technologies based on the use of biomass and solar sources (PV panels and solar thermal collectors);
5. The preliminary design of all included components of the plant (the tank, solar PV field, solar thermal collectors, and pyrogisifier) in order to obtain a grid-independent system and reduce conventional fossil fuels' adoption;
6. The development of a suitable energy, economic, and environmental model to evaluate the performance of the plant with respect to the conventional system based on a conventional natural gas boiler and national grid.

This paper was partially published in the Proceedings of the 12th International Conference on Smart Cities and Green ICT Systems—SMARTGREENS [31].

2. Method

In this section, the method adopted to develop this work will be described. Here, the greenhouse model and its validation vs. the literature data will be reported. Then, this model will be integrated into a comprehensive simulation model including the investigated renewable technologies according to the investigated layout (Figure 1). The section also includes some details of the modelling of the main components, such as the solar thermal collector and PV panel fields, and the main economic and energy indexes evaluated to perform the technoeconomic analysis. The flowchart of the adopted approach is reported in Figure 2. This follows the bullet list reported in the aim of the work subsection.

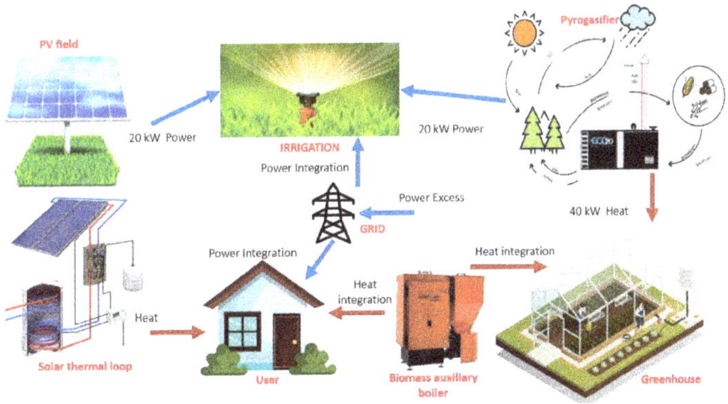

Figure 1. Layout.

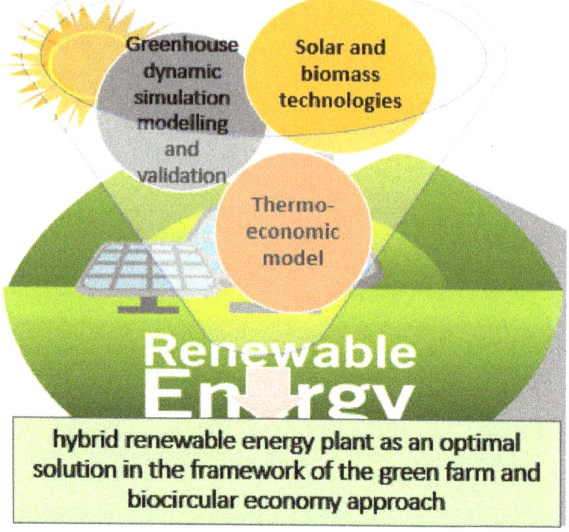

Figure 2. Flowchart of the proposed approach.

2.1. Layout

Figure 1 shows the layout investigated in this work.

The polygeneration system proposed for the greenhouse integrates a photovoltaic field, a pyrogasification unit, a solar thermal collector field, and a biomass auxiliary heater.

The pyrogasifier is fed with the residuals of the agricultural crops and wood to produce both electricity for the user and heat for the greenhouse. Power is supplied to the user also by means of solar PV panels which also provide electricity for the irrigation pumps. In addition, solar thermal collectors are adopted to provide heat to meet the demand for heating and domestic hot water by the user. Moreover, the heat produced by means of the solar thermal collectors is used for the heating system of the corn dryer unit. In case the heat produced is greater than the user and greenhouse demand, a water tank is used for storage purposes. Otherwise, in case of scarce solar radiation, the pyrogasifier is switched off, or simply during night hours, an auxiliary biomass-fed boiler is used for meeting the heating demands. The polygeneration system proposed is also connected to the electric grid in case of scarce energy production with respect to the user demand.

2.2. Model

The modelling of this plant was developed utilizing the well-known dynamic simulation software TRNSYS 18 (Transient System Simulation Tool, Solar Energy Laboratory, Univ. of Wisconsin-Madison, Madison, WI, USA). The software is based on a specific library of components, suitable to accurately simulate several energy units. The models of the TRNSYS components were validated and judged reliable [32]. For the sake of shortness, the TRNSYS types used to simulate the components included in the plant are reported in Table 1.

TRNSYS software is very accurate for the estimation of the energy demand of buildings [33], and it can be considered a reliable tool for the validation of in-house building simulation models [34–36]. However, its application can also be suitable for the simulation of greenhouses, as reported in ref. [23]. The next subsection includes the description of the greenhouse model and its validation, the model of the PV field (Type 94) and the storage tank (Type 4), and the thermo-economic equations used to evaluate the plant's performance. For all the other components, a detailed explanation of the mathematical modelling is reported in ref. [32]. Note that the simulation modelling steps included the following steps: (i) the development of the greenhouse model; (ii) the validation of the greenhouse model using the literature data; (iii) the integration of the greenhouse model into the whole multi-energy system plant model; and (iv) the development of a thermo-economic model and an environmental model for the evaluation of the performance of the proposed plant with respect to the reference plant.

Table 1. TRNSYS types.

Type 1b	Solar thermal collectors
Type 94	Photovoltaic panels
Type 109	Weather conditions
Type 48	Inverter regulator
Type 4c	Thermal storage tank
Type 114	Circulation pump
Type 6	Biomass auxiliary boiler
Type 641	Humidification system
Type 77	Ground modelling

2.2.1. Greenhouse Model

Type 56 was selected to model the greenhouse. By taking into account the 3D geometry of the building, defined in the Google SketchUp TRNSYS3d plug-in [37], type 56 determines the dynamic energy demand by considering both the effects of the weather conditions (i.e., solar radiation, ambient temperature, humidity, etc.) and the thermophysical proprieties of the envelope. Ventilation, infiltration, and internal heat gain were also simulated. In Figure 3, the greenhouse geometry investigated in this research is reported. In reference [38], the validation of type 56 is reported. Note that considering the radiative properties of the surfaces as a function of the wavelength, type 56 considers a detailed model for the radiation calculation in the greenhouse, considering a complex model for the view factors' calculation.

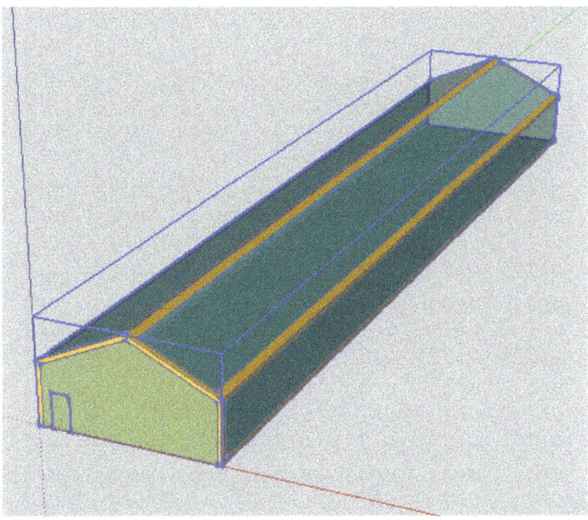

Figure 3. Geometric model of the investigated greenhouse.

The model calculates the temperature of the surfaces and the radiate flows transmitted by the glazing surfaces and emitted by the surfaces.

Validation of the model of the greenhouse was carried out considering the greenhouse model developed in TRNSYS according to ref. [23], where all of the assumptions to redevelop the model are reported.

In Figure 4, the monthly average daily heating requirement obtained both by ref. [23] and our model are summarized.

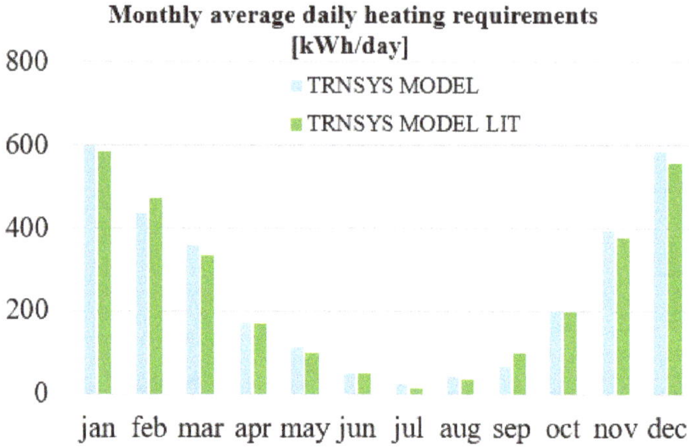

Figure 4. Model validation.

2.2.2. PV Panels Model

The "four parameters" model is used to simulate the PV panels' field. The four parameters in this model are: (i) $I_{L,ref}$, the photocurrent of the module under the reference condition; (ii) $I_{o,ref}$, the diode reverse saturation current under the reference condition; (iii) γ, the empirical PV curve-fitting parameter; and (iv) R_s the module series resistance.

The model assumes that the slope of the IV curve under the short-circuit condition is zero.

$$\left(\frac{dI}{dV}\right)_{v=0} = 0 \quad (1)$$

These parameters were calculated by type 94 from the manufacturer's data to obtain the IV curve for each time step. Assuming Rs and γ to be constant, the circuit current-voltage equation is:

$$I = I_{L,ref}\frac{G_T}{G_{T,ref}} - I_o\left[\exp\left(\frac{q}{\gamma k T_c}(V + IR_s)\right) - 1\right] \quad (2)$$

where $G_{T,ref}$ is the reference solar irradiance, equal to 1000 W/m^2, and G_T is the total incident solar irradiance on the PV field.

I_o is the diode reverse saturation current, depending on temperature, according to Equation (3):

$$\frac{I_o}{I_{o,ref}} = \left(\frac{T_c}{T_{c,ref}}\right)^3 \quad (3)$$

To know I_0, the PV current was calculated according to Newton's method. By means of an iterative routine, the current (I_{mpp}) and the voltage (V_{mpp}) at the point of maximum power along the IV curve were calculated.

For handling the system, the following three equations, which are functions of $I_{L,ref}$, $I_{o,ref}$, and γ (and of the open-circuit, short circuit, and maximum power conditions), are obtained:

$$I_{L,ref} \approx I_{sc,ref} \tag{4}$$

$$\gamma = \frac{q\left(V_{mp,ref} - V_{oc,ref} + I_{mp,ref}R_s\right)}{kT_{c,ref}\ln\left(1 - \frac{I_{mp,ref}}{I_{sc,ref}}\right)} \tag{5}$$

$$I_{o,ref} = \frac{I_{sc,ref}}{\exp\left(\frac{qV_{oc,ref}}{\gamma kT_{c,ref}}\right)} \tag{6}$$

To determine the last unknown parameter, a further equation, i.e., the analytical derivate of voltage with respect to the temperature under the reference open-circuit condition, is needed:

$$\frac{\partial V_{oc}}{\partial T_c} = \mu_{voc} = \frac{\gamma k}{q}\left[\ln\left(\frac{I_{sc,ref}}{I_{o,ref}}\right) + \frac{T_c\mu_{isc}}{I_{sc,ref}} - \left(3 + \frac{q\varepsilon}{\frac{\gamma}{N_s}kT_{c,ref}}\right)\right] \tag{7}$$

The manufacture's specification about the open-circuit temperature was equaled to this analytical value. A search routine was employed iteratively to calculate the equivalent open-circuit characteristics.

2.2.3. Storage Tank Model

According to the model implemented in TRNSYS type 4 [32], the tank was virtually divided into sub-volumes (N), where the fluid was considered fully mixed. In unsteady state conditions, the temperature T_i is calculated by mass and energy balances, for each volume i, according to the following equation:

$$M_i c_p \frac{dT_i}{d\vartheta} = \alpha_i \dot{m}_h c_p (T_h - T_i) + \beta_i \dot{m}_L c_p (T_L - T_i) + UA_i (T_{amb} - T_i) \tag{8}$$

where

T_h is the temperature of the fluid entering the tank from the heat source;

T_L is the temperature of the fluid replacing that extracted to supply the load.

Note that α_i is equal to 1 if the i-th sub-volume is the bottom volume of the tank; otherwise, it is 0. β_i. In contrast, it is equal to 0 if the i-th sub-volume corresponds to the bottom of the tank; otherwise, it is 1.

To the right side of Equation (8), the following term has to be added:

$$\begin{aligned} &+\gamma_i c_p (T_{i-1} - T_i) \quad if \gamma_i > 0 \\ &+\gamma_i c_p (T_i - T_{i+1}) \quad if \gamma_i < 0 \end{aligned} \tag{9}$$

γ_i is the control function defined by the following equation:

$$\gamma_i = \dot{m}_{Heat} \sum_{j=1}^{i-1} \alpha_j - \dot{m}_L \sum_{j=i+1}^{N} \beta_j \tag{10}$$

2.2.4. Thermo-Economic Model

The thermo-economic model is necessary to determine the environmental impact and the economic feasibility of the polygeneration system model proposed. The environmental analysis is based on the calculation of the primary energy saving (*PES*). This index is calculated as the percentage difference between the primary energy consumption in the proposed system (PS) with respect to a reference system (RS). In this case, the RS is the one in which the electricity demand is met by the grid and the thermal energy demand is met with a conventional gas boiler. The efficiency of these systems is 46% and 90% ($\eta_{el,grid}$ and η_{boiler}), respectively.

$$PES = \frac{PE_{RS} - PE_{PS}}{PE_{RS}} = \frac{\left(\frac{E_{th,boiler}}{\eta_{boiler}} + \frac{E_{el,fromGRID}}{\eta_{el,grid}}\right)_{RS} - \left[\frac{E_{el,fromGRID} - E_{el,toGRID}}{\eta_{el,grid}}\right]_{PS}}{\left(\frac{E_{th,boiler}}{\eta_{boiler}} + \frac{E_{el,fromGRID}}{\eta_{el,grid}}\right)_{RS}} \quad (11)$$

Together with primary energy consumption, the CO_2 equivalent emissions also need to be calculated. To do so, the electricity withdrawn from the grid and the primary energy consumed by the boiler were multiplied by their respective CO_2 emissions factors, shown in Table 4.

The equivalent CO_2 emissions difference was evaluated according to Equation (12):

$$\Delta CO_2 = \left(E_{el,fromGRID} f_{el} + \frac{E_{th,boiler}}{\eta_{boiler}} f_{NG}\right)_{RS} - \left[\left(E_{el,fromGRID} - E_{el,toGRID}\right) f_{el}\right]_{PS} \quad (12)$$

The economic analysis is instead based on the calculation of the yearly operating cost saving ΔC of the PS with respect to the RS. In this analysis, the unit purchasing cost of electricity from the grid $c_{el,fromGRID}$ and natural gas c_{NG}, for the RS, and the purchasing/selling $c_{el,toGRID}$ of the electricity from/to the grid for PS, are considered. In the PS, the unit purchasing cost of the biomass c_{wood} for the auxiliary wood-chip heater is also considered. $c_{bio,pyr}$ is the unit purchasing cost for the biomass (wood and agricultural crops) which feed the pyrogasifier. Moreover, the maintenance cost M was considered.

$$\begin{aligned}\Delta C = &\left(E_{el,fromGRID} c_{el,fromGRID} + V_{NG} c_{NG}\right)_{RS} - \left(M + M_{bio,pyr} c_{bio,pyr} + M_{bio,boiler} c_{bio,boiler}\right)_{PS} \\ &- \left(E_{el,fromGRID} c_{el,fromGRID} - E_{el,toGRID} c_{el,toGRID}\right)_{PS}\end{aligned} \quad (13)$$

To calculate the payback time of the proposed solution, the capital costs were also considered. More details about this term are shown in the Case Study section.

2.3. Case Study

The model of the greenhouse was applied to a suitable case study located in Castelvolturno (Naples, South of Italy). The main features of the greenhouse are reported in Table 2. In Table 3, the design data of the proposed plant are also summarized. The plant is designed to produce the electricity for the buildings close to the greenhouse and the related irrigation pumps and to produce the thermal energy both for the greenhouse heating and the domestic hot water and space heating energy demand of the user. The thermo-economic and environmental assumptions for the analysis of the PS with respect to the RS are summarized in Table 4. Figure 5 reports the thermal energy demand of the greenhouse; Figure 6 reports the power and heat load of the user.

Table 2. Greenhouse features.

Area	**450 m^2 (9 m × 50 m)**
Max height	5 m
Slope of the roof	30°
Air change infiltration	0.5 1/h
Ventilation	0.1 m/s
Artificial lightning	30 W/m^2
Day/night humidification rate for evapotranspiration	21.5/3.6 g/h
Heating temperature	20 °C
Materials	Plastic cover, steel structure, and chalk/clay floor

Table 3. Design data of the proposed plant.

Rated power PV field A	15 kW
Rated power PV field B	5 kW
Slope PV field A	0°
Slope solar thermal field/PV field B	30°
Area solar thermal field	28 m^2
Rated power pyrogasifier	20 kW
Rated thermal flow rate pyrogasifier	40 kW
Equivalent oper. hours pyrogasifier	7500 h/y
Efficiency curve coeff. solar collector	
a_0	0.785
a_1	1.03 W/m^2K
a_2	0.0033 W/m^2K^2

Table 4. Thermo-economic and environmental parameters.

Data	Value
Pyrogasifier cost	EUR 150 k
Biomass auxiliary boiler cost	EUR 10 k
Ordinary maint. pyrogasifier	3%/y
Extraordinary maint. pyrogasifier	EUR 5 k/2 y
Maint. biomass auxiliary boiler	2.50%
Unit cost of purchased biomass	EUR 0.12/kg
Unit cost of self-produced biomass	EUR 0.07/kg
Lower heating value of wood-chips	EUR 4/kg
Unit cost of PV field	EUR 1800/kW
Maint. PV field	2%
Unit cost solar thermal field	EUR 400/m
Maint. solar thermal field	2.5%
Lifetime proposed system	20 y
Discount rate	5%
CO_2 emission factor for electricity	0.48 $kgCO_2/kWh$
CO_2 emission factor for primary energy	0.20 $kgCO_2/kWh$

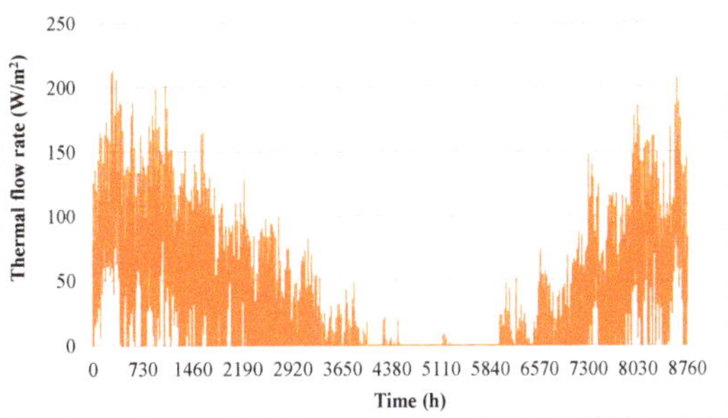

Figure 5. Thermal energy demand of the greenhouse.

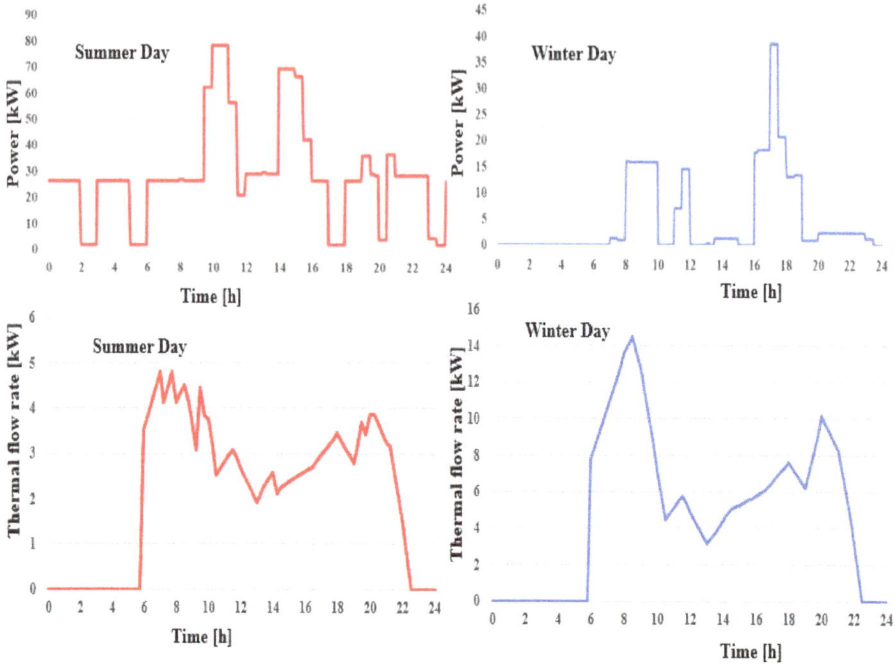

Figure 6. The heat (space heating and DHW) and power demand of the user (irrigation and offices' demand).

It is clearly shown how the thermal energy demand is very high during the winter months due to the cold temperatures. Over 50% of all yearly heating demand is concentrated in the coldest winter months. The power load is mainly due to the irrigation pumps (with a rated power of 55 kW). Note the high peak value during the summer day, considering that the irrigation time range is wider in the summer days. In addition, electric consumption due to technologies and offices is negligible with respect to irrigation. The thermal flow rate during the winter day reaches the peak value of 14 kW at 8 a.m., higher than the summer one, considering that the winter thermal flow rate is due to the space heating, DHW purposes, and corn drying.

3. Results

In this section, the results of the dynamic simulations performed for one year of operation will be presented according to different time bases: hourly and yearly bases. In particular, the results of the energy, economic, and environmental analysis will also be reported when the PS is compared with the RS. In addition, an economic analysis will be presented considering the purchasing costs before and after the energy crisis.

Figure 7 shows the trends of temperature of greenhouse and outdoor air without the heating system. Note that heating of the greenhouse quickly occurs during the central hours of the day and that the greenhouse temperature follows the same trend of the ambient temperature. The heating of the greenhouse occurs because the rays of the sun enter through the glass of the greenhouse, due to the particularly high transmission coefficients of the glass itsef. However, the infrared radiation emitted by the ground cannot be transmitted through some materials, such as glass, guaranteeing a higher temperature than the outdoor air temperature.

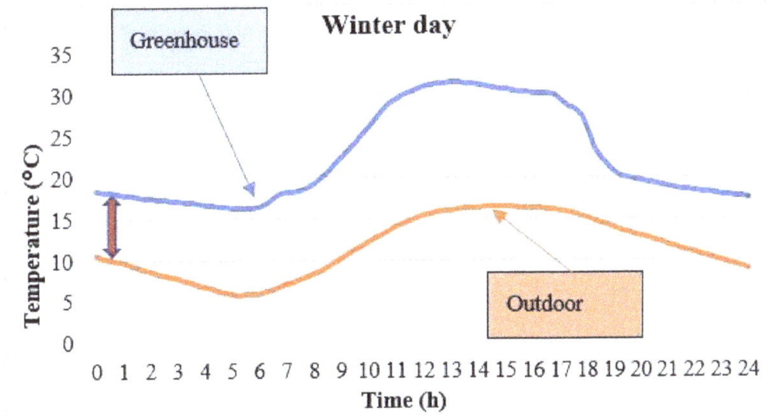

Figure 7. Temperature difference between the greenhouse and the outdoor air.

A heated greenhouse allows for an increase in the yield of the crop so that some crops can be cultivated even in the winter months. Conversely, in the case of greenhouses with heating systems, the trend of the thermal flow rates represented in Figure 8 can be observed. The heating demand of the greenhouse by the heating system occurs only if the greenhouse temperature is lower than 20 °C, mainly when radiation is absent or for cold ambient temperatures. During these hours, the thermal losses by the construction materials are high. Due to the transmitted solar radiation, the thermal energy demand is null from 11 a.m. to 4 p.m. because the greenhouse temperature is higher than 20 °C, although this was a winter day.

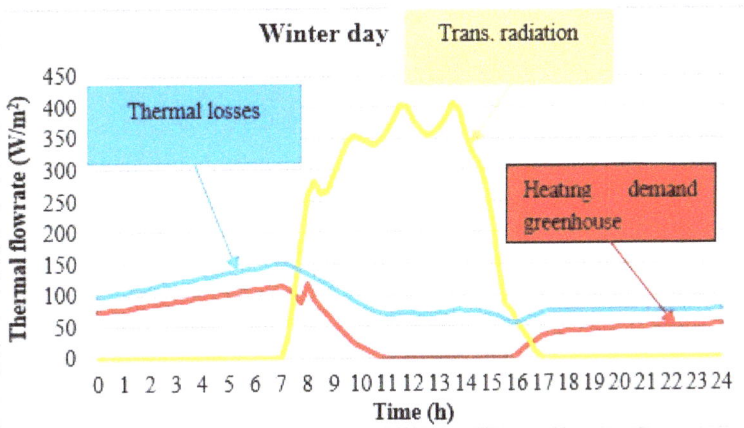

Figure 8. The transmitted solar radiation, thermal losses, and heating demand of the greenhouse.

In the central hours of the day, although the pyrogasifier is switched off to carry out the ordinary maintenance of the unit, the integrations are null due to high solar production. Simultaneously, the greenhouse heat demand decreases. During the night, the pyrogasifier is unable to cover the demand. Therefore, the boiler is needed. This integration increases in the early morning hours when the boiler has to heat the greenhouse and provide heat to the user due to the low solar thermal production, see Figure 9.

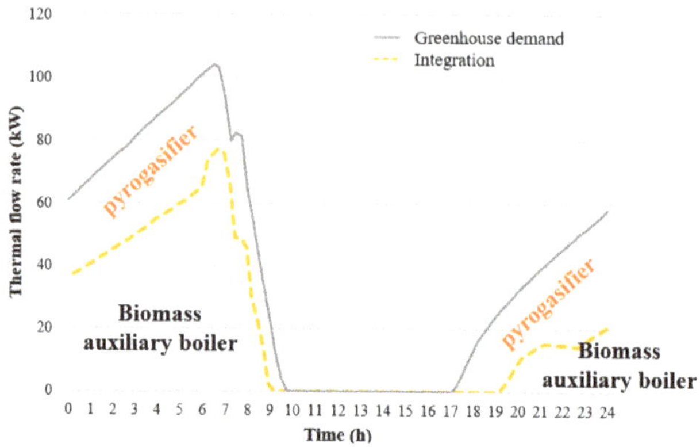

Figure 9. Thermal flow rate: greenhouse heating demand and biomass auxiliary production and the pyrogasifier.

Figures 10 and 11 report the powers of the electric loop without and with the pyrogasifier, respectively. As previously noted, the winter demand is clearly lower than the summer one due to the scarce use of irrigation pumps. PV production without a pyrogasifier guarantees very low integrations and particularly high excesses to the grid in the central hours. During the winter day, the production is almost higher than the demand. Without the pyrogasifier for a typical summer day, given the high consumption, excesses are null and important integrations were obtained. In this scenario, the plant appears to be undersized.

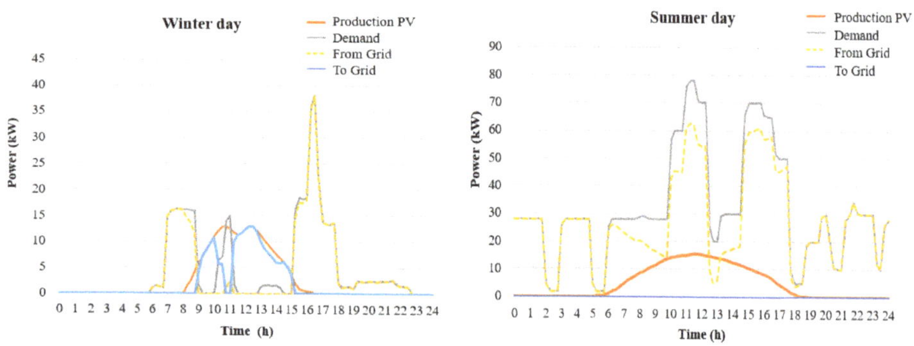

Figure 10. Power production of the PV panels, the total power demand, and power from/to the grid.

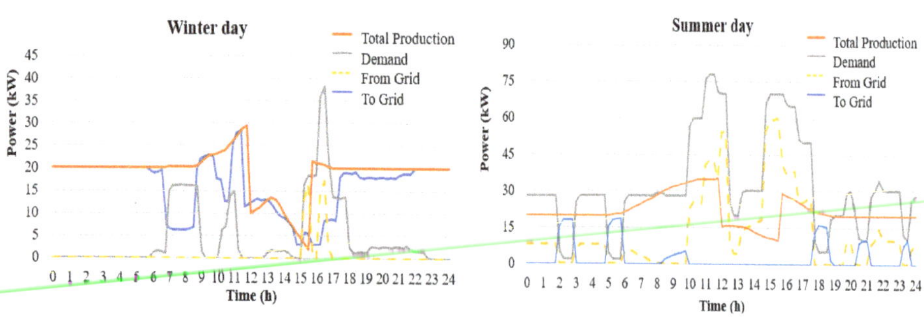

Figure 11. The total power production of the PV panels and the pyrogasifier, the total power demand, and the power from/to the grid.

The power production of the pyrogasifier Is not dependent on the weather conditions and it is constant and significant (Figure 11). Both of the systems, the pyrogasifier and the PV panels, are able to reduce the integrations of electricity from the grid; although, the higher power demand in the summer season is due to the irrigation pumps than in the winter season. Note that during the central hours of the day, the pyrogasifier is switched off to carry out the ordinary maintenance of the unit, and the electricity is only provided by the PV panels. Note that the electricity is delivered to the grid mainly during the night and late afternoon hours. This mainly occurs because the irrigation pumps operate during the central hours, doubling the power consumption.

By considering the monthly electric energy ratio between the self-consumption energy and the electric demand (Figure 12), it was determined that in the colder months, this ratio was almost about 90%, while during the summer months, it decreased to 58% although there was higher PV production. This is due to the high summer demand being about 10 times higher than the winter one. From this point of view, the use of energy storage would be a useful solution to reduce the energy delivered to the grid mainly during the winter months.

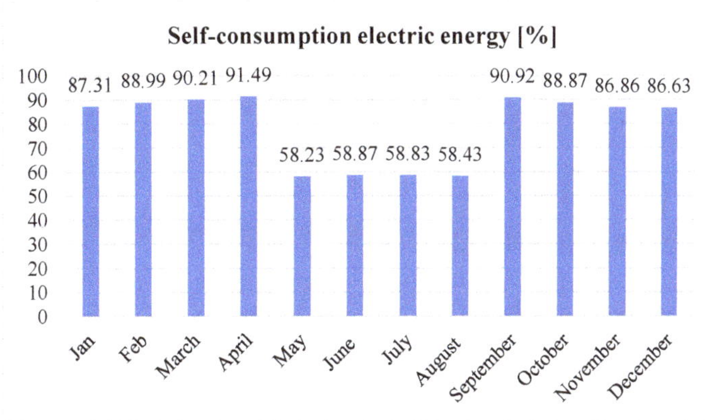

Figure 12. Monthly self-consumption of electric energy.

The yearly results of the energy analysis are summarized in Figure 13 and Table 5. The electricity integration from the grid is about 35% of the total electric energy demand, whereas thermal self-consumption is 63% of the total thermal energy demand and 57% of the total production.

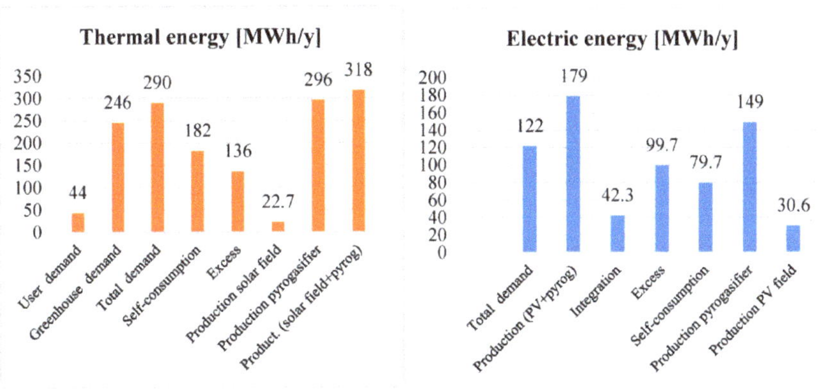

Figure 13. Yearly energy results.

Table 5. Yearly energy and environmental results.

Primary energy RS	587.04 MWh/y
Primary energy PS	−124.67 MWh/y
PES (primary energy saving)	121%
CO_2 emissions RS	122.93 t/y
CO_2 emissions PS	−25.73
Avoided CO_2 emissions	121%

Thermal integration is 37% of the total demand and the electric self-consumption is 65% of the total electric energy demand. The electric production of the solar field covers only 17% of the total electric energy production, a result that confirms the small size of the PV field with respect to the pyrogasifier.

The proposed system acquires a reduction of 149 t/y (Table 5). The primary energy saving of 121% is due to the high amount of electric energy delivered to the grid, equal to 55% of the total electricity production. During the winter months, due to the lower electric energy demand when the irrigation pumps operate for only few hours per day, the electric self-consumption with respect to the demand reaches a high value, also 90%. Therefore, the excess of electricity reduces during the summer months although there is higher PV production. Considering the increase in the purchasing costs before and after the energy crisis, EUR 0.70 vs. EUR 1.58/Sm^3 for natural gas and EUR 0.19 vs. EUR 0.66/kWh, the economic indexes, reported in Table 6, clearly improve, with a simple payback period, decreasing from 6.7 to 1.7 years.

Table 6. Economic results.

Adopted Purchasing Costs	Post Crisis	Pre Crisis
ΔC	122 k EUR/y	31 k EUR/y
SPB	1.7 y	6.7 y
DPB	1.8 y	8.4 y
NPV	1317 k EUR	177 k EUR
PI	6.4	0.86

4. Conclusions

This work proposed a dynamic simulation model of a polygeneration system based on solar thermal collectors and biomasses to meet the energy demand of a farm including a residential user and a greenhouse. To analyze the performance of the system, an economic, environmental, and energy analysis of the polygeneration system was also proposed. This renewable-based polygeneration system integrates photovoltaic panels, solar thermal collectors, a biomass-fed pyrogasifier, and a biomass auxiliary boiler. Therefore, dynamic and thermo-economic analyses are fundamental to observe the dynamic matching of renewable energy produced with the users demand and to assess the economic feasibility of the solution proposed.

To this scope, the model was developed in a TRNSYS environment, where a detailed model for the greenhouse was integrated. The greenhouse model was validated against data available from research works in the literature to accurately evaluate the power and heating demands of the greenhouse. The model was applied to a case study in Castelvolturno (Naples, South of Italy) where a greenhouse with a total heating demand of 246 MWh/year was considered. In addition to the greenhouse, a domestic user was considered, leading to an overall heating and power demand by the end user of 289 MWh/year and 122 MWh/year, respectively. The main results of the work can be summarized as follows:

- The thermal self-consumption is 63% of the total thermal energy demand (182 MWh/year)
- The electric production covers more than 65% of the electric consumption (79 MWh/year).
- Considering the purchasing energy costs before and after the energy crisis, the performed economic analysis obtained substantially different simple payback values, decreasing from 6.7 to 1.7 years.
- The energy and environmental analysis showed how much the implementation of green systems connected to a circular economy reduces the emissions, with −148.66 tons of CO_2/year, and the exploitation of fossil fuels, with −711.7 MWh/year of primary energy.

Future research topics regard the implementation of a suitable electric storage system in order to increase the grid-independency as well as the development of a detailed dynamic simulation model to simulate the performance of the pyrogasifier unit. In fact, in the current work, the assumption regarding the pyrogasifier unit operation expects that it constantly produces both heat and power, without taking into account a suitable dynamic simulation model. This limitation could be overcome in a future research study.

Author Contributions: Conceptualization, F.C. and M.V.; methodology, F.L.C. and M.V.; software, L.C. and F.L.C.; validation, F.L.C. and M.V.; formal analysis, L.C.; investigation, L.C. and F.L.C.; resources, L.C. and F.L.C.; data curation, M.V.; writing—original draft preparation, M.V.; writing—review and editing, F.C.; visualization, M.V.; supervision, F.C.; project administration, F.C. and M.V.; funding acquisition, none. All authors have read and agreed to the published version of the manuscript.

Funding: This research received no external funding.

Informed Consent Statement: Informed consent was obtained from all subjects involved in the study.

Data Availability Statement: The data presented in this study are available on request from the corresponding author.

Conflicts of Interest: The authors declare no conflict of interest.

References

1. Rahman, A.; Farrok, O.; Haque, M. Environmental impact of renewable energy source based electrical power plants: Solar, wind, hydroelectric, biomass, geothermal, tidal, ocean, and osmotic. *Renew. Sustain. Energy Rev.* **2022**, *161*, 112279. [CrossRef]
2. Elahi, E.; Khalid, Z.; Zhang, Z. Understanding farmers' intention and willingness to install renewable energy technology: A solution to reduce the environmental emissions of agriculture. *Appl. Energy* **2022**, *309*, 118459. [CrossRef]
3. Zahedi, A.R.; Mirnezami, S.A. Experimental analysis of biomass to biodiesel conversion using a novel renewable combined cycle system. *Renew. Energy* **2020**, *162*, 1177–1194. [CrossRef]
4. Calise, F.; Cappiello, F.L.; Cimmino, L.; Napolitano, M.; Vicidomini, M. Dynamic Simulation and Thermoeconomic Analysis of a Novel Hybrid Solar System for Biomethane Production by the Organic Fraction of Municipal Wastes. *Energies* **2023**, *16*, 2716. [CrossRef]
5. Cappiello, F.L.; Cimmino, L.; Napolitano, M.; Vicidomini, M. Thermoeconomic Analysis of Biomethane Production Plants: A Dynamic Approach. *Sustainability* **2022**, *14*, 5744. [CrossRef]
6. Calise, F.; Cappiello, F.L.; Cimmino, L.; d'Accadia, M.D.; Vicidomini, M. Integration of photovoltaic panels and solar collectors into a plant producing biomethane for the transport sector: Dynamic simulation and case study. *Heliyon* **2023**, *9*, e14681. [CrossRef]
7. Calise, F.; Cappiello, F.L.; Cimmino, L.; D'accadia, M.D.; Vicidomini, M. Dynamic analysis and investigation of the thermal transient effects in a CSTR reactor producing biogas. *Energy* **2023**, *263*, 126010. [CrossRef]
8. Saleem, M. Possibility of utilizing agriculture biomass as a renewable and sustainable future energy source. *Heliyon* **2022**, *8*, e08905. [CrossRef]
9. Shen, Y.; Linville, J.L.; Ignacio-de Leon, P.A.A.; Schoene, R.P.; Urgun-Demirtas, M. Towards a sustainable paradigm of waste-to-energy process: Enhanced anaerobic digestion of sludge with woody biochar. *J. Clean. Prod.* **2016**, *135*, 1054–1064. [CrossRef]
10. Kothari, R.; Tyagi, V.; Pathak, A. Waste-to-energy: A way from renewable energy sources to sustainable development. *Renew. Sustain. Energy Rev.* **2010**, *14*, 3164–3170. [CrossRef]
11. Brunner, P.H.; Rechberger, H. Waste to energy—Key element for sustainable waste management. *Waste Manag.* **2015**, *37*, 3–12. [CrossRef] [PubMed]
12. Xue, J. Photovoltaic agriculture—New opportunity for photovoltaic applications in China. *Renew. Sustain. Energy Rev.* **2017**, *73*, 1–9. [CrossRef]
13. Chantasiriwan, S. Solar-aided power generation in biomass power plant using direct steam generating parabolic trough collectors. *Energy Rep.* **2021**, *8*, 641–648. [CrossRef]

14. Calise, F.; Cappiello, F.L.; D'accadia, M.D.; Vicidomini, M. Concentrating photovoltaic/thermal collectors coupled with an anaerobic digestion process: Dynamic simulation and energy and economic analysis. *J. Clean. Prod.* **2021**, *311*, 127363. [CrossRef]
15. Azam, M.M.; Eltawil, M.A.; Amer, B.M. Thermal analysis of PV system and solar collector integrated with greenhouse dryer for drying tomatoes. *Energy* **2020**, *212*, 118764. [CrossRef]
16. Okakwu, I.; Alayande, A.; Akinyele, D.; Olabode, O.; Akinyemi, J. Effects of total system head and solar radiation on the techno-economics of PV groundwater pumping irrigation system for sustainable agricultural production. *Sci. Afr.* **2022**, *16*, e01118. [CrossRef]
17. Xu, Z.; Lu, J.; Xing, S. Thermal performance of greenhouse heating with loop heat pipe solar collector and ground source heat pump. *Results Eng.* **2022**, *15*, 100626. [CrossRef]
18. Kut, P.; Pietrucha-Urbanik, K. Most Searched Topics in the Scientific Literature on Failures in Photovoltaic Installations. *Energies* **2022**, *15*, 8108. [CrossRef]
19. Hu, G.; You, F. Renewable energy-powered semi-closed greenhouse for sustainable crop production using model predictive control and machine learning for energy management. *Renew. Sustain. Energy Rev.* **2022**, *168*, 112790. [CrossRef]
20. Singh, G.; Singh, P.P.; Lubana, P.P.S.; Singh, K. Formulation and validation of a mathematical model of the microclimate of a greenhouse. *Renew. Energy* **2006**, *31*, 1541–1560. [CrossRef]
21. Fitz-Rodríguez, E.; Kubota, C.; Giacomelli, G.A.; Tignor, M.E.; Wilson, S.B.; McMahon, M. Dynamic modeling and simulation of greenhouse environments under several scenarios: A web-based application. *Comput. Electron. Agric.* **2010**, *70*, 105–116. [CrossRef]
22. Ahamed, S.; Guo, H.; Tanino, K. A quasi-steady state model for predicting the heating requirements of conventional greenhouses in cold regions. *Inf. Process. Agric.* **2018**, *5*, 33–46. [CrossRef]
23. Ahamed, M.S.; Guo, H.; Tanino, K. Modeling heating demands in a Chinese-style solar greenhouse using the transient building energy simulation model TRNSYS. *J. Build. Eng.* **2020**, *29*, 101114. [CrossRef]
24. Dong, S.; Ahamed, S.; Ma, C.; Guo, H. A Time-Dependent Model for Predicting Thermal Environment of Mono-Slope Solar Greenhouses in Cold Regions. *Energies* **2021**, *14*, 5956. [CrossRef]
25. Klemm, C.; Vennemann, P. Modeling and optimization of multi-energy systems in mixed-use districts: A review of existing methods and approaches. *Renew. Sustain. Energy Rev.* **2020**, *135*, 110206. [CrossRef]
26. Calise, F.; Cappiello, F.L.; Vicidomini, M.; Song, J.; Pantaleo, A.M.; Abdelhady, S.; Shaban, A.; Markides, C.N. Energy and Economic Assessment of Energy Efficiency Options for Energy Districts: Case Studies in Italy and Egypt. *Energies* **2021**, *14*, 1012. [CrossRef]
27. Calise, F.; Cappiello, F.L.; D'Accadia, M.D.; Vicidomini, M. Thermo-economic optimization of a novel hybrid renewable trigeneration plant. *Renew. Energy* **2021**, *175*, 532–549. [CrossRef]
28. Calise, F.; Daccadia, M.D.; Vicidomini, M.; Ferruzzi, G.; Vanoli, L. Design and Dynamic Simulation of a Combined System Integration Concentrating Photovoltaic/Thermal Solar Collectors and Organic Rankine Cycle. *Am. J. Eng. Appl. Sci.* **2015**, *8*, 100–118. [CrossRef]
29. Calise, F.; Cappiello, F.L.; Vicidomini, M.; Petrakopoulou-Robinson, F. Water-energy nexus: A thermoeconomic analysis of polygeneration systems for small Mediterranean islands. *Energy Convers. Manag.* **2020**, *220*, 113043. [CrossRef]
30. Calise, F.; Cappiello, F.L.; D'accadia, M.D.; Vicidomini, M. Thermo-Economic Analysis of Hybrid Solar-Geothermal Polygeneration Plants in Different Configurations. *Energies* **2020**, *13*, 2391. [CrossRef]
31. Calise, F.; Cimmino, L.; Cappiello, F.L. Dynamic Simulation and Energy, Economic and Environmental Analysis of a Greenhouse Supplied by Renewable Energy Sources. In Proceedings of the 12th International Conference on Smart Cities and Green ICT Systems—SMARTGREENS, Prague, Czech Republic, 26–28 April 2023.
32. Klein, S.A.B.W.; Mitchell, J.W.; Duffie, J.A.; Duffie, N.A.; Freeman, T.L. *Solar Energy Laboratory, TRNSYS. A Transient System Simulation Program*; University of Wisconsin: Madison, WI, USA, 2006.
33. Calise, F.; Cappiello, F.L.; D'accadia, M.D.; Vicidomini, M. Dynamic simulation, energy and economic comparison between BIPV and BIPVT collectors coupled with micro-wind turbines. *Energy* **2020**, *191*, 116439. [CrossRef]
34. Buonomano, A.; Palombo, A. Building energy performance analysis by an in-house developed dynamic simulation code: An investigation for different case studies. *Appl. Energy* **2014**, *113*, 788–807. [CrossRef]
35. Buonomano, A.; Calise, F.; Palombo, A.; Vicidomini, M. Transient analysis, exergy and thermo-economic modelling of façade integrated photovoltaic/thermal solar collectors. *Renew. Energy* **2019**, *137*, 109–126. [CrossRef]
36. Calise, F.; Dâaccadia, M.D.; Libertini, L.; Quiriti, E.; Vicidomini, M. Dynamic Simulation and Optimum Operation Strategy of a Trigeneration System Serving a Hospital. *Am. J. Eng. Appl. Sci.* **2016**, *9*, 854–867. [CrossRef]
37. Murray, M.C.; Finlayson, N.; Kummert, M.; Macbeth, J. Live Energy Trnsys–Trnsys Simulation within Google Sketchup. In Proceedings of the Eleventh International IBPSA Conference, Glasgow, UK, 27–30 July 2009.
38. Voit, P.; Lechner, T.; Schuler, M.T. Common EC validation procedure for dynamic building simulation programs—Application with TRNSYS. In Proceedings of the Conference of International Simulation Societies, Zürich, Switzerland, 22–25 August 1994.

applied sciences

MDPI

Article

ForecastNet Wind Power Prediction Based on Spatio-Temporal Distribution

Shurong Peng [1,2], Lijuan Guo [2], Haoyu Huang [2], Xiaoxu Liu [1] and Jiayi Peng [3,*]

1 Sino-German College of Intelligent Manufacturing, Shenzhen Technology University, Shenzhen 518118, China; peng_sr@126.com (S.P.); liuxiaoxu@sztu.edu.cn (X.L.)
2 School of Electrical and Information Engineering, Changsha University of Science and Technology, Changsha 410114, China; guolijuan@stu.csust.edu.cn (L.G.); 23205061121@stu.csust.edu.cn (H.H.)
3 State Grid Zhuzhou Power Supply Company, State Grid Hunan Power Company, Zhuzhou 412011, China
* Correspondence: zzdw_pengjy@163.com

Abstract: The integration of large-scale wind power into the power grid threatens the stable operation of the power system. Traditional wind power prediction is based on time series without considering the variability between wind turbines in different locations. This paper proposes a wind power probability density prediction method based on a time-variant deep feed-forward neural network (ForecastNet) considering a spatio-temporal distribution. First, the outliers in the wind turbine data are detected based on the isolated forest algorithm and repaired through Lagrange interpolation. Then, based on the graph attention mechanism, the features of the proximity node information of the individual wind turbines in the wind farm are extracted and the input feature matrix is constructed. Finally, the wind power probability density prediction results are obtained using the ForecastNet model based on three different hidden layer variants. The experimental results show that the ForecastNet model with a hidden layer as a dense network based on the attention mechanism (ADFN) predicts better. The average width of the prediction intervals at achieved confidence levels for all interval coverage is reduced by 34.19%, 35.41%, and 35.17%, respectively, when compared to the model with the hidden layer as a multilayer perceptron. For different categories of wind turbines, ADFN also achieves relatively narrow interval average widths of 368.37 kW, 315.87 kW, and 299.13 kW, respectively.

Keywords: time-variant deep feed-forward neural network; probability density prediction; spatio-temporal distribution

Citation: Peng, S.; Guo, L.; Huang, H.; Liu, X.; Peng, J. ForecastNet Wind Power Prediction Based on Spatio-Temporal Distribution. *Appl. Sci.* **2024**, *14*, 937. https://doi.org/10.3390/app14020937

Academic Editor: Maria Vicidomini

Received: 14 November 2023
Revised: 18 January 2024
Accepted: 19 January 2024
Published: 22 January 2024

1. Introduction

With energy and environmental issues becoming more and more prominent [1], global energy is moving towards low-carbon, clean, safe, and efficient development [2]. In recent years, the wind energy industry has been developing rapidly around the world. According to statistics, the global newly installed capacity of wind power reached 93.6 GW in 2021. However, wind power generation is random and fluctuating due to climate conditions. Therefore, when large-scale wind power is integrated into the grid, it can jeopardize the safe and stable operation of the grid [3,4]. Wind power prediction helps to optimize the grid integration of wind power, reduce operating costs, and promote the development of energy systems.

Existing wind power prediction methods are categorized into ultra-short-term prediction [5,6], short-term prediction [7,8] and mid-to-long-term prediction [9,10], with multiple prediction time dimensions. Commonly used wind power prediction models include physical [11–14] and statistical [15–18] methods. Physical methods utilize weather forecast data and related geographic information, and the models are generally more complex. Moreover, the weather forecast data will affect the prediction accuracy to a certain extent. They are often used for the mid-to-long-term prediction of new wind farms or wind farms with incomplete data. Statistical methods use learning algorithms to analyze historical data and establish the intrinsic connections between historical data. They are commonly used in short-term and ultra-short-term prediction.

Of the many forecasting methods, deep learning [19] algorithms are the most widely used, mainly including time series methods [20] and artificial intelligence algorithms [21]. Common artificial intelligence algorithms include random forests [22], support vector machines [23], extreme learning machines [24], and artificial neural networks [25]. Long- and short-term memory neural networks [26,27] are widely used in wind power prediction due to their excellent properties in analyzing time-series data. Convolutional neural networks (CNN) [28], which have achieved good performance in image processing, have also been applied to wind power prediction. Zhou et al. [29] proposed a combination of RANSAC (random sample consensus) noise screening and the Seq2Seq-Attention-BiGRU model to enhance prediction accuracy. Zhang et al. [30] proposed a novel hybrid prediction method involving individual prediction model training, model ensemble, and error correction considering temporal correlation. Chen et al. [31] proposed a variational mode decomposition-gate recurrent unit network prediction mode to enhance the accuracy of ultra-short wind power forecasting. Although these studies have achieved more accurate wind power prediction results, they lacked research on the spatial characteristics of the wind power data.

Most wind power prediction studies predict the sum of the power of all turbines in the region, taking into account the temporal characteristics. However, the location and contextual information of wind turbines are neglected, and consideration of the variability of turbine power output at different spatial locations is lacking. The studies show that considering both temporal and spatial characteristics of wind turbine data can help improve the accuracy of wind power prediction. Yu et al. [32] proposed a spatiotemporal wind power forecasting model, which utilized a GAT (graph attention network), GRU (gated recursion unit), and GAT-TCN (temporal convolutional network) as the main prediction methods. Zhang et al. [33] proposed a wind power prediction method based on spatiotemporal correlations considering the influences of wind speed, wind direction, and temperature. Wang et al. [34] proposed a short-term wind power forecasting method based on feature clustering and correlation analysis, improving forecasting accuracy through data feature clustering, variable correlation analysis, and building forecasting models.

Comprehensively analyzing existing studies, this paper analyzes the spatial distribution and dynamic contextual information of wind turbines. In order to facilitate the study of the spatial characteristics of wind power data and better analyze the location information, this paper predicts wind power from the perspective of wind turbines rather than wind farms. The dataset used in this paper includes data such as the angle of the wind received by each turbine, the environmental temperatures around the different wind turbines, the internal temperatures of the turbine nacelles, and the orientation of each turbine nacelle. These data are used to model the spatial correlation between the wind turbines. Predictions for each wind turbine are obtained considering the spatial and temporal distribution of the turbines, and then the predictions for the different turbines are summed to obtain the final prediction. Specifically, this paper makes the following contributions:

- We detect outliers based on the isolated forest algorithm and repair them using the Lagrange interpolation method.
- We extract the information about the neighboring nodes of individual turbines in a wind farm based on the graph attention mechanism and construct the input feature matrix.
- We use a time-variant deep feed-forward neural network (ForecastNet) model to obtain the wind power probability density prediction results based on three different hidden layer variants.

The rest of the paper is structured as follows: Section 2 presents the theoretical knowledge of the method proposed in this paper. Section 3 presents the experiments using the method proposed in this paper and analyzes the experimental results. Section 4 provides the conclusions of this paper.

2. Materials and Methods

2.1. Analysis and Processing of Wind Power Data

2.1.1. Wind Power Anomaly Data Processing

Wind farm data mainly includes the spatial distribution of wind turbines, as well as dynamic background factors such as the temperature, weather, and internal state of turbines. Wind power data will inevitably have anomalies and missing information in the process of collection, transmission, processing, and storage. The existence of these anomalous data can reduce the efficiency of the model training and the accuracy of the prediction [35]. The anomalous data need to be handled without interfering too much with the original data.

The isolated forest algorithm quickly separates the outliers and vacancies in the data from the normal data. Based on the length of the data path to score the data to determine the degree of abnormality, the higher the score, the higher the degree of abnormality of the data. By setting a certain proportion of abnormal data, the algorithm can quickly distinguish between abnormal and normal data, and the abnormal score of the isolated forest algorithm is calculated as shown in Equation (1).

$$S(x) = 2^{-\frac{E(h(x))}{c(x)}}, \quad (1)$$

where $S(x)$ is the outlier score of samples x, with a value range of [0,1]. The larger the value, the more likely the sample is to be labeled as an outlier. $h(x)$ is the path length of the sample in the tree, $h(x) = \ln(x) + \zeta$, and ζ is Euler's constant. $E(h(x))$ is the mean value of the path length of the sample x in the tree. $c(x)$ is the average search path length of the binary tree constituted by a dataset containing x samples.

The Lagrange interpolation method is used to repair the abnormal data detected by the isolated forest, as shown in Equation (2).

$$L(x) = \sum_{i=0}^{n} y_i \prod_{j=0, j \neq i} \frac{x - x_i}{x_i - x_j}, \quad (2)$$

where x_j corresponds to wind speed and y_j corresponds to wind power. Figure 1 shows the wind speed–power plot for the detection of anomalous data. Figure 1a shows the results before detection and repair, and Figure 1b shows the results after detection and repair. The orange dots are the anomalous and repaired data, respectively, and the blue dots are the normal data.

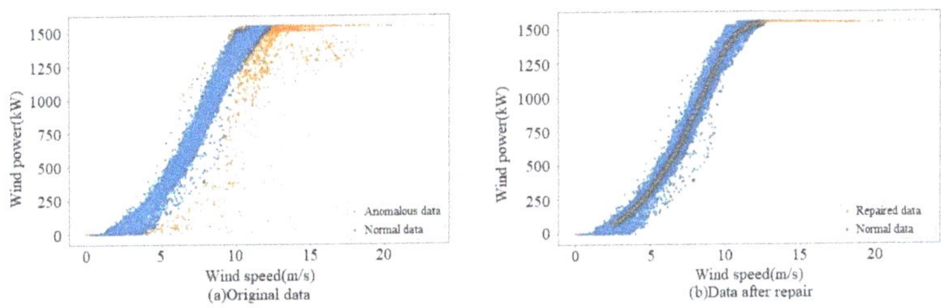

Figure 1. Detection and repair of wind power outliers.

2.1.2. Feature Clustering of Wind Power Data

Wind power output is closely related to factors in the weather environment. Generally, the feature with the strongest correlation with wind power is wind speed. Figure 2 shows the Pearson correlation heat map of wind power data features. From Figure 2, it can be seen that the features that have the greatest degree of influence on wind turbine power output are wind speed (Wspd), the angle between the wind direction and the position of the turbine nacelle (Wdir), and the pitch angle of blades 1–3 (Pab1–Pab3).

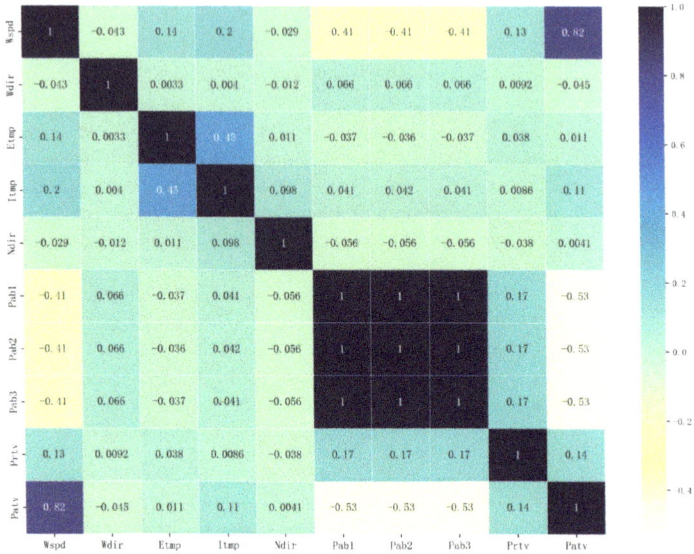

Figure 2. Characteristic correlation heat map.

In order to improve the intrinsic connection between wind power data and better explore the spatial and temporal distribution state of wind power, the K-sums algorithm is used to segmentally cluster wind turbines at different locations [36]. Based on the features related to wind speed, wind turbines distributed in different spatial locations are differentiated. In this paper, the wind direction factor is added on the basis of wind speed. The wind speed and direction of the natural wind in the duration of time is constantly changing, and the atmosphere has a certain inertia and incompressible mobility in a small-time scale range, which leads to periodic changes in wind speed and direction within a certain range. This is manifested by the direction of the wind force and the angle between the wind turbine oscillating back and forth within a certain range [37].

The wind turbines are clustered into two classes based on wind speed characteristics, defined as the class that receives higher wind speeds and the class that receives lower wind speeds. In fact, due to geographic location and environmental factors, low output units still exist in the high wind speed range. For this reason, the high wind speed category will be clustered for the second time using wind direction characteristics such as Wdir and Pab as input variables, and finally, three types of wind turbines will be obtained. Type 0 is for high wind speeds and high output units, type 1 is for high wind speeds and low output units, and type 2 is for low wind speeds and low output units. The results of the K-sums algorithm clustering for visualizing the location of the wind farm are shown in Figure 3.

The spatial categorization of the three types of turbines is more concentrated, and the existence of some outliers is due to the different topography and ground roughness at different turbine locations. The gap between the turbines in the upper right corner of Figure 3 is exactly a railroad.

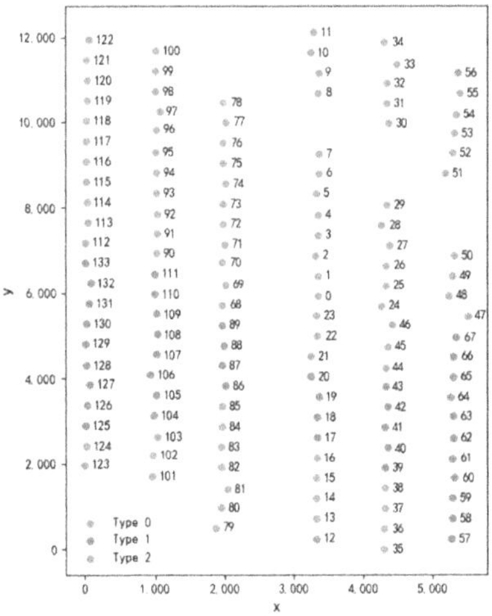

Figure 3. Results of K-sums clustering.

2.2. Wind Power Prediction Model

2.2.1. Euclidean Distance and Differential Distance

In order to obtain the spatio-temporal distribution characteristics of wind turbines, the spatial distance perception function G_d is constructed based on the Euclidean distance, and the differential distance perception function G_s is constructed based on the differential similarity. G_d can reflect the explicit neighboring relationship between wind turbines, and G_s can reflect the invisible neighboring relationship. We calculate the Euclidean distance between two nodes, take the K nodes with the closest distance as neighboring nodes, and form the set $N_d(i)$ to obtain the spatial matrix $A(i,j)$. The expression of $A(i,j)$ is shown in Equation (3). We calculate the differential similarity $Sim(i,j)$ between two nodes, which is calculated as in Equation (4), and the closest K nodes are taken as the differential proximity nodes to form the set $N_s(i)$.

$$A(i,j) = \begin{cases} 1, j \in N_d(i) \\ 0, j \notin N_d(i) \end{cases}, \tag{3}$$

$$Sim(i,j) = \sum_{t=1}^{T} \left(\left(x_{i,w}^{t} - x_{i,w}^{t-1} \right) \cdot \left(x_{j,w}^{t} - x_{j,w}^{t-1} \right) \right), \tag{4}$$

where $x_{i,w} \in R^{T\times 1}$ denotes the wind speed sequence of the i-th wind turbine.

For wind turbines, both $N_d(i)$ and $N_s(i)$ are combined to aggregate wind speed information from two neighboring turbines and merged as input features to improve the performance of the wind power prediction.

2.2.2. Graph Attention

In order to establish spatial and temporal correlations between wind turbines in different geographic locations, an attention-based spatio-temporal graph network [38] is introduced into the time series prediction model. In order to improve the training efficiency and prediction accuracy of the model, the feature information needs to be filtered, and the information with low importance is ignored [39].

Graph attention networks process information by calculating the weights of the information and weighting the information according to certain weights to aggregate the information. Specifically, the attention score of the information is calculated using query vector (Query), key-value vector (Key), and value vector (Value) [40]. The formula is shown in Equation (5):

$$Attention(Q, K, V) = \sum_{i}^{L} < Q, K_i > V_i, \tag{5}$$

where Q is the feature vector of the current node, K is the feature vector of the neighboring node, and V is the feature vector of the neighboring node after mapping the weight value W. An attention score is obtained by performing an inner product calculation on the feature vector of a node itself and the feature vectors of the neighboring nodes, and the score is weighted with the node's feature vectors after a normalization operation. Specifically:

1. The central target node and neighbor node attention scores are calculated.

The formula for calculating the value of attention on node i based on the features of node j is shown in Equation (6), where e_{ij} is the attention score between node i and node j, a represents the correlation calculation function between nodes, h_j is the output vector of node i, and W is the weight.

$$e_{ij} = a(Wh_i, Wh_j). \tag{6}$$

2. The weight scores are activated using the activation function.

$$e_{i,j} = Leaky\mathrm{ReLU}(a^T[Wh_i||Wh_j]). \tag{7}$$

The calculation formula is shown in Equation (7), where $||$ is the vector vertical splicing operation, which splices mapped column vectors; $Leaky\mathrm{ReLU}$ is the activation function; a is the vector to be learnt; and a^T is the transpose of vector a.

3. Weight normalization

The sum of all the weights should be 1; therefore, the attention score is normalized. The weights are normalized for all neighboring node attention values of node i using the softmax function. The calculation formula is shown in Equation (8).

$$a_{ij} = \frac{\exp(Leaky\mathrm{ReLU}(e_{ij}))}{\sum_{k \in N_i} \exp(Leaky\mathrm{ReLU}(e_{ik}))}. \tag{8}$$

4. Information aggregation

The graph attention layer adds the node's feature information and neighbor node's feature information according to a certain weight coefficient, performs feature extraction to form a new node to represent the feature, and outputs the new node feature as a result. The calculation formula is shown in Equation (9):

$$h_i' = \sigma(\sum_{j \in N_i} a_{ij} Wh_j), \tag{9}$$

where h_i' is the new node features, $\sigma(\cdot)$ is the activation function, and a_{ij} is the effect of node i features on the node j attention score.

The features of the new node obtained according to the parameter W will be different in dimension from the features of the original node, and the parameter W will map the information of the original node to the new space.

5. Multi-head attention mechanisms

The multi-head attention mechanism introduces multiple attention mechanisms to aggregate information, and each attention mechanism is able to focus on different features to enhance the expression ability of the attention layer [41]. The multi-head attention mechanism splices the outputs of multiple nodes into column vectors to obtain the final new node features, and the new node features are calculated using the Equation (10):

$$h_i' = ||_{k=1}^{K} \sigma(\sum_{j=N_i} a_{ij}^k W^k h_j), \tag{10}$$

where $||$ denotes the vector splicing operation, a_{ij}^k represents the normalized value of the attention score calculated by the k-th attention mechanism, and W^k is the weight matrix of the linear transformation.

Using the graph attention mechanism to deal with the pair of spatial feature information, combined with the multi-head attention mechanism, different attention score weights are calculated for the target node and its neighboring nodes [42]. It is beneficial to improve the model's representation of spatial dimensions and reduce the risk of overfitting.

2.2.3. ForecastNet Model

ForecastNet [43] is a multi-layer feed-forward neural network model that is commonly used to perform multi-step time series forecasting. Its network structure is shown in Figure 4, where each neuron can receive signals from the previous layer of neurons, resulting in a multilayer structure. This neural network structure is commonly used for processing time series data because it captures trends and periodicities in time series, as well as other complex dynamic features.

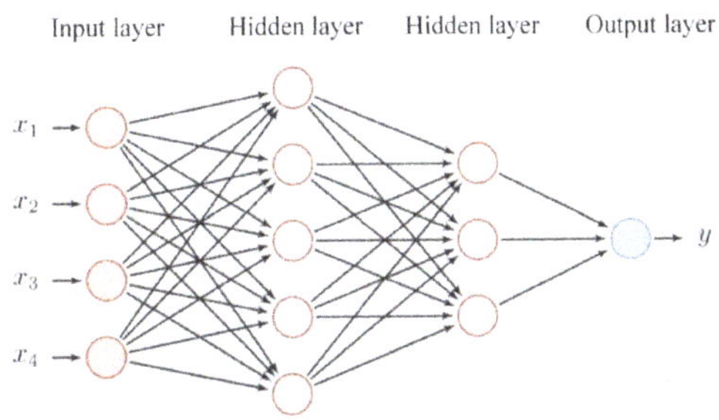

Figure 4. Multilayer feedforward neural network structure.

ForecastNet has time-varying and interleaved output properties among model neurons. The former improves the problem of gradient vanishing during the training process of recurrent neural network (RNN) and CNN models. The latter improves the problem of gradient explosion and gradient vanishing that occurs during neural network training. The root cause of gradient explosion and gradient vanishing is the reuse of chaining laws in neural network gradient computation. The interleaved output properties of ForecastNet are shown in Figure 5, where the upper row represents the hidden layer network, the lower row represents the output layer, $a^{[l]}$ is the output vector of the l-th layer, $z^{[l]} = W^{[l]T} a^{[l-1]} b^{[l]}$, $W^{[l]}$ is the weight matrix of the l-th layer, and $b^{[l]}$ is the bias parameter matrix. It breaks down the chain multiplication chain in the chain rule into the sum of multiple terms. The factor accumulation process is more stable than the multiplication process, which can significantly reduce the depth of the network while alleviating the gradient explosion and gradient vanishing problems.

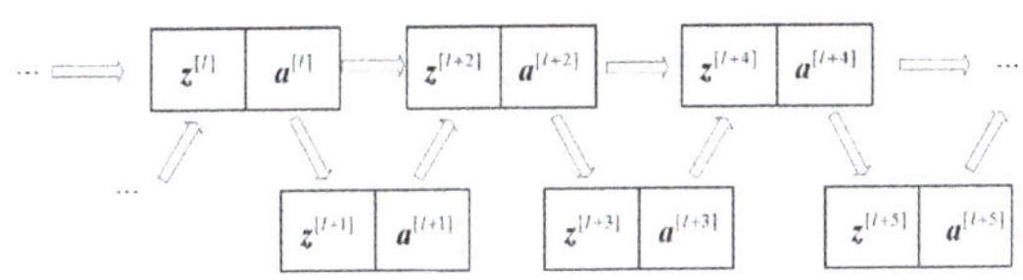

Figure 5. The interleaved output properties of ForecastNet.

The input layer of ForecastNet is univariate or a set of multivariate inputs; the hidden layers are different forms of feed-forward neural networks, such as common back propagation (BP) networks, radial basis function (RBF) networks, etc. The architecture of each hidden layer can be heterogeneous or identical, and the parameters of each hidden layer are independent of each other, which are used to simulate the dynamic characteristics of the time series. Different variants of the ForecastNet model can be obtained by using different feed-forward networks in the hidden layer:

6. ForecastNet model with a multilayer perceptron (MLP) as the hidden layer (MLPFN)

A multilayer perceptron is a special form of a fully connected neural network. Its main difference from a fully connected network is its hidden layer. The hidden layer can improve the MLP's expressive ability for the network, thus improving its ability to solve complex prediction or classification problems. Figure 6 shows a schematic diagram of the ForecastNet hidden layer using an MLP structure, where dense is the fully connected layer and *h* represents the number of neuron nodes in each hidden layer. The hidden layer and the output layer of the MLP network are fully connected layers. Each layer has 24 ReLU neuron units, where the neuron nodes are fully connected.

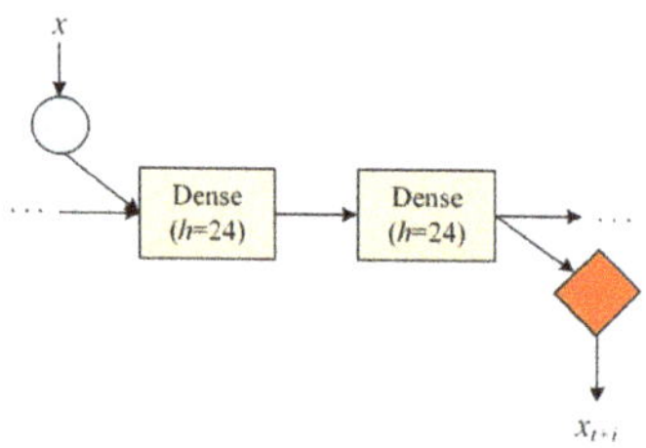

Figure 6. Hidden layer structure of MLPFN.

7. ForecastNet model with CNN as the hidden layer (CNNFN)

CNN is a kind of artificial neural network. Its structure is mainly composed of three parts: the convolution layer, pooling layer, and dense layer, as shown in Figure 7. The main role of the convolution layer is to extract the features, where *f* is the number of convolution kernels and *k* is the size of the convolution kernel. The pooling layer is used for down sampling, where *s* is the filling of the pooling layer and *p* is the step size of the pooling layer. The dense layer is mainly used for feature classification, where *h* is the number of hidden layers of the dense layer, which is composed of 24 ReLU neurons.

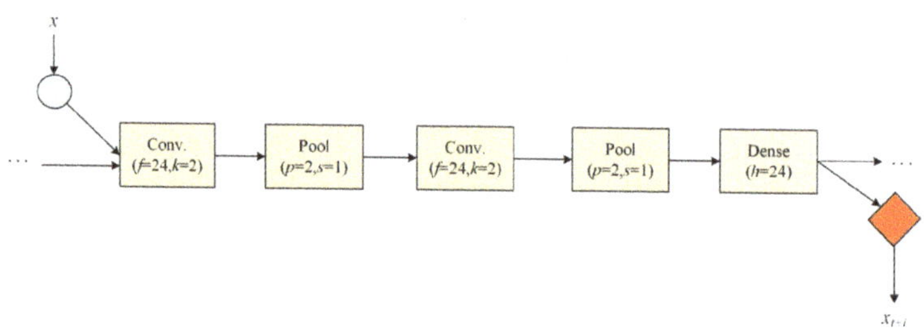

Figure 7. Hidden layer structure of CNNFN.

8. ForecastNet model with the hidden layer as a dense network based on the attention mechanism (ADFN)

The model is shown in Figure 8.

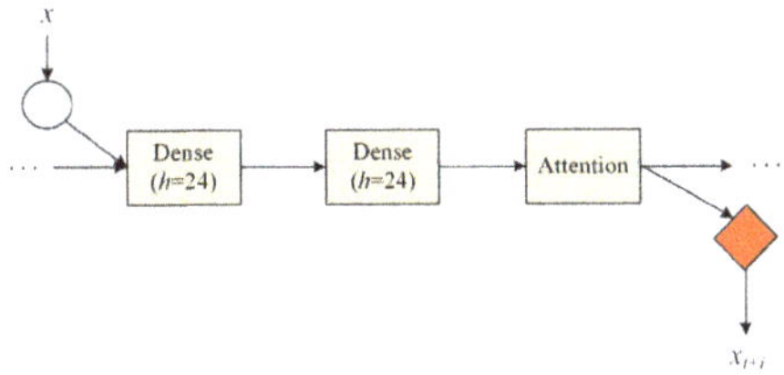

Figure 8. Hidden layer structure of ADFN.

In this paper, the ForecastNet model is used as the basis for wind power spatio-temporal distribution prediction research by using a MLP, CNN, and the dense network based on the attention mechanism in its hidden layer. Each output in the output layer is a prediction result, and a linear output can be used to obtain a deterministic prediction result, or a Gaussian mixture density can be used to establish a probability density output.

2.2.4. ForecastNet Methodology

There is a certain difference in the output power of each wind turbine, and it is not only a large amount of work to build a model for each wind turbine, but also a tedious and complicated model training process. Therefore, the prediction is based on the average wind power. Specifically, the higher the active power (Patv) data of the wind turbines, the closer the output of the turbines to the theoretical output, which means that it is more appropriate to select the turbines with high wind speeds in the k-sums clustering. The data from the N wind turbines with the highest output and high spatio-temporal correlation are selected as the historical data for training the model. Predictions were made using these data as the output of other turbines, and the average of the predictions was used as the prediction for the other turbines in the wind plant.

The new node data obtained from the data processed by the graph attention network mentioned in Section 2.2.2 contains not only the wind turbine feature data and historical wind power data of its own node but also the feature data of the neighboring nodes. These new node data are used as inputs to the ForecastNet model to build a prediction model and perform multi-step predictions. Each step predicts the data for the next 12 time points, and the prediction results obtained from the previous step are merged with the historical features before the next moment prediction. Together, they are used as inputs to the next model. This allows for more accurate learning of wind power trends over time while also fully considering the correlation of wind power data between inputs and outputs and between outputs at different moments. Following the multi-step prediction steps shown in Figure 9, a multi-step prediction of wind power is performed to obtain the prediction results at multiple future points in time. Adding a linear output model to the ForecastNet output layer allows for obtaining deterministic prediction results, and adding a Gaussian mixture density module [44] allows for obtaining probability density prediction results. The specific process framework diagram is shown in Figure 10.

2.2.5. Evaluation Metrics

Common error evaluation metrics such as root mean squared error (RMSE), mean absolute error (MAE), etc., are only applicable to deterministic prediction results and cannot be applied to probabilistic prediction results. In interval probabilistic prediction, the prediction interval coverage is a major metric for evaluating uncertain predictions. The formula is shown in Equation (11):

$$R_{cover} = \frac{c}{v} \times 100\%, \quad (11)$$

where R_{cover} is the interval coverage, c is the number of true values in the test set that fall within the prediction interval, and v is the total number of true values in the test set.

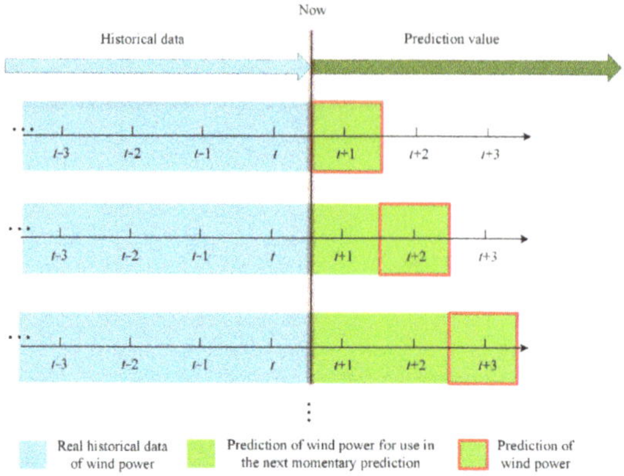

Figure 9. Schematic of multi-step prediction.

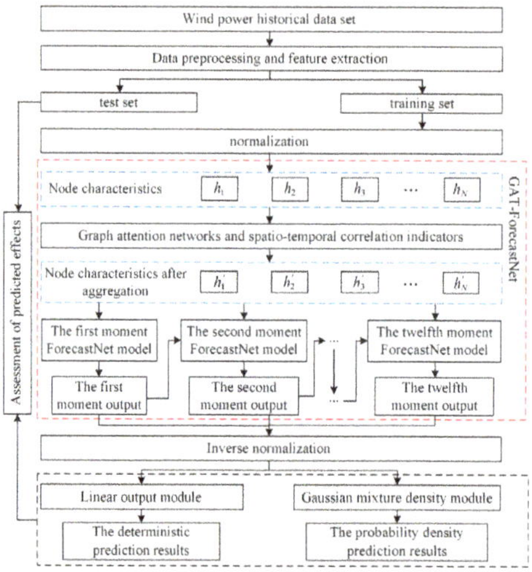

Figure 10. A wind power prediction framework based on spatio-temporally distributed ForecastNet.

Interval coverage indicates the proportion of real values included in the prediction interval. A higher coverage indicates a more accurate prediction. Under a given confidence level, the coverage of the prediction interval cannot be lower than the requirement of the confidence level; otherwise, the prediction result is unreliable. In wind power prediction, interval coverage is one of the important metrics used to assess the prediction accuracy. However, relying solely on the interval coverage is insufficient to evaluate quality of the interval predictions; the interval width metric is also essential. The interval average width metric describes the average width of the prediction interval, which is usually used to assess the accuracy and effectiveness of interval prediction. When the interval coverage is higher, the average width of the interval should be smaller, indicating that the model can provide a more accurate range of intervals in the prediction process. Relatively speaking, the model prediction results of the interval coverage are high-quality, but the average width of the interval is too large may have certain defects. The calculation of the average width of the interval metric is shown in Equation (12):

$$\delta_{mean} = \frac{1}{v}\sum_{t=1}^{v} \frac{(U_t - L_t)}{(Y_{\max} - Y_{\min})}, \tag{12}$$

where Y_{max} is the maximum value of wind power in the test set at the prediction time t, Y_{min} is the minimum value of wind power in the test set at the prediction time t, U_t is the upper boundary of the prediction interval, and L_t is the lower boundary of the prediction interval.

The larger the average interval width indicator, the larger the power interval obtained from the prediction, and the more valid the information it can provide. Combining the two metrics, interval average width and interval coverage, to jointly judge the uncertain prediction results is of practical significance.

3. Results and Discussion

In this paper, we utilize the data from the open-source dataset 2022 Long yuan Power Group Co., Beijing, China wind farm to verify the effectiveness of the proposed method.

The proximity nodes are identified based on the spatial perception function and differential distance perception function. The data of the neighboring turbine nodes are aggregated based on the graph attention model to improve the model effect. Predictions are made using the ForecastNet model with different implicit layers, predicting data at 12 time points in the future at a time. Rolling predictions are used to determine the power generation from wind turbines in the next two days. The deterministic prediction results are obtained using linear output at the output layer, and the interval prediction results of wind power are obtained using a Gaussian mixture density network for the probability density output.

Wind energy fluctuates over time, and there are differences in the variability at different locations in the same wind farm. A spatio-temporal graph is constructed based on the physical spatial location of individual wind turbines, and the data of the model are reconstructed based on the established relationships between nodes and edges. The strong temporal correlation of the time series model is taken into account while also fully considering the characteristics of the wind turbines in terms of their physical location in space.

3.1. Analysis of Deterministic Prediction Results

Deterministic prediction values can be obtained by using linear output in the output layer of ForecastNet. The RMSE and MAE metric values are calculated separately for the ForecastNet model and the other three prediction models: support vector regression (SVR), K-nearest neighbor (KNN), and light gradient-boosting machine (LightGBM). Tables 1 and 2 show the RMSE and MAE metrics for selecting the top 1–5 proximate wind turbine prediction results with the highest power outputs.

Table 1. RMSE metrics for wind turbine prediction results.

Models	Top = 1	Top = 2	Top = 3	Top = 4	Top = 5
ForecastNet	55.930	59.595	62.336	53.447	65.332
SVR	114.660	109.169	120.044	100.730	106.931
KNN	121.807	120.008	137.426	106.321	121.102
LightGBM	82.583	95.353	106.340	92.396	96.301

Table 2. MAE metrics for wind turbine prediction results.

Models	Wind Turbine 1	Wind Turbine 2	Wind Turbine 3	Wind Turbine 4	Wind Turbine 5
ForecastNet	48.828	43.145	53.664	44.771	62.602
SVR	96.914	93.318	103.126	92.814	110.663
KNN	105.121	100.702	112.003	113.630	129.361
LightGBM	69.743	73.088	80.355	75.230	90.505

Figure 11 demonstrates the trend in RMSE evaluation metrics for selecting different numbers of wind turbines with the highest power output (THPO) in the multi-step prediction process.

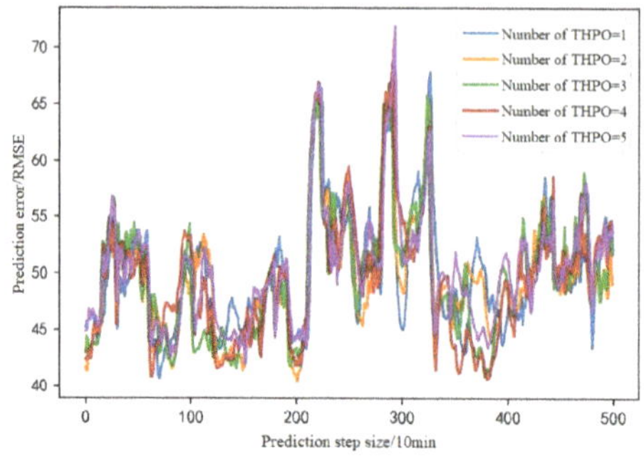

Figure 11. Multi-step prediction error curves for different numbers of THPO.

From Figure 11, it can be concluded that ForecastNet obtains similar error distributions when predicting proximity wind turbines. It indicates that the ForecastNet model achieves relative feature aggregation in dealing with the proximity of wind turbines, obtaining similar error outputs. The RMSE within the short-term prediction is controlled to be less than 60, and its prediction trend is relatively good and stable. However, after 200 predictions, a large error fluctuation occurs, and then the error fluctuation tends to be more stable. The reason for this may be that the wind speed and other characteristics have fluctuated more during this prediction. Overall, most of the prediction steps have good error performance, and from the data in the Figure 11 and Tables 1 and 2, it can be seen that better prediction results can be achieved by selecting the four wind turbines with the highest power outputs for prediction.

3.2. Analysis of Probability Density Prediction Results

In order to verify the prediction effect of ForecastNet on the uncertainty of the spatio-temporal distribution of wind power, the hidden layer of ForecastNet is improved. Different hidden layer connection methods such as MLP, CNN, and attention mechanism fully connected network are chosen to analyze their effects on the prediction effect. The rolling forecast is used to predict the wind power, and each time the data of the next 12 time points are predicted, the multi-step prediction is performed to predict the distribution of wind power in the next 2 days. The prediction results of different prediction steps in the multi-step prediction process were randomly selected, and the prediction intervals at different confidence levels (80%, 85%, and 90%) are shown in Figure 12.

As can be seen from the Figure 12, the prediction intervals of each probabilistic prediction model cover most of the real values more accurately, and the trends in the upper and lower limits of the prediction intervals are also consistent with the trend in the wind power. The wind power prediction intervals obtained at different confidence levels are different, and the larger the confidence level, the wider the given prediction interval. The specific prediction evaluation indexes are shown in Table 3.

Different variants can be obtained by changing the hidden layer structure of ForecastNet. As can be seen from Table 3, the highest coverage of prediction intervals at different confidence levels is achieved by the CNNFN model, with interval coverage of reliability metrics at 90%, 85%, and 80%, and confidence levels of 100%, 98.95%, and 95.32%, respectively. However, interval coverage that is too high may cause the average width of the intervals to rise. The average width of the intervals of the CNNFN model is also the highest among the three models, which is 72.11 kW, 67.75 kW, and 63.31 kW at 90%, 85%, and 80% confidence levels, respectively. The wider the average width of the intervals in the case of certainty of the interval coverage, the less practical significance of the interval prediction.

Both the ADFN model and the MLPFN model obtained relatively good prediction results, and the predicted interval coverages obtained are up to the pre-specified confidence level. At confidence levels of 90%, 85%, and 80%, the interval coverage of the MLPFN model is higher than that of the ADFN model by 4.11%, 4.28%, and 3.59%, respectively. However, the average width of the intervals of the ADFN is narrower, which is reduced by 34.19%, 35.41%, and 35.17% compared with that of the MLPFN model, respectively, achieving better prediction results. Combining the above analyses, it can be concluded that the prediction effects of the ADFN and MLPFN models are better than that of the CNNFN model. The ADFN model achieves a narrower interval prediction width based on the attention mechanism at the confidence level of interval coverage, and it is able to better fit the trend in wind power. The CNNFN model, on the other hand, obtained 100% interval coverage at a higher confidence level, making the prediction intervals less meaningful in practice.

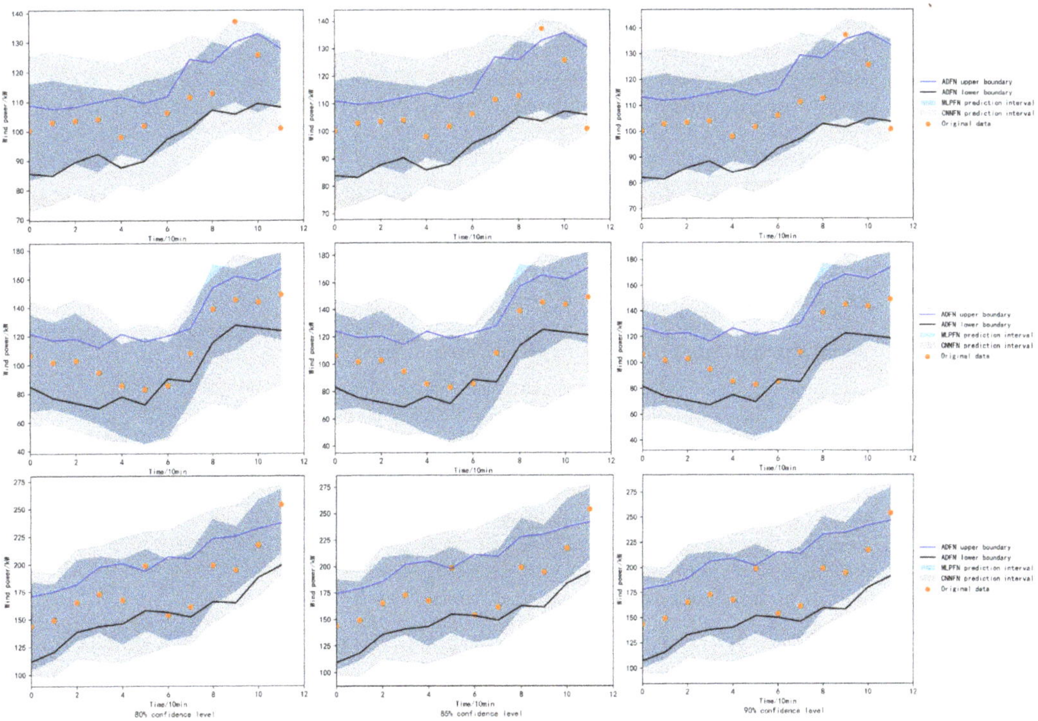

Figure 12. Comparison of wind power prediction intervals for different ForecastNet models.

Table 3. Comparison of evaluation indexes for the different models.

Confidence Levels %	Models	Interval Coverage/%	Average Width of the Interval/kW
	CNNFN	95.32	63.31
80	MLPFN	87.26	50.46
	ADFN	83.67	32.71
	CNNFN	98.95	67.75
85	MLPFN	90.89	54.36
	ADFN	86.61	35.11
	CNNFN	100.0	72.11
90	MLPFN	94.54	59.95
	ADFN	90.43	39.45

In addition, predictions are made for the three different classes of wind turbines obtained from the previous clustering. Figures 13–15 demonstrate the probability density predictions of the ADFN model over the next two days. The more common probabilistic prediction models selected for comparison at the 90% confidence level are LSTM (combination of long short-term memory network and Gaussian mixture density model) and the QRGBM model (gradient booster using quantile output). Table 4 shows the corresponding prediction results.

As can be seen from Table 4, the turbines in the high wind speed and high output units and the turbines in the high wind speed and low output units achieve the given confidence level when uncertainty prediction is made for turbines in different clusters at the 90% confidence level. The highest interval coverage is obtained by QRGBM in the prediction results, with 99.50%, 88.13%, and 97.01%, respectively. However, the interval width of QRGBM is wider than the other models, reaching 452.58 kW, 358.46 kW, and 329.57 kW, respectively. The overly wide interval width will lead to less valid information available in the prediction results. The LSTM and ADFN models have similar interval coverages. For type 0 (high wind speed and high output units), the ADFN model obtains a slightly narrower average width of intervals, but it is wider for other types, and its average width of intervals increases by 5.23% and 4.34% compared to LSTM model. ADFN, based on the attention mechanism, with a small increase in interval coverage, also increases the average width of intervals by a certain amount, with better prediction results. For type 2 (low wind speed and low output units), the interval coverages of all three models do not reach the confidence level, and the prediction effects are not good.

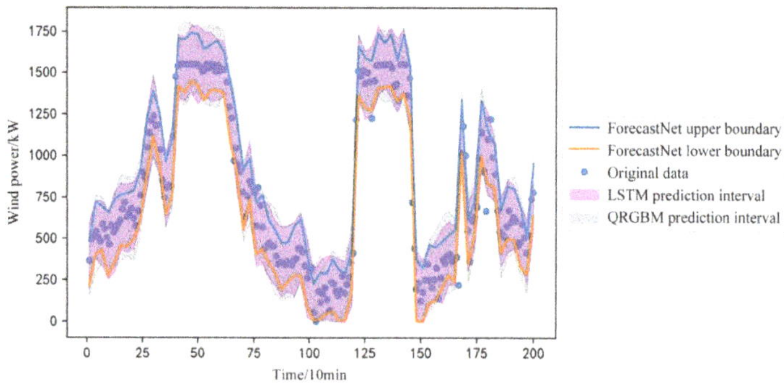

Figure 13. Comparison of interval prediction results for different models for turbine #1.

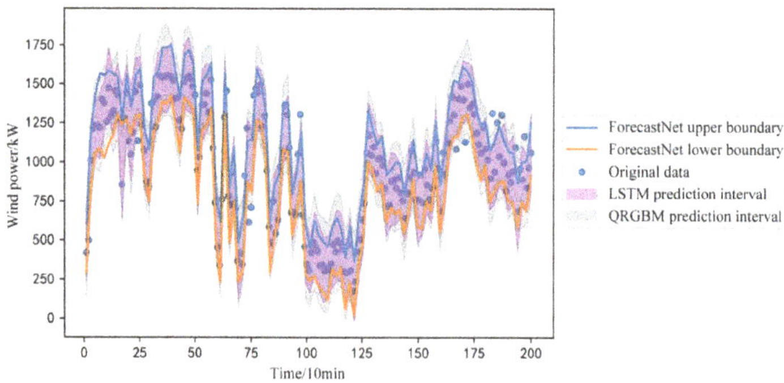

Figure 14. Comparison of interval prediction results for different models for turbine #96.

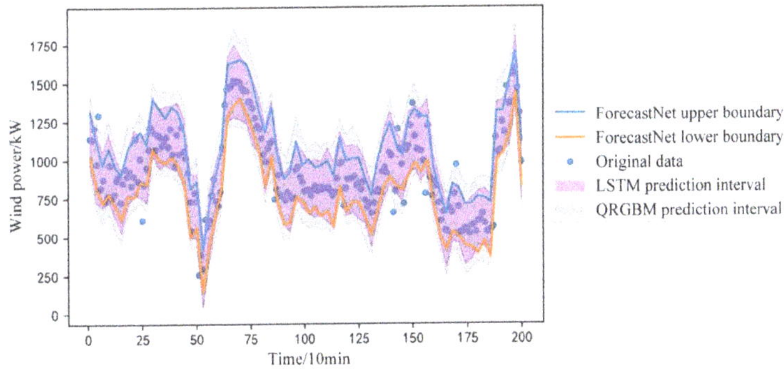

Figure 15. Comparison of interval prediction results for different models for turbine #128.

Table 4. Comparison of forecast indicators for different wind turbines.

Wind Turbine Serial Number	Models	Interval Coverage/%	Average Width of the Interval/kW
#1 (Type 0)	LSTM	94.52	373.43
	ADFN	95.36	368.37
	QRGBM	99.50	452.58
#96 (Type 2)	LSTM	83.92	300.16
	ADFN	84.13	315.87
	QRGBM	88.13	358.46
#128 (Type 1)	LSTM	94.43	286.68
	ADFN	95.06	299.13
	QRGBM	97.01	329.57

Furthermore, this paper randomly selected the probability density function of the ADFN prediction model at different wind turbines and different time prediction points from the multi-step prediction, as shown in Figure 16.

The straight line perpendicular to the x-coordinate axis in Figure 16 is the true value at that prediction moment, and the curve is the probability density distribution of the corresponding model. From Figure 16, it can be seen that the probability density curve is complete and smooth, and there is no missing, very high, or very low values. The curve is also not too broad or too narrow, indicating that the prediction effect of the algorithm is appropriate. Most of the real values in the prediction results of the ADFN model fall near the highest probability points of the probability density curve, such as when $t = 167$, $t = 110$, $t = 100$, and $t = 0$. This indicates that the algorithm predicts with a high accuracy. When the actual wind power value is in the vicinity of the crest of the probability density curve, it indicates that the real value of wind power falls in the high probability interval given by the interval prediction, and the prediction error of these moments is small. When $t = 20$, the real value of wind power deviates from the center of the probability curve, and when $t = 30$, the deviation is even further away. This indicates that the prediction error at these moments is large, and the prediction interval may not even cover the real value of wind power. In the probability density curve results obtained from the test set, the prediction results are reliable if most of the actual values are close to the center range of the curve and only a small number of actual values deviate farther. If most of the actual values are deviated or even far away from the position of the peak of the probability density curve, the obtained probability density prediction results are unreliable. In summary, the ADFN prediction model is reliable in predicting the probability density of different wind turbines.

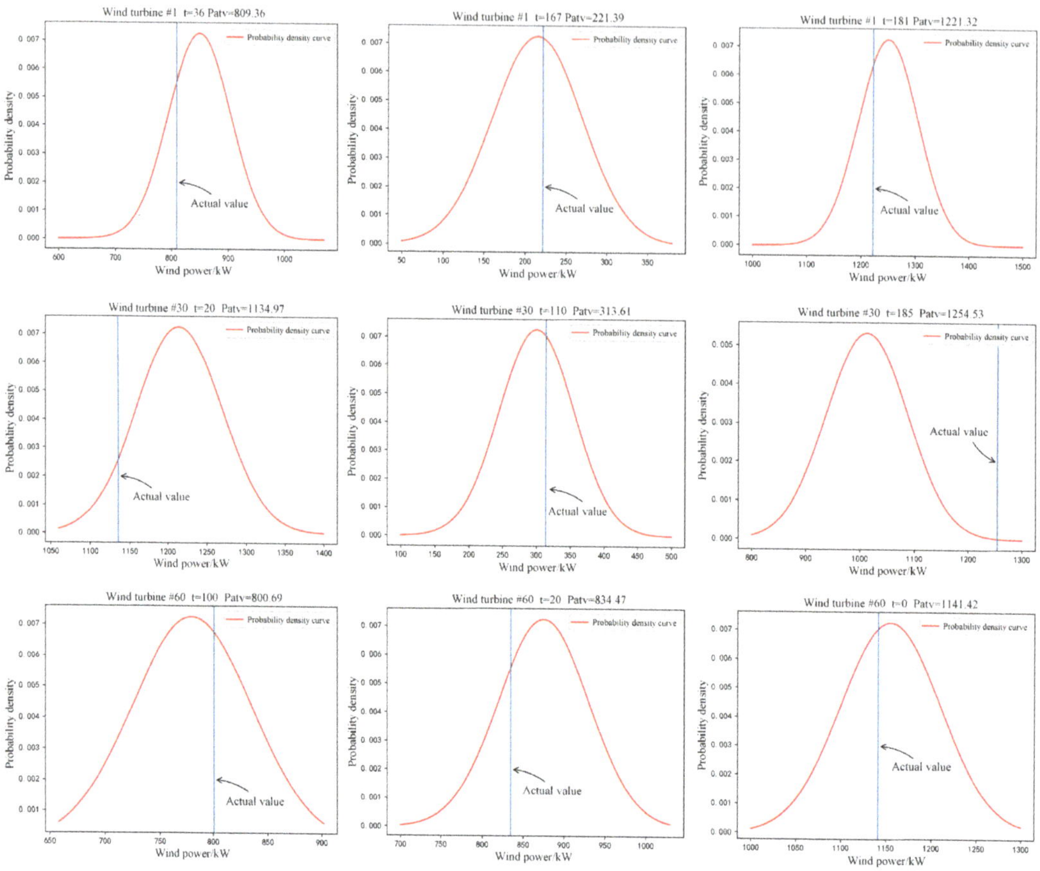

Figure 16. Probability density function of wind power at different moments.

4. Conclusions

This paper focuses on predicting the spatio-temporal distribution of wind power from different turbines in wind farms and proposes a ForecastNet prediction model based on the spatio-temporal distribution. In order to simplify the prediction, the data of N wind turbines with the highest output power in the wind farm are selected to train the model, where the determination of the N value is based on the deterministic prediction results. Then, the uncertainty prediction is performed on the wind power data. The prediction results of ForecastNet variant models based on different hidden layers are compared with those of common prediction models. The results show that the ADFN model has better prediction results when considering the reliability index and acuity index. In addition, the prediction model in this paper has the following features:

- The model uses Euclidean distance and difference distance combined with a graph attention network to aggregate spatio-temporal information on input features.
- In order to avoid the gradient problem when the neural network model predicts wind power in the short term, the time-varying characteristics and interleaved output characteristics of ForecastNet are used to improve the prediction effect.
- The model can predict the probability density curve of wind power at future moments, which can provide more effective information for grid decision makers.

Author Contributions: Conceptualization, S.P., X.L. and J.P.; methodology, S.P. and J.P.; software, L.G. and J.P.; validation, L.G., H.H. and J.P.; writing—original draft preparation, L.G. and J.P.; writing—review and editing, L.G., H.H. and J.P.; supervision, S.P. and X.L. All authors have read and agreed to the published version of the manuscript.

Funding: This research was supported in part by National Natural Science Foundation of China (grant number: No. 62003218) and in part by the Stable Support Projects for Shenzhen Higher Education Institutions (grant number: No. 20220717223051001).

Institutional Review Board Statement: Not applicable.

Informed Consent Statement: Not applicable.

Data Availability Statement: All the data used in this paper can be obtained by contacting the authors of this study.

Conflicts of Interest: Author Jiayi Peng was employed by the company State Grid Zhuzhou Power Supply Company. The remaining authors declare that the research was conducted in the absence of any commercial or financial relationships that could be construed as a potential conflict of interest.

Abbreviations

ForecastNet	Deep feed-forward neural network
Wspd	Wind speed
Wdir	The angle between the wind direction and the position of turbine nacelle
Etmp	Environmental temperature
Itmp	Turbine nacelle internal temperature
Ndir	Nacelle direction
Pab 1	Pitch angle of blade 1
Pab 2	Pitch angle of blade 2
Pab 3	Pitch angle of blade 3
Prtv	Reactive power
Patv	Active power
CNN	Convolutional neural network
MLP	Multilayer perceptron
MLPFN	ForecastNet model with multilayer perceptron as hidden layer
CNNFN	ForecastNet model with CNN as hidden layer
ADFN	ForecastNet model with hidden layer as a dense network based on the attention mechanism
RMSE	Root mean squared error
MAE	Mean absolute error
SVR	Support vector regression
KNN	K nearest neighbor
LightGBM	Light gradient boosting machine
THPO	Wind turbines with the highest power output

References

1. Li, J.; Liu, K.; Yang, Z.X.; Qu, Y. Evolution and impacting factors of global renewable energy products trade network: An empirical investigation based on ERGM model. *Sustainability* **2023**, *15*, 8701. [CrossRef]
2. Lu, Y.; Liu, X.; Li, H.J.; Wang, H.R.; Kong, J.J.; Zhong, C.; Cui, M.L.; Li, Y.; Sun, X.Q.; Xuan, J.D.; et al. What is the impact of the renewable energy power absorption guarantee mechanism on China's green electricity market? *Energies* **2023**, *16*, 7434. [CrossRef]
3. Hu, S.; Xiang, Y.; Zhang, H.C.; Xie, S.Y.; Li, J.H.; Gu, C.H.; Sun, W.; Liu, J.H. Hybrid forecasting method for wind power integrating spatial correlation and corrected numerical weather prediction. *Appl. Energy* **2021**, *293*, 116951. [CrossRef]
4. Chen, W.H.; Zhou, H.T.; Cheng, L.S.; Xia, M. Prediction of regional wind power generation using a multi-objective optimized deep learning model with temporal pattern attention. *Energy* **2023**, *278*, 127942. [CrossRef]
5. Yang, M.; Guo, Y.F.; Huang, Y.T. Wind power ultra-short-term prediction method based on NWP wind speed correction and double clustering division of transitional weather process. *Energy* **2023**, *282*, 128947. [CrossRef]
6. Peng, C.; Zhang, Y.Q.; Zhang, B.W.; Song, D.; Lyu, Y.; Tsoi, A. A novel ultra-short-term wind power prediction method based on XA mechanism. *Appl. Energy* **2023**, *351*, 121905. [CrossRef]
7. Ying, H.M.; Deng, C.H.; Xu, Z.H.; Huang, H.X.; Deng, W.S.; Yang, Q.L. Short-term prediction of wind power based on phase space reconstruction and BiLSTM. *Energy Rep.* **2023**, *9*, 474–482. [CrossRef]

8. Xu, P.H.; Zhang, M.Y.; Chen, Z.H.; Wang, B.Q.; Cheng, C.; Liu, R.F. A deep learning framework for day ahead wind power short-term prediction. *Appl. Sci.* **2023**, *13*, 4042. [CrossRef]
9. Cai, Z.; Dai, S.; Ding, Q.; Zhang, J.L.; Xu, D.; Li, Y.X. Gray wolf optimization-based wind power load mid-long term forecasting algorithm. *Comput. Electr. Eng.* **2023**, *109*, 108769. [CrossRef]
10. Han, S.; Qiao, Y.H.; Yan, J.; Liu, Y.Q.; Li, L.; Wang, Z. Mid-to-long term wind and photovoltaic power generation prediction based on copula function and long short term memory network. *Appl. Energy* **2019**, *239*, 181–191. [CrossRef]
11. Guo, N.Z.; Shi, K.Z.; Li, B.; Qi, L.W.; Wu, H.H.; Zhang, Z.L.; Xu, J.Z. A physics-inspired neural network model for short-term wind power prediction considering wake effects. *Energy* **2022**, *261*, 125208. [CrossRef]
12. Yakoub, G.; Mathew, S.; Leal, J. Intelligent estimation of wind farm performance with direct and indirect 'point' forecasting approaches integrating several NWP models. *Energy* **2023**, *263*, 125893. [CrossRef]
13. Zhang, Y.; Li, Y.T.; Zhang, G.Y. Short-term wind power forecasting approach based on Seq2Seq model using NWP data. *Energy* **2020**, *213*, 118371. [CrossRef]
14. Liu, C.Y.; Zhang, X.M.; Mei, S.W.; Zhou, Q.Y.; Fan, H. Series-wise attention network for wind power forecasting considering temporal lag of numerical weather prediction. *Appl. Energy* **2023**, *336*, 120815. [CrossRef]
15. Wang, J.N.; Zhu, H.Q.; Cheng, F.; Zhou, C.; Zhang, Y.J.; Xu, H.L.; Liu, M.H. A novel wind power prediction model improved with feature enhancement and autoregressive error compensation. *J. Clean. Prod.* **2023**, *420*, 138386. [CrossRef]
16. Li, D.Q.; Yu, X.D.; Liu, S.L.; Dong, X.; Zang, H.Z.; Xu, R. Wind power prediction based on PSO-Kalman. *Energy Rep.* **2022**, *8*, 958–968. [CrossRef]
17. Zhang, W.Q.; Lin, Z.; Liu, X.L. Short-term offshore wind power forecasting—A hybrid model based on Discrete Wavelet Transform (DWT), Seasonal Autoregressive Integrated Moving Average (SARIMA), and deep-learning-based Long Short-Term Memory (LSTM). *Renew. Energy* **2022**, *185*, 611–628. [CrossRef]
18. Le Goff Latimier, R.; Le Bouedec, E.; Monbet, V. Markov switching autoregressive modeling of wind power forecast errors. *Electr. Power Syst. Res.* **2020**, *189*, 106641. [CrossRef]
19. Zhen, Z.; Liu, J.M.; Zhang, Z.Y.; Wang, F.; Chai, H.; Yu, Y.L.; Lu, X.X.; Wang, T.Q.; Lin, Y.Z. Deep learning based surface irradiance mapping model for solar PV power forecasting using sky image. *IEEE Trans. Ind. Appl.* **2020**, *56*, 3385–3396. [CrossRef]
20. Yang, M.; Chen, X.X.; Du, J.; Cui, Y. Ultra-short-term multistep wind power prediction based on improved EMD and reconstruction method using run-length analysis. *IEEE Access* **2018**, *6*, 31908–31917. [CrossRef]
21. Lu, H.F.; Ma, X.; Huang, K.; Azimi, M. Prediction of offshore wind farm power using a novel two-stage model combining kernel-based nonlinear extension of the Arps decline model with a multi-objective grey wolf optimizer. *Renew. Sustain. Energy Rev.* **2020**, *127*, 109856. [CrossRef]
22. Lahouar, A.; Ben Hadj Slama, J. Hour-ahead wind power forest based on random forest. *Renew. Energy* **2017**, *109*, 529–541. [CrossRef]
23. Zhao, Y.Y.; Song, H.Y.; Guo, Y.J.; Zhao, L.J.; Sun, H.X. Super short term combined power prediction for wind power hydrogen production. *Energy Rep.* **2022**, *8*, 1387–1395. [CrossRef]
24. Wang, X.Y.; Li, J.; Shao, L.; Liu, H.L.; Ren, L.; Zhu, L.H. Short-term wind power prediction by an extreme learning machine based on an improved hunter–prey optimization algorithm. *Sustainability* **2023**, *15*, 991. [CrossRef]
25. Ogliari, E.; Guilizzoni, M.; Giglio, A.; Pretto, S. Wind power 24-h ahead forecast by an artificial neural network and an hybrid model: Comparison of the predictive performance. *Renew. Energy* **2021**, *178*, 1466–1474. [CrossRef]
26. Wei, J.Q.; Wu, X.J.; Yang, T.M.; Jiao, R.H. Ultra-short-term forecasting of wind power based on multi-task learning and LSTM. *Int. J. Electr. Power Energy Syst.* **2023**, *149*, 109073. [CrossRef]
27. Xiao, Z.X.; Tang, F.; Wang, M.Y. Wind power short-term forecasting method based on LSTM and multiple error correction. *Sustainability* **2023**, *15*, 3798. [CrossRef]
28. Liu, X.; Yang, L.X.; Zhang, Z.J. Short-term multi-step ahead wind power predictions based on a novel deep convolutional recurrent network method. *IEEE Trans. Sustain. Energy* **2021**, *12*, 1820–1833. [CrossRef]
29. Zhou, G.Y.; Hu, G.F.; Zhang, D.X.; Zhang, Y. A novel algorithm system for wind power prediction based on RANSAC data screening and Seq2Seq-Attention-BiGRU model. *Energy* **2023**, *283*, 128986. [CrossRef]
30. Zhang, S.Y.; Liu, M.B.; Liu, M.D.; Lei, Z.X.; Zeng, G.H.; Chen, Z.R. Day-ahead wind power prediction using an ensemble model considering multiple indicators combined with error correction. *Appl. Soft. Comput.* **2023**, *148*, 110873. [CrossRef]
31. Chen, H.P.; Wu, H.; Kan, T.Y.; Zhang, J.H.; Li, H.L. Low-carbon economic dispatch of integrated energy system containing electric hydrogen production based on VMD-GRU short-term wind power prediction. *Int. J. Electr. Power Energy Syst.* **2023**, *154*, 109420. [CrossRef]
32. Yu, C.Q.; Yan, G.X.; Yu, C.M.; Zhang, Y.; Mi, X.W. A multi-factor driven spatiotemporal wind power prediction model based on ensemble deep graph attention reinforcement learning networks. *Energy* **2023**, *263*, 126034. [CrossRef]
33. Zhang, J.A.; Liu, D.; Li, Z.J.; Han, X.; Liu, H.; Dong, C.; Wang, J.Y.; Liu, C.Y.; Xia, Y.P. Power prediction of a wind farm cluster based on spatiotemporal correlations. *Appl. Energy* **2021**, *302*, 117568. [CrossRef]
34. Wang, Y.J.; Wang, J.D.; Cao, M.; Li, W.X.; Yuan, L.; Wang, N. Prediction method of wind farm power generation capacity based on feature clustering and correlation analysis. *Electr. Power Syst. Res.* **2022**, *212*, 108634. [CrossRef]
35. Li, Y.; Xiao, Z.Q.; Nie, S.S.; Cao, J.W.; Hua, H.C. Review of research on generative adversarial network and its application in new energy data quality. *South. Power Syst. Technol.* **2020**, *14*, 25–33. [CrossRef]

36. Xu, W.J.; Wu, S.L.; Wang, F.Y.; Lin, L.; Li, G.J.; Zhang, Z. Large-scale post-disaster user distributed coverage optimization based on multi-agent reinforcement learning. *J. Commun.* **2022**, *43*, 1–16.
37. Xu, B.K. Fan yaw control optimization strategy based on clustering algorithm and wind direction prediction. *J. Shandong Electr. Power Coll.* **2017**, *20*, 50–54.
38. Ni, Q.J.; Wang, Y.H.; Yuan, J.Y. Adaptive scalable spatio-temporal graph convolutional network for PM2.5 prediction. *Eng. Appl. Artif. Intell.* **2023**, *126*, 107080. [CrossRef]
39. Yin, L.S.; Wu, Y.Y. Traffic flow combination prediction model based on improved VMD-GAT-GRU. *J. Electron. Meas. Instrum.* **2022**, *36*, 62–72. [CrossRef]
40. Feng, B.; Zhang, Y.W.; Tang, X.; Guo, C.X.; Wang, J.J.; Yang, Q.; Wang, H.F. Power equipment defect record text mining based on BiLSTM-attention neural network. *Proc. CSEE* **2020**, *40*, 1–10. [CrossRef]
41. Wei, J.; Zhao, H.T.; Liu, D.N.; Jia, H.P.; Wang, X.Y.; Zhang, H.; Liu, Z. Short-term power load forecasting method by attention-based CNN-LSTM. *J. North China Electr. Power Univ. (Nat. Sci. Ed.)* **2021**, *48*, 42–47.
42. Li, J.Q.; Hu, X.D.; Qin, J.R.; Zhang, C.Z. Ultra-short term wind power prediction based on multi-head attention and CNN model. *Electr. Power Sci. Eng.* **2022**, *38*, 34–40.
43. Dabrowski, J.J.; Zhang, Y.; Rahman, A. ForecastNet: A Time-Variant Deep Feed-Forward Neural Network Architecture for Multi-step-Ahead Time-Series Forecasting. *Int. Conf. Neural Inf. Process.* **2020**, *12534*, 579–591. [CrossRef]
44. Zheng, W.Q.; Peng, X.G.; Lu, D.; Zhang, D.; Liu, Y.; Lin, Z.H.; Lin, L.X. Composite quantile regression extreme learning machine with feature selection for short-term wind speed forecasting: A new approach. *Energy Convers. Manag.* **2017**, *151*, 737–752. [CrossRef]

MDPI
St. Alban-Anlage 66
4052 Basel
Switzerland
www.mdpi.com

Applied Sciences Editorial Office
E-mail: applsci@mdpi.com
www.mdpi.com/journal/applsci

www.ingramcontent.com/pod-product-compliance
Lightning Source LLC
Chambersburg PA
CBHW040834250926
51850CB00007BA/1309
9783725807277